The Argument of Mathematics

LOGIC, EPISTEMOLOGY, AND THE UNITY OF SCIENCE

VOLUME 30

Logic, Epistemology, and the Unity of Science aims to reconsider the question of the unity of science in light of recent developments in logic. At present, no single logical, semantic or methodological framework dominates the philosophy of science. However, the editors of this series believe that formal techniques like, for example, independence friendly logic, dialogical logics, multimodal logics, game theoretic semantics and linear logics, have the potential to cast new light on basic issues in the discussion of the unity of science.

This series provides a venue where philosophers and logicians can apply specific technical insights to fundamental philosophical problems. While the series is open to a wide variety of perspectives, including the study and analysis of argumentation and the critical discussion of the relationship between logic and the philosophy of science, the aim is to provide an integrated picture of the scientific enterprise in all its diversity.

For further volumes:
http://www.springer.com/series/6936

Andrew Aberdein • Ian J. Dove

Editors

The Argument
of Mathematics

 Springer

Editors
Andrew Aberdein
Department of Humanities
   and Communication
Florida Institute of Technology
Melbourne, FL, USA

Ian J. Dove
Department of Philosophy
University of Nevada, Las Vegas
Las Vegas, NV, USA

ISBN 978-94-007-6533-7   ISBN 978-94-007-6534-4 (eBook)
DOI 10.1007/978-94-007-6534-4
Springer Dordrecht Heidelberg New York London

Library of Congress Control Number: 2013941250

Printed on acid-free paper

Springer is part of Springer Science+Business Media (www.springer.com)

Preface

Foundational questions and the status of proof have long been central to philosophy of mathematics, although mathematicians do a lot more than just prove results. However, most mathematical practice may still be understood in terms of argument. Thus philosophy of mathematics needs an account of argument. Argumentation theory seems like a good place to look. The intersection of the two, although largely unexplored, has the potential to be hugely fruitful. This book brings together important pioneering work from a variety of sources with newly commissioned articles to provide an overview of the topic.

Our first editorial venture on this theme was a special issue of the journal *Foundations of Science* published in 2009. We are grateful to Diederik Aerts, the editor, for his support. Two of the articles from that collection are republished here, and the authors of two others have contributed new chapters, however we should like to thank all of the contributors: Jody Azzouni; Edwin Coleman; Bart Van Kerkhove and Jean Paul Van Bendegem; David Sherry; Zenon Kulpa; Matthew Inglis and Juan Pablo Mejía-Ramos; Alison Pease, Alan Smaill, Simon Colton and John Lee. We are grateful to our editors at Springer, Lucy Fleet, Ties Nijssen and Christi Lue.

Edinburgh, Scotland   Andrew Aberdein
Las Vegas, Nevada   Ian J. Dove
December 2012

Contents

Contributors

Andrew Aberdein Department of Humanities and Communication, Florida Institute of Technology, Melbourne, FL, USA

Jesse Alama Faculty of Science and Technology, Center for Artificial Intelligence, Universidade Nova de Lisboa, Caparica, Portugal

Jesús Alcolea Banegas Departament de Lògica i Filosofia de la Ciència, Facultat de Filosofia i CC. de l'Educació, Universitat de València, València, Spain

Patrick Allo Center for Logic and Philosophy of Science, Vrije Universiteit Brussel, Brussels, Belgium

Paul Bartha Department of Philosophy, University of British Columbia, Vancouver, Canada

Paola Cantù CNRS & CEPERC, Université Aix-Marseille, Aix-en-Provence Cedex 1, France

Simon Colton Department of Computing, Imperial College, London, UK

Ian J. Dove Department of Philosophy, University of Nevada, Las Vegas, Las Vegas, NV, USA

Michel Dufour Department of Communication, Sorbonne Nouvelle, Paris Cedex 05, France

Richard L. Epstein Advanced Reasoning Forum, Socorro, NM, USA

James Franklin School of Mathematics and Statistics, University of New South Wales, Sydney, Australia

Matthew Inglis Mathematics Education Centre, Loughborough University, Loughborough, UK

Reinhard Kahle CENTRIA and DM, FCT, Universidade Nova de Lisboa, Caparica, Portugal

Christine Knipping Universität Bremen, AG Didaktik, Bremen, Germany

Erik C.W. Krabbe Faculteit Wijsbegeerte, University of Groningen, Groningen, The Netherlands

Brendan Larvor Department of Philosophy, School of Humanities, University of Hertfordshire, Hatfield, UK

John Lee School of Informatics, University of Edinburgh, Edinburgh, UK

Juan Pablo Mejía-Ramos Graduate School of Education, Rutgers University, New Brunswick, NJ, USA

Alison Pease Department of Computing, Imperial College, London, UK

School of Electronic Engineering and Computer Science, Queen Mary, University of London, London, UK

Lawrence H. Powers Department of Philosophy, Wayne State University, Detroit, MI, USA

David Reid Universität Bremen, AG Didaktik, Bremen, Germany

School of Education, Acadia University, Wolfville, NS, Canada

Alan Smaill School of Informatics, University of Edinburgh, Edinburgh, UK

Jean Paul Van Bendegem Center for Logic and Philosophy of Science, Vrije Universiteit Brussel, Brussels, Belgium

Bart Van Kerkhove Center for Logic and Philosophy of Science, Vrije Universiteit Brussel, Brussels, Belgium

Chapter 1
Introduction

Andrew Aberdein and Ian J. Dove

Our goal in this book is to explore the relationship between argumentation theory and the philosophy of mathematical practice. By 'argumentation theory' we intend the study of reasoning and argument, and especially those aspects not addressed (or not addressed well) by formal deduction. Such work goes back at least to Aristotle, who invented formal logic but wrote more extensively about informal reasoning. The great success of formal logic in the nineteenth and twentieth centuries led to an eclipse of informal techniques, but a revival began in the 1950s with the publication of two classic works: (Perelman and Olbrechts-Tyteca, 1969) and (Toulmin, 1958). These pioneers initiated a thriving research tradition with particular strengths in Canada (e.g. Johnson, 2000; Walton et al., 2008) and the Netherlands (e.g. Van Eemeren and Grootendorst, 2004). The philosophy of mathematical practice diverges from mainstream philosophy of mathematics in the emphasis it places on what the majority of working mathematicians actually do, rather than on mathematical foundations, an issue most mathematicians ignore (rightly or wrongly). This leads to a closer relationship with history and sociology of mathematics, mathematics education, and mathematics itself. Important milestones include (Pólya, 1954) and (Lakatos, 1976), both of which pay close attention to mathematical argumentation, although neither work was directly informed by argumentation theory. (For overviews of each work, see Newell, 1983, and Larvor, 1998, respectively.) In the last decade philosophy of mathematical practice has been

A. Aberdein (✉)
Department of Humanities and Communication, 150 West University Boulevard, Florida Institute of Technology, Melbourne, FL 32901-6975, USA
e-mail: aberdein@fit.edu

I.J. Dove
Department of Philosophy, University of Nevada, Las Vegas, 4505 Maryland Parkway, Box 455028, Las Vegas, NV 89154, USA
e-mail: ian.dove@unlv.edu

A. Aberdein and I.J. Dove (eds.), *The Argument of Mathematics*, Logic, Epistemology, and the Unity of Science 30, DOI 10.1007/978-94-007-6534-4_1,
© Springer Science+Business Media Dordrecht 2013

developed further by many authors, including contributors to (Hersh, 2006) and (Mancosu, 2008), but the potential argumentation theory holds for this work has mostly been overlooked. This collection is designed to remedy that oversight.

1.1   What Are Mathematical Arguments?

Conventional wisdom amongst both philosophers of mathematics and argumentation theorists holds that formal logic, whatever its shortcomings in the analysis of everyday reasoning, is well-adapted to mathematical reasoning and, therefore, that there is no role for informal logic in mathematics. The chapters in the first part of this book directly challenge this assumption.

James Franklin is a mathematician, a philosopher, and the author of numerous works on a diverse array of subjects. His chapter is a revised version of one of the earliest philosophical accounts of non-deductive mathematical practice, originally published in the *British Journal for the Philosophy of Science* in 1987. He argues for an experimental approach in mathematics, with illustrations drawn from the varieties of evidence for (and against) the Riemann Hypothesis, heuristic arguments for Goldbach's Conjecture and a cumulative case argument regarding the classification of finite groups. The variety of non-deductive inferences common in mathematical practice has unfortunately received little consideration since the first publication of Franklin's original paper. (Notable exceptions include, Baker, 2007, 2009.) Franklin concludes by advocating a Bayesian approach to mathematical reasoning. Independently (of Franklin and each other) such an approach has been pursued in both philosophy of mathematics (Corfield, 2003) and argumentation theory (Hahn and Oaksford, 2007).

Erik C. W. Krabbe is a philosopher who has published on an impressive number of topics. He has conducted influential research in both formal and informal logic; in the former category, Krabbe's work is foundational for dialogue logic (see, for example, Krabbe, 1985). In the first of two chapters included in this volume, Krabbe argues that some types of mathematical proof are open to analysis and evaluation by the techniques of argumentation theory. An early version of this chapter was published in Dutch in 1991, and a revised version appeared in English in 1997.

Jesús Alcolea Banegas was one of the first philosophers to apply Stephen Toulmin's work to mathematical proof. Toulmin was a pioneer of argumentation theory who developed an influential account of the 'layout' of an argument. This was primarily intended as a characterization of informal, non-mathematical reasoning, although he does include the occasional mathematical example (Toulmin, 1958; Toulmin et al., 1979). Alcolea's chapter, which was first published in Catalan in 1998, shows how the Toulmin layout may profitably be applied to mathematical and metamathematical arguments. This treatment has proved influential, both in philosophy (see, for example, Aberdein, 2005, 2007) and mathematics education (for example, Inglis et al., 2007).

Michel Dufour is a philosopher who mostly researches scientific communication, including scientific popularization and didactical practices, and combines this topic with argumentation, including that of mathematics (see, for example, Dufour, 2011). His chapter takes on two highly influential works in argumentation theory, both of which expressly deny that mathematical proofs are arguments: (Perelman and Olbrechts-Tyteca, 1969) and (Johnson, 2000). (Somewhat coincidentally, both editors have also challenged Johnson's position: (Dove, 2007) and (Aberdein, 2011). Neither of us tackles Perelman.) Dufour proceeds to discuss the circumstances in which arguments arise in the vicinity of proofs, whether or not those proofs are themselves arguments.

1.2 Argumentation as a Methodology for Studying Mathematical Practice

Once we have established that argumentation theory has something to say about mathematics, what use can we make of it? The chapters in the second part offer an impressively diverse set of answers to this question, covering the history of mathematics, mathematics education and, perhaps surprisingly, formal proof verification.

Paola Cantù is a philosopher and historian of mathematics who has also published works on argumentation theory. However, her chapter is the first of her publications to pursue both interests simultaneously. She makes careful use of the concept of an 'argumentation scheme', a stereotypical pattern of reasoning designed to capture informal, defeasible arguments in a similar fashion to the way inference rules capture formal, deductive arguments (Walton et al., 2008, is the most extensive survey to date). She demonstrates how this methodology, and in particular the argumentation scheme for Value-Based Practical Reasoning, sheds new light on the debate over the introduction of ideal elements in nineteenth and early twentieth century mathematics.

Matthew Inglis and Juan Pablo Mejía-Ramos are educational psychologists who have made extensive use of the Toulmin layout as the basis for empirical enquiry into the efficacy of mathematical proofs in securing conviction (Inglis et al., 2007; Inglis and Mejía-Ramos, 2009a,b). This work builds on earlier applications of the layout in mathematics education research (beginning with Krummheuer, 1995). In their chapter, originally published in *Research in Mathematics Education* in 2008, Inglis and Mejía-Ramos show how careful use of the layout can bring to light hitherto concealed ambiguities in the characterization of how persuasive a subject finds a proof. This work has important implications for the design of empirical research into the teaching of proof.

Christine Knipping and David Reid are also mathematics education researchers and their work also reflects the influence of Toulmin's layout (Knipping, 2003, 2008; Reid and Knipping, 2010). However, they differ from Inglis and Mejía-Ramos in their focus on the macro-structure of argumentation. They generalize Toulmin's layout from the consideration of individual argument steps (or whole arguments considered as single steps) to complex networks of steps. This permits a qualitative comparison of large-scale argumentation structures: in their chapter they compare two structure types they characterize as 'source' and 'spiral' structures. This is a novel area of research which should have considerable interest to scholars of argumentation in general, and not just that conducted in mathematics classrooms.

Jesse Alama and Reinhard Kahle are logicians with an interest in the formal verification of proofs. This may seem to be the least promising territory for the application of argumentation theory to mathematics. However, as their chapter demonstrates, topics that might be thought of as essentially informal reappear in the computer-assisted, formal setting. These include the question of 'obviousness': at what stage of explication has an inference become so detailed that no further justification is required?

1.3   Mathematics as a Testbed for Argumentation Theory

The previous part explored the usefulness of argumentation theory in the understanding of mathematical practice. The chapters of the next part demonstrate that the relationship is reciprocated: mathematics is a valuable testbed for argumentation theory. Mathematicians are peculiarly reflective about their own reasoning practice. This makes mathematics a useful source of well-documented instances of a wide range of reasoning practices, many of them of much more general application.

Lawrence H. Powers is the originator of a novel and provocative thesis about fallacies, the 'One Fallacy Theory' (Powers, 1995; see also Zuckero, 2003). Fallacy theorists are traditionally 'splitters': from Aristotle's *Sophistical Refutations* onwards, the standard approach has been a catalogue of diverse sources of error. Powers, by contrast, is a 'lumper': he holds that all reputed fallacies are fallacies of equivocation, or no fallacy at all. Mathematical fallacies might seem an exception to this approach, but as he explains in his chapter, which first appeared in the *Proceedings of the Fourth Conference of the International Society for the Study of Argumentation* in 1999, his theory can accommodate them too.

Krabbe's second chapter, which was first published in *Argumentation* in 2008, applies the analytic techniques of pragma-dialectics to mathematical proofs. Pragma-dialectics was pioneered by the communication theorists Frans van Eemeren and Rob Grootendorst (Van Eemeren and Grootendorst, 2004, is a consolidation of their views). It is an influential approach to argumentation based around an idealized concept of a critical discussion. This gives rise to an explicit code of conduct — a set of rules that discussants must follow, or their arguments will be fallacious. In later work, Van Eemeren extends pragma-dialectics to recognize that arguers do not just seek a rational outcome to their disputes, they set out to win (Van Eemeren and Houtlosser, 2002). He calls the process of balancing these two objectives strategic maneuvering. Krabbe shows how strategic maneuvering may be extended to encompass other objectives, and how mathematical fallacies may be understood as failed attempts to accommodate the tension between them.

Paul Bartha is a philosopher working mainly in philosophy of science and decision theory. His recent book on analogical reasoning (Bartha, 2010) is notable for the attention he pays to mathematical cases. In his chapter, Bartha summarizes this research, showing how a general account of analogy may be refined to cover arguments for the plausibility of mathematical conjectures.

Brendan Larvor is a philosopher of mathematics whose work pays particular attention to mathematical practice. Notably, he is the author of a monograph (Larvor, 1998) on Imre Lakatos, one of the first philosophers of mathematics to attend to the structure of mathematical arguments (in Lakatos, 1976). More recently, Larvor's work has turned more explicitly to argumentation theory (see, for example Larvor, 2012). In his chapter he looks at visual argumentation, a difficult problem for both philosophy of mathematics and argumentation theory. Each discipline has tackled the problem largely in ignorance of the efforts of the other; Larvor argues that argumentation theorists in particular would benefit from closer interaction.

1.4 An Argumentational Turn in the Philosophy of Mathematics

The last part consists of chapters defending attention to mathematical argumentation as the basis for new perspectives on the philosophy of mathematics.

Richard L. Epstein is the author of numerous works in both formal logic and critical thinking (for example, Epstein and Kernberger, 1998; Epstein, 2006). In his chapter he develops an account of mathematics as grounded in human experience. The argumentational character of mathematical proof is central to this account. He includes an appendix contrasting his own views with those of several illustrious predecessors, ultimately finding a kindred spirit in John Stuart Mill.

Ian Dove's chapter, which first appeared in *Foundations of Science* in 2009, offers a variety of evidence for the utility of argumentation theory not just in the analysis of mathematical practice but also in the resolution of some of the philosophical debates to which that practice has given rise. He pays particular attention to Jody Azzouni's (2004, 2009) 'derivation indicator' view of informal mathematical practice, arguing that, despite initial appearances, it is compatible with a central role for argumentation. Dove also surveys further open questions that remain to be answered by theorists of mathematical argumentation.

Alison Pease, Alan Smaill, Simon Colton and John Lee have backgrounds in artificial intelligence research, and a specific interest in its implications for mathematical cognition. This has led them to pay particularly close attention to the work of Lakatos (for example, in Pease et al., 2005; Pease, 2007). Their chapter, also first published in *Foundations of Science* in 2009, explores how resources from artificial intelligence, and especially the computational modelling of mathematical reasoning, may be utilized to further develop both argumentation theory and the philosophy of mathematical practice.

Patrick Allo, Jean Paul van Bendegem and Bart van Kerkhove are philosophers working primarily in logic and the philosophy of science. Van Bendegem and Van Kerkhove in particular have a long-standing interest in mathematical practice (see, for example, Van Bendegem, 1988; Van Kerkhove, 2005). More recently their work has begun to integrate themes from the study of argumentation (Van Bendegem and Van Kerkhove, 2009). In their chapter they develop an account of mathematical justification from the perspective of computer science and epistemic logic. The result is a fascinating combination of insights drawing on research in artificial intelligence, distributed systems theory and new media studies.

Andrew Aberdein is a philosopher of mathematics who also writes on informal logic. Much of his recent work has focused on the intersection of argumentation theory and mathematics, especially the use of Toulmin diagrams (2005, 2006), argumentation schemes (2013) and fallacy theory (2010) to assess mathematical reasoning. His paper offers an account of mathematical reasoning as comprising two parallel structures, one inferential and one argumentational. The uniquely mathematical character of the reasoning proceeds from the rigour of the inferential structure. However, much actual mathematical practice takes place in the argumentational structure. If this account is correct, the understanding of mathematics can only properly be addressed when *both* of the parallel structures are accounted for.

References

Aberdein, A. (2005). The uses of argument in mathematics. *Argumentation, 19*(3), 287–301.

Aberdein, A. (2006). Managing informal mathematical knowledge: Techniques from informal logic. In J. M. Borwein & W. M. Farmer (Eds.), *MKM 2006*, Vol. 4108 in *LNAI* (pp. 208–221). Berlin: Springer.

Aberdein, A. (2007). The informal logic of mathematical proof. In B. Van Kerkhove & J. P. Van Bendegem (Eds.), *Perspectives on mathematical practices: Bringing together philosophy of mathematics, sociology of mathematics, and mathematics education* (pp. 135–151). Dordrecht: Springer.

Aberdein, A. (2010). Observations on sick mathematics. In B. Van Kerkhove, J. P. Van Bendegem, & J. De Vuyst (Eds.), *Philosophical perspectives on mathematical practice* (pp. 269–300). London: College Publications.

Aberdein, A. (2011). The dialectical tier of mathematical proof. In F. Zenker (Ed.), *Argumentation: Cognition and community. Proceedings of the 9th international conference of the Ontario Society for the Study of Argumentation (OSSA)*. Windsor, ON: OSSA.

Aberdein, A. (2013). Mathematical wit and mathematical cognition. *Topics in Cognitive Science, 5*(2), 231–250.

Azzouni, J. (2004). The derivation-indicator view of mathematical practice. *Philosophia Mathematica, 12*(2), 81–105.

Azzouni, J. (2009). Why do informal proofs conform to formal norms? *Foundations of Science, 14*(1–2), 9–26.

Baker, A. (2007). Is there a problem of induction for mathematics? In M. Leng, A. Paseau, & M. Potter (Eds.), *Mathematical knowledge* (pp. 59–73). Oxford: Oxford University Press.

Baker, A. (2009). Non-deductive methods in mathematics. *Stanford Encyclopedia of Philosophy*. http://plato.stanford.edu/entries/mathematics-nondeductive/. Cited 20 Mar 2011.

Bartha, P. (2010). *By parallel reasoning: The construction and evaluation of analogical arguments.* New York: Oxford University Press.

Corfield, D. (2003). *Towards a philosophy of real mathematics.* Cambridge: Cambridge University Press.

Dove, I. J. (2007). On mathematical proofs and arguments: Johnson and Lakatos. In F. H. Van Eemeren & B. Garssen (Eds.), *Proceedings of the sixth conference of the International Society for the Study of Argumentation* (Vol. 1, pp. 346–351). Amsterdam: Sic Sat.

Dufour, M. (2011). Didactical arguments and mathematical proofs. In F. H. van Eemeren, B. Garssen, D. Godden, & Mitchell, G. (Eds.), *Proceedings of the 7th conference of the International Society for the Study of Argumentation* (pp. 390–397). Amsterdam: Rozenberg/Sic Sat.

Epstein, R. L. (2006). *Classical mathematical logic.* Princeton, NJ: Princeton University Press.

Epstein, R. L., & Kernberger, C. ([2005] 1998). *Critical thinking.* Belmont, CA: Wadsworth.

Hahn, U., & Oaksford, M. (2007). The rationality of informal argumentation: A Bayesian approach to reasoning fallacies. *Psychological Review, 114*(3), 704–732.

Hersh, R. (Ed.). (2006). *18 Unconventional essays about the nature of mathematics.* New York: Springer.

Inglis, M., & Mejía-Ramos, J. P. (2009a). The effect of authority on the persuasiveness of mathematical arguments. *Cognition and Instruction, 27*(1), 25–50.

Inglis, M., & Mejía-Ramos, J. P. (2009b). On the persuasiveness of visual arguments in mathematics. *Foundations of Science, 14*(1–2), 97–110.

Inglis, M., Mejía-Ramos, J. P., & Simpson, A. (2007). Modelling mathematical argumentation: The importance of qualification. *Educational Studies in Mathematics, 66*(1), 3–21.

Johnson, R. H. (2000). *Manifest rationality: A pragmatic theory of argument.* Mahwah, NJ: Lawrence Erlbaum Associates.

Knipping, C. (2003). *Beweisprozesse in der Unterrichtspraxis: Vergleichende Analysen von Mathematikunterricht in Deutschland und Frankreich.* Hildesheim: Franzbecker Verlag.

Knipping, C. (2008). A method for revealing structures of argumentations in classroom proving processes. *ZDM Mathematics Education, 40*, 427–441.

Krabbe, E. C. W. (1985). Formal systems of dialogue rules. *Synthese, 63*, 295–328.

Krummheuer, G. (1995). The ethnography of argumentation. In P. Cobb & H. Bauersfeld (Eds.), *The emergence of mathematical meaning: Interaction in classroom cultures* (pp. 229–269). Hillsdale, NJ: Lawrence Erlbaum Associates.

Lakatos, I. (1976). *Proofs and refutations: The logic of mathematical discovery* (edited by J. Worrall & E. Zahar). Cambridge: Cambridge University Press.

Larvor, B. (1998). *Lakatos: An introduction.* London: Routledge.

Larvor, B. (2012). How to think about informal proofs. *Synthese, 187*, 715–730.

Mancosu, P. (Ed.). (2008). *The philosophy of mathematical practice.* Oxford: Oxford University Press.

Newell, A. (1983). The heuristic of George Polya and its relation to artificial intelligence. In R. Groner, M. Groner, & W. F. Bischof (Eds.), *Methods of heuristics* (pp. 195–243). Hillsdale, NJ: Lawrence Erlbaum Associates.

Pease, A. (2007). *A computational model of Lakatos-style reasoning.* PhD thesis, School of Informatics, University of Edinburgh. Online at http://hdl.handle.net/1842/2113. Cited 5 Feb 2012.

Pease, A., Colton, S., Smaill, A., & Lee, J. (2005). Modelling Lakatos's philosophy of mathematics. In L. Magnani & R. Dossena (Eds.), *Computing, philosophy and cognition: Proceedings of the European computing and philosophy conference (ECAP 2004)* (pp. 57–85). London: College Publications.

Perelman, C., & Olbrechts-Tyteca, L. (1969). *The new rhetoric: A treatise on argumentation.* Notre Dame, IN: University of Notre Dame Press.

Pólya, G. (1954). *Mathematics and plausible reasoning* (2 Vols.). Princeton, NJ: Princeton University Press.

Powers, L. H. (1995). The one fallacy theory. *Informal Logic, 17*(2), 303–314.

Reid, D., & Knipping, C. (2010). *Proof in mathematics education: Research, learning and teaching.* Rotterdam: Sense.

Toulmin, S. (1958). *The uses of argument.* Cambridge: Cambridge University Press.

Toulmin, S., Rieke, R., & Janik, A. (1979). *An introduction to reasoning.* London: Macmillan.

Van Bendegem, J. P. (1988). Non-formal properties of real mathematical proofs. In J. Leplin, A. Fine, & M. Forbes (Eds.), *PSA: Proceedings of the biennial meeting of the philosophy of science association* (vol. 1: Contributed Papers, pp. 249–254). East Lansing, MI: Philosophy of Science Association.

Van Bendegem, J. P., & Van Kerkhove, B. (2009). Mathematical arguments in context. *Foundations of Science, 14*(1–2), pp. 45–57.

Van Eemeren, F. H., & Grootendorst, R. (2004). *A systematic theory of argumentation: The pragma-dialectical approach.* Cambridge: Cambridge University Press.

Van Eemeren, F. H., & Houtlosser, P. (2002). Strategic manoeuvring in argumentative discourse: A delicate balance. In Van Eemeren, F. H. & Houtlosser, P. (Eds.), *Dialectic and rhetoric: The warp and woof of argumentation analysis* (pp. 131–159). Amsterdam: Kluwer.

Van Kerkhove, B. (2005). Aspects of informal mathematics. In Sica, G. (Ed.), *Essays on the foundations of mathematics and logic* (pp. 268–351). Monza: Polimetrica International Scientific Publisher.

Walton, D. N., Reed, C., & Macagno, F. (2008). *Argumentation schemes.* Cambridge: Cambridge University Press.

Zuckero, M. (2003). Three potential problems for Powers' One-Fallacy Theory. *Informal Logic, 23*(2), 285–292.

Part I
What Are Mathematical Arguments?

Chapter 2
Non-deductive Logic in Mathematics: The Probability of Conjectures

James Franklin

2.1 Introduction

Mathematicians often speak of conjectures as being confirmed by evidence that falls short of proof. For their own conjectures, evidence justifies further work in looking for a proof. Those conjectures of mathematics that have long resisted proof, as Fermat's Last Theorem did and the Riemann Hypothesis still does, have had to be considered in terms of the evidence for and against them. It is not adequate to describe the relation of evidence to hypothesis as "subjective", "heuristic" or "pragmatic" there must be an element of what it is objectively rational to believe on the evidence, that is, of non-deductive logic. Mathematics is therefore (among other things) an experimental science.

The occurrence of non-deductive logic, or logical probability, or the rational support for unproved conjectures, in mathematics is however an embarrassment. It is embarrassing to mathematicians, used to regarding deductive logic as the only real logic. It is embarrassing for those statisticians who wish to see probability as solely about random processes or relative frequencies: surely there is nothing probabilistic about the truths of mathematics? It is a problem for philosophers who believe that induction is justified not by logic but by natural laws or the "uniformity of nature": mathematics is the same no matter how lawless nature may be. It does not fit well with most philosophies of mathematics. It is awkward even for proponents of non-deductive logic. If non-deductive logic deals with logical relations weaker than entailment, how can such relations hold between the necessary truths of mathematics?

Work on this topic was therefore rare in the mid-twentieth century "classical" period in the philosophy of science and mathematics. The recent turning of attention

J. Franklin (✉)
School of Mathematics and Statistics, University of New South Wales, Sydney 2052, Australia
e-mail: j.franklin@unsw.edu.au

A. Aberdein and I.J. Dove (eds.), *The Argument of Mathematics*, Logic, Epistemology, and the Unity of Science 30, DOI 10.1007/978-94-007-6534-4_2,
© Springer Science+Business Media Dordrecht 2013

11

in philosophy of mathematics towards mathematical practice has produced a number of examinations of experimental mathematics (Franklin, 1987; Fallis, 1997; Brown, 1999, Ch. 10; Fallis, 2000; Corfield, 2003, Ch. 5; Lehrer Dive, 2003; Van Kerkhove and Van Bendegem, 2008; Baker, 2009; Dove, 2009; brief earlier remarks in Kolata, 1976) but these have mostly not discussed in depth the theoretical issues raised. Many of these works were inspired by one important earlier contribution, the pair of books by the mathematician George Pólya on *Mathematics and Plausible Reasoning* (1954). Despite their excellence, these books of Pólya's had been little noticed by mathematicians, and even less by philosophers. Undoubtedly that is largely because of Pólya's unfortunate choice of the word "plausible" in his title—"plausible" has a subjective, psychological ring to it, so that the word is almost equivalent to "convincing" or "rhetorically persuasive". Arguments that happen to persuade, for psychological reasons, are rightly regarded as of little interest in mathematics and philosophy. Pólya made it clear, however, that he was not concerned with subjective impressions, but with what degree of belief was *justified* by the evidence (Pólya, 1954, I, 68).

Non-deductive logic deals with the support, short of entailment, that some propositions give to others. If a proposition has already been proved true, there is of course no longer any need to consider non-conclusive evidence for it. Consequently, non-deductive logic will be found in mathematics in those areas where mathematicians consider propositions which are not yet proved. These are of two kinds. First there are those that any working mathematician deals with in his preliminary work before finding the proofs he hopes to publish, or indeed before finding the theorems he hopes to prove. The second kind are the long-standing conjectures which have been written about by many mathematicians but which have resisted proof.

It is obvious on reflection that a mathematician must use non-deductive logic in the first stages of his work on a problem. Mathematics cannot consist just of conjectures, refutations and proofs. Anyone can generate conjectures, but which ones are worth investigating? Which ones are relevant to the problem at hand? Which can be confirmed or refuted in some easy cases, so that there will be some indication of their truth in a reasonable time? Which might be capable of proof by a method in the mathematician's repertoire? Which might follow from someone else's theorem? Which are unlikely to yield an answer until after the next review of tenure? The mathematician must answer these questions to allocate his time and effort. But not all answers to these questions are equally good. To stay employed as a mathematician, he must answer a proportion of them well. But to say that some answers are better than others is to admit that some are, on the evidence he has, more reasonable than others, that is, are rationally better supported by the evidence. That is to accept a role for non-deductive logic.

The area where a mathematician must make the finest discriminations of this kind—and where he might, in theory, be guilty of professional negligence if he makes the wrong decisions—is as a supervisor advising a prospective Ph.D. student. It is usual for a student beginning a Ph.D. to choose some general field of mathematics and then to approach an expert in the field as a supervisor. The supervisor then selects a problem in that field for the student to investigate. In mathematics, more than in any other discipline, the initial choice of problem is the crucial event in the Ph.D.-gathering process. The problem must be

1. unsolved at present
2. not being worked on by someone who is likely to solve it soon

but most importantly

3. tractable, that is, probably solvable, or at least partially solvable, by 3 years' work at the Ph.D. level

It is recognised that of the enormous number of unsolved problems that have been or could be thought of, the tractable ones form a small proportion, and that it is difficult to discern which they are. The skill in non-deductive logic required of a supervisor is high. Hence the advice to Ph.D. students not to worry too much about what field or problem to choose, but to concentrate on finding a good supervisor.

It is also clear why it is hard to find Ph.D. problems that are also
4. interesting

It is not possible to dismiss these non-deductive techniques as simply "heuristic" or "pragmatic" or "subjective". Although those are correct descriptions as far as they go, they give no insight into the crucial differences among techniques, namely, that some are more reasonable and consistently more successful than others. "Successful" can mean "lucky", but "consistently successful" cannot. "If you have a lot of lucky breaks, it isn't just an accident", as Groucho Marx said (Chandler, 1999, 560). Many techniques can be heuristic, in the sense of leading to the discovery of a true result, but we are especially interested in those which give reason to believe the truth has been arrived at, and justify further research. Allocation of effort on attempted proofs may be guided by many factors, which can hence be called "pragmatic", but those which are likely to lead to a completed proof need to be distinguished from those, such as sheer stubbornness, which are not. Opinions on which approaches are likely to be fruitful in solving some problem may differ, and hence be called "subjective", but the beginning graduate student is not advised to pit his subjective opinion against the experts' without good reason. Damon Runyon's observation on horse-racing applies equally to courses of study: "The race is not always to the swift, nor the battle to the strong, but that's the way to bet" (Fadiman, 1955, 794). An example where the experts agreed on their opinion and were eventually proved right is the classification of finite simple groups, described in Sect. 2.4.

It is true that similar remarks could be made about any attempt to see rational principles at work in the evaluation of hypotheses, not just those in mathematical research. In scientific investigations, various inductive principles obviously produce results, and are not simply dismissed as pragmatic, heuristic or subjective. Yet it is common to suppose that they are not principles of *logic*, but work because of natural laws (or the principle of causality, or the regularity of nature). This option is not available in the mathematical case. Mathematics is true in all worlds, chaotic or regular. Any principles governing the relationship between hypothesis and evidence in mathematics can only be logical.

2.2   Evidence for (and Against) the Riemann Hypothesis

In modern mathematics, it is usual to cover up the processes leading to the construction of a proof, when publishing it—naturally enough, since once a result is proved, any non-conclusive evidence that existed before the proof is no longer of interest. That was not always the case. Euler, in the eighteenth century, regularly published conjectures which he could not prove, with his evidence for them. He used, for example, some daring and obviously far from rigorous methods to conclude that the infinite sum

$$1 + \frac{1}{4} + \frac{1}{9} + \frac{1}{16} + \frac{1}{25} + \cdots \tag{2.1}$$

(where the numbers on the bottom of the fractions are the successive squares of whole numbers) is equal to the *prima facie* unlikely value $\frac{\pi^2}{6}$. Finding that the two expressions agreed to seven decimal places, and that a similar method of argument led to the already proved result

$$1 - \frac{1}{3} + \frac{1}{5} - \frac{1}{7} + \frac{1}{9} - \frac{1}{11} + \cdots = \frac{\pi}{4} \tag{2.2}$$

Euler concluded, "For our method, which may appear to some as not reliable enough, a great confirmation comes here to light. Therefore, we shall not doubt at all of the other things which are derived by the same method" (Pólya, 1954, I, 18–21). He later proved the result. A translation of another of Euler's publications devoted to presenting "such evidence … as might be regarded as almost equivalent to a rigorous demonstration" of a proposition is given as a chapter in Pólya's books (1954, I, 91–98).

   Even today, mathematicians occasionally mention in print the evidence that led to a theorem. Since the introduction of computers, and even more since the recent use of symbolic manipulation software packages, it has become possible to collect large amounts of evidence for certain kinds of conjectures. (Many examples in Borwein and Bailey, 2004; Borwein et al., 2004; Müller and Neunhöffer, 1987; some comments on experimental mathematics of this kind in Epstein et al., 1992; philosophical examination in Baker, 2008). A few mathematicians argue that in some cases, it is not worth the excessive cost of achieving certainty by proof when "semi-rigorous" checking will do (Zeilberger, 1993).

   At present, it is usual to delay publication until proofs have been found. This rule is broken only in work on those long-standing conjectures of mathematics which are believed to be true but have so far resisted proof. The most notable of these, which stands since the proof of Fermat's Last Theorem as the Everest of mathematics, is the Riemann Hypothesis.

   Riemann stated in a celebrated paper of 1859 (Riemann, 1974) that he thought it "very likely" that

   All the roots of the Riemann zeta function (with certain trivial exceptions) have real part equal to $\frac{1}{2}$.

Table 2.1 Hand calculations of roots of the Riemann zeta function

Date	Worker	Number of roots found to have real part $\frac{1}{2}$
1903	Gram	15
1914	Bäcklund	79
1925	Hutchinson	138
1935/6	Titchmarsh	1,041

This is the still unproved Riemann Hypothesis. The Riemann zeta function is defined on positive whole numbers $s > 1$ by the formula

$$\zeta(s) = \frac{1}{1^s} + \frac{1}{2^s} + \frac{1}{3^s} + \ldots \tag{2.3}$$

(Thus for example $\zeta(2) = 1 + \frac{1}{4} + \frac{1}{9} + \frac{1}{16} + \ldots$, which is $\frac{\pi^2}{6}$, as mentioned above.) The definition can be extended to the entire complex plane: $\zeta(s)$ is the unique complex function, analytic except at $s = 1$, which agrees with the above formula on the positive integers greater than 1. It is found that $\zeta(s)$ has obvious ("trivial") zeros at the negative even integers. The Riemann Hypothesis is that all the (infinitely many) other zeros have real part equal to $\frac{1}{2}$. For the present purpose an understanding of complex functions is not necessary: it is only important that this is a simple universal proposition like "all ravens except Texan ones are black". It is also true that the infinitely many non-trivial roots of the Riemann zeta function have a natural order, so that one can speak of "the first million roots". (Accounts in Edwards, 1974; Derbyshire, 2003, Ch. 5; Sabbagh, 2002; du Sautoy, 2003.)

Once it became clear that the Riemann Hypothesis would be very hard to prove, it was natural to look for evidence of its truth (or falsity). The simplest kind of evidence would be ordinary induction: Calculate as many of the roots as possible and see if they all have real part $\frac{1}{2}$. This is in principle straightforward (though in practice computational mathematics is difficult, since one needs to devise subtle algorithms which save as much calculation as possible, so that the results can go as far as possible). Such numerical work was begun by Riemann and was carried on later with the results in Table 2.1.

"Broadly speaking, the computations of Gram, Bäcklund and Hutchinson contributed substantially to the plausibility of the Riemann Hypothesis, but gave no insight into the question of why it might be true" (Edwards, 1974, 97). The next investigations were able to use electronic computers; the results are shown in Table 2.2 (Brent et al., 1982; Gourdon, 2004).

It is one of the largest inductions in the world.

Besides this simple inductive evidence, there are some other reasons for believing that Riemann's Hypothesis is true (and some reasons for doubting it). In favour, there are:

1. Hardy proved in 1914 that infinitely many roots of the Riemann zeta function have real part $\frac{1}{2}$ (Edwards, 1974, 226–9). This is quite a strong consequence

Table 2.2 Computer calculations of roots of the Riemann zeta function

Date	Worker	Number of roots found to have real part $\frac{1}{2}$
1956	Lehmer	25,000
1958	Meller	35,337
1966	Lehman	250,000
1968	Rosser, Yohe and Schoenfeld	3,500,000
1979	Brent	81,000,001
1986	Te Riele, van de Lune et al.	1,500,000,001
2004	Gourdon	10^{13}

of Riemann's Hypothesis, but is not sufficient to make the Hypothesis highly probable, since if the Riemann Hypothesis is false, it would not be surprising if the exceptions to it were rare.

2. Riemann himself showed that the Hypothesis implied the "prime number theorem", then unproved. This theorem was later proved independently. This is an example of the general non-deductive principle that non-trivial consequences of a proposition support it.

3. Also in 1914, Bohr and Landau proved a theorem roughly expressible as "Almost all the roots have real part very close to $\frac{1}{2}$." More exactly, "For any $\delta > 0$, all but an infinitesimal proportion of the roots have real part within δ of $\frac{1}{2}$." This result "is to this day the strongest theorem on the location of the roots which substantiates the Riemann hypothesis" (Edwards, 1974, 193).

4. Studies in number theory revealed areas in which it was natural to consider zeta functions analogous to Riemann's zeta function. In some famous and difficult work, André Weil proved that the analogue of Riemann's Hypothesis is true for these zeta functions (Weil, 1948), and his related conjectures for an even more general class of zeta functions were proved to widespread applause in the 1970s. "It seems that they provide some of the best reasons for believing that the Riemann hypothesis is true—for believing, in other words, that there is a profound and as yet uncomprehended number-theoretic phenomenon, one facet of which is that the roots ρ all lie on Re $s = \frac{1}{2}$" (Edwards, 1974, 298).

5. Finally, there is the remarkable "Denjoy's probabilistic interpretation of the Riemann hypothesis" (Edwards, 1974, 268–269). If a coin is tossed n times, then of course we expect about $\frac{1}{2}n$ heads and $\frac{1}{2}n$ tails. But we do not expect *exactly* half of each. We can ask, then, what the average deviation from equality is. The answer, as was known by the time of Bernoulli, is $\sqrt{n}$. One exact expression of this fact is:

For any $\varepsilon > 0$, with probability one the number of heads minus the number of tails in n tosses grows less rapidly than $n^{\frac{1}{2}+\varepsilon}$.

Now we form a sequence of "heads" and "tails" by the following rule: Go along the sequence of numbers and look at their prime factors. If a number has two or more prime factors equal (i.e., is divisible by a square), do nothing. If not, its prime factors must be all different; if it has an even number of prime factors,

Table 2.3 "Head" and "tail" sequence from the factors of integers

2	3	4	5	6	7	8	9	10	11	12	13	14	15	16	17	...
2	3	2^2	5	2×3	7	2^3	3^2	2×5	11	$2^2\times3$	13	2×7	3×5	2^4	17	...
T	T		T	H	T			H		T		T	H	H		T	...

write "heads". If it has an odd number of prime factors, write "tails". Table 2.3 shows the beginning of the sequence.

The resulting sequence is of course not "random" in the sense of "probabilistic", since it is totally determined. But it is "random" in the sense of "patternless" or "erratic"; such sequences are common in number theory, and are studied by the branch of the subject called misleadingly "probabilistic number theory" (Tenenbaum, 1995). From the analogy with coin tossing, it is likely that

For any $\varepsilon > 0$, the number of heads minus the number of tails in the first n "tosses" in this sequence grows less rapidly than $n^{\frac{1}{2}+\varepsilon}$.

This statement is equivalent to Riemann's Hypothesis. Edwards comments, in his book on the Riemann zeta function,

One of the things which makes the Riemann Hypothesis so difficult is the fact that there is no plausibility argument, no hint of a reason, however unrigorous, why it should be true. This fact gives some importance to Denjoy's probabilistic interpretation of the Riemann hypothesis which, though it is quite absurd when considered carefully, gives a fleeting glimmer of plausibility to the Riemann hypothesis (Edwards, 1974, 268).

Not all of the probabilistic arguments bearing on the Riemann Hypothesis are in its favour. In the balance against, there are the following arguments:

1. Riemann's paper is only a summary of his researches, and he gives no reasons for his belief that the Hypothesis is "very likely". No reasons have been found in his unpublished papers. Edwards does give an account, however, of facts which Riemann knew, which would naturally have seemed to him evidence of the Hypothesis. But the facts in question are true only of the early roots; there are some exceptions among the later ones. Edwards concludes:

   The discoveries ... completely vitiate any argument based on the Riemann-Siegel formula and suggest that, unless some basic cause is operating which has eluded mathematicians for 110 years, occasional roots ρ off the line [i.e., with real part not $\frac{1}{2}$] are altogether possible. In short, although Riemann's insight was stupendous it was not supernatural, and what seemed "probable" to him in 1859 might seem less so today (Edwards, 1974, 166).

   This is an example of the non-deductive rule given by Pólya, "Our confidence in a conjecture can only diminish when a possible ground for the conjecture is exploded" (Pólya, 1954, II, 20).
2. Although the calculations by computer did not reveal any counterexamples to the Riemann Hypothesis, Lehmer's and later work did unexpectedly find values which it is natural to see as "near counterexamples" (Edwards, 1974, 175–9, further in Ivić, 2003). An extremely close one appeared near the 13,400,000th

root. It is partly this that prompted the calculators to persevere in their labours, since it gave reason to believe that if there were a counterexample it would probably appear soon. So far it has not, despite the distance to which computation has proceeded, so the Riemann Hypothesis is not so undermined by this consideration as appeared at first.

3. Perhaps the most serious reason for doubting the Riemann Hypothesis comes from its close connections with the prime number theorem The theorem states that the number of primes less than x is (for large x) approximately equal to the integral

$$\int_2^x \frac{dt}{\log t} \qquad (2.4)$$

If tables are drawn up for the number of primes less than x and the values of this integral, for x as far as calculations can reach, then it is always found that the number of primes less than x is actually *less* than the integral. On this evidence, it was thought for many years that this was true for all x. Nevertheless Littlewood proved that this is false. While he did not produce an actual number for which it is false, it appears that the first such number is extremely large—well beyond the range of computer calculations. Edwards comments

> In the light of these observations, the evidence for the Riemann hypothesis provided by the computations of Rosser *et al.* ... loses all its force.

That seems too strong a conclusion, since the degree of relevance of Littlewood's discovery to the Riemann Hypothesis is far from clear. But it does give some reason to suspect that there may be a very large counterexample to the Hypothesis even though there are no small ones.

It is plain, then, that there is much more to be said about the Riemann Hypothesis than, "It is neither proved nor disproved." Without non-deductive logic, though, nothing more can be said.

2.3   Goldbach's Conjecture

The situation with Goldbach's Conjecture, possibly the easiest to state of the classic unsolved problems of mathematics, is similar. Based on a letter of 1742 from Goldbach, Euler conjectured that every even number (except 2) is the sum of two primes.

The conjecture is still neither proved nor disproved and it is believed that a proof is not close. There is a simple heuristic argument that the larger the number, the more ways it can be made up of smaller numbers, so the easier it should be to write it as the sum of two primes; but there seems to be no way of converting that into a deductive argument. Computer verification for individual numbers is possible and there is a distributed computing project that has checked the Conjecture for even numbers up to and beyond 10^{18} (Wang, 2002; discussed from the point of view of experimental methods in Echeverría, 1996 and Baker, 2007).

Various consequences of it have been proved (Pólya, 1954, II, 210), and, remarkably, connections have appeared between Goldbach's Conjecture and the Riemann Hypothesis. Hardy and Littlewood proved in 1924 that a generalisation of the Riemann Hypothesis and a certain estimate implied that *most* even integers are the sum of two primes. Vinogradov in 1937 showed that every sufficiently large odd integer is the sum of three primes, and these methods were soon adapted to show Hardy and Littlewood's result without any assumptions. In 1948 Renyi found that every even number is the sum of a prime and an "almost prime" (a number with few prime factors) (Renyi, 1962). Linnik showed in 1952 that the Riemann Hypothesis itself implied a proposition relevant to Goldbach's Conjecture (Linnik, 1952). Results on the problem are still sometimes found, but there do not seem to have been dramatic advances in the last 50 years.

2.4   The Classification of Finite Groups

A last mathematical example of the central role of non-deductive inference is provided by the classification of finite simple groups, one of the great co-operative efforts of modern pure mathematics. As a case study, it has the merit that the non-deductive character of certain aspects was admitted rather explicitly by the principals. That was so because of the size of the project. Since so many people were involved, living in different continents and working over some years, it was necessary to present partial findings in print and at conferences, with explanations as to how these bore on the overall results hoped for.

Groups are one of the basic abstract entities of mathematics, having uses in describing symmetry, in classifying the kinds of curved surfaces and in many other areas. To read the following it is only necessary to know:

1. A group consists of finitely or infinitely many members; the number of members of a finite group is called its order.
2. Any group is composed, in a certain sense, of "simple" groups. ("Simple", like "group", is a technical term; "simple" groups are not in any sense uncomplicated or easy to understand but are so-called because they are not composed of smaller groups.)

A fundamental question is then: how many different finite simple groups are there? And what is the order of each? It is these questions that were attacked by the classification of finite groups project.

The project proper covered the 20 years from 1962 to 1981 inclusive. Groups had been studied in the nineteenth and early twentieth centuries, and various finite simple groups were found. It was discovered that most of them fell into a number of infinite families. These families were quite well described by the mid-1950s, with some mopping-up operations later. There were, however, five finite simple groups left over from this classification, called the Mathieu groups after their discoverer in the 1860s. Around 1960 it was not known whether any more should be expected, or, if not, how much work it might take to prove that these were the only possible simple groups.

The field was opened up by the celebrated theorem of Feit and Thompson in 1963 ("a moment in the evolution of finite group theory analogous to the emergence of fish onto dry land" Solomon, 2001). The theorem stated:

> The order of any finite simple group is an even number.

Though the result is easy to state and understand, their proof required an entire 255-page issue of the *Pacific Journal of Mathematics*. This theorem is a consequence of the full classification result (since if one knew all the finite simple groups, one could easily check that the order of each of them was even). It thus appeared that if the full classification could be found at all it would be a vast undertaking.

The final step in the answer was announced as completed in February, 1981. The full proof is spread over some 300–500 journal papers, taking up somewhere between 5,000 and 10,000 pages (Gorenstein, 1982, 1; "cleaned-up" version in Gorenstein et al., 2005). Of interest is the logical situation as the proof developed, particularly the increasing confidence—justified as it happened—that the workers in the field had in the answer long before the end was reached.

It turned out that the five Mathieu groups were not the only "sporadic" groups, as groups outside the infinite families came to be called. The first new one was discovered by Zvonimir Janko in Canberra (Janko, 1966), and excitement ran high as researchers applied many methods and discovered more. The final tally of sporadic groups stands at 26. These "discoveries" had in many cases a strong non-deductive aspect, as explained by Daniel Gorenstein of Rutgers, who became the father figure of the project and leading expert on how it was progressing:

> Another aspect of sporadic group theory makes the analogy with elementary particle theory even more apt. In a number of cases (primarily but not exclusively those in which computer calculations were ultimately required) "discover" did not include the actual construction of a group—all that was established was strong evidence for the existence of a simple group G satisfying some specified set of conditions X. The operative metamathematical group principle is this: if the investigation of an arbitrary group G having property X does not lead to a contradiction but rather to a "compatible" internal subgroup structure, then there exists an actual group with property X. In all cases, the principle has been vindicated; however, the interval between discovery and construction has varied from a few months to several years (Gorenstein, 1982, 3–4).

Michael Aschbacher, another leader of the field in the 1970s, distinguished three stages for any new group: discovery, existence and uniqueness.

> I understand a sporadic group to be discovered when a sufficient amount of self-consistent information about the group is available ... Notice that under this definition the group can be discovered before it is shown to exist ... Of course the group is said to exist when there is a proof that there exists some finite simple group satisfying P ... (Aschbacher, 1980, 6–7).

Some groups attracted more suspicion than others; for example that discovered by Richard Lyons was for some time habitually denoted Ly? and spoken of in such terms as, "If this group exists, it has the following properties . . ." (Tits, 1971, 204). Lyons entitled his original paper 'Evidence for the existence of a new finite simple group' (Lyons, 1972). A similar situation arose with another of the later groups, discovered by O'Nan. His paper, 'Some evidence for the existence of a new simple group', was devoted to finding "some properties of the new simple group G, whose existence is pointed at by the above theorems" (O'Nan, 1976, 422).

The rate of discovery of new sporadic groups slowed after 1970 and attention turned to the problem of showing that there were no more possible. At a conference at the University of Chicago in 1972 Gorenstein laid out a 16-point program for completing the classification (Gorenstein, 1979). It was thought over-optimistic at the time but immense strides were soon made by Aschbacher, Glauberman and others, more or less following Gorenstein's program.

> The turning point undoubtedly occurred at the 1976 summer conference in Duluth, Minnesota. The theorems presented there were so strong that the audience was unable to avoid the conclusion that the full classification could not be far off. From that point on, the practicing finite group theorists became increasingly convinced that the "end was near"—at first within five years, then within two years, and finally momentarily. Residual skepticism was confined largely to the general mathematical community, which quite reasonably would not accept at face value the assertion that the classification theorem was "almost proved" (Gorenstein, 1982, 5–6).

Notice that "almost proved" indeed does not mean anything in deductive logic. With hindsight, one can say that a theorem was almost proved when most of the steps in the proof were found; but before a proof is complete, there can only be good non-deductive reason to believe that a sequence of existing steps will constitute most of a future proof.

By the time of the conference at Durham, England in 1978 (described in its Proceedings as on "the classification of simple groups, a programme which is now almost complete") optimism ran even higher. At that stage existence and uniqueness had been proved for 24 of the sporadic groups, leaving two "for which considerable evidence exists" (Collins, 1980, 21). One of these was successfully dealt with in 1980 ("four years after Janko's initial evidence for such a sporadic group" Gorenstein, 1982, 110), and attention focussed on the last one, known as the "Monster" because of its immense size (order about 10^{54}).

> That the search for sporadic groups was not totally haphazard can be seen from the remarkable simultaneous realization by Fischer in West Germany and Griess in the United States in 1974 that there might be a simple group having a covering group . . . (Gorenstein, 1982, 92).

Consequences of the existence of this group were then studied:

> Soon after the initial "discovery", Griess, Conway and Norton noticed that every nontrivial irreducible character of a group G of type F, has degree at least 196,883 and very likely such a group G must have a character of this exact degree. Indeed, on this assumption, Fischer, Livingstone and Thorne eventually computed the full character table of such a group G (Gorenstein, 1982, 126–7).

Aschbacher, lecturing at Yale in 1978, said:

> When the Monster was discovered it was observed that, if the group existed, it must contain
> two new sporadic groups (the groups denoted by F_3 and F_5 in Table 2.2) whose existence
> had not been suspected up to that time. That is, these groups were discovered as subgroups
> of the Monster. Since that time the groups F_3 and F_5 have been shown to exist. This is
> analogous to the situation in the physical sciences where a theory is constructed which
> predicts certain physical phenomena that are later verified experimentally. Such verification
> is usually interpreted as evidence that the theory is correct. In this case, I take the existence
> of F_3 and F_5 to be very good evidence that the Monster exists ... My belief is that there are
> at most a few groups yet to be discovered. If I were to bet, I would say no more (Aschbacher,
> 1980, 13–15).

Gorenstein's survey article of 1978 contains perhaps the experts' last sop to
deductivism, the thesis that all logic is deductive. He wrote:

> At the present time the determination of all finite simple groups is very nearly complete.
> Such an assertion is obviously presumptuous, if not meaningless, since one does not speak
> of theorems as "almost proved" (Gorenstein, 1979, 50–51).

To the deductivist, the fact that most steps in a proposed proof are completed is no
reason to believe that the rest will be. Undeterred, however, Gorenstein went on to
say:

> The complete proof, when it is obtained, will run to well over 5,000 journal pages!
> Moreover, it is likely that at the present time more than 80% of those pages exist ... The
> assertion that the classification is nearly complete is really a prediction that the presently
> available techniques will be sufficient to deal with the problems still outstanding. In its
> support, we cite the fact that, with two exceptions, all open questions are open because no
> one has yet examined them and not because they involve some intrinsic difficulty.

A year after the Durham conference, the experts assembled again at Santa Cruz,
California, in a mood of supreme confidence. Gorenstein's survey opened with the
remark:

> My aim here is to present a brief outline of the classification of the finite simple groups,
> now rapidly nearing completion (Gorenstein, 1980, 3).

Another contributor to the conference began his talk:

> Now that the problem of classifying finite simple groups is probably close to completion ...
> (Hunt, 1980).

What concern remained was less about the completion of the project than about
what to do next; the editor of the conference proceedings began by commenting, "In
the last year or so there have been widespread rumors that group theory is finished,
that there is nothing more to be done" (Mason, 1980, xii). The *New York Times Week
in Review* (June 22, 1980) headlined an article 'A School of Theorists Works Itself
Out of a Job.'

The confidence proved justified. Griess was able to show the existence of the
Monster, and finally, in 1981, Simon Norton of Cambridge University completed
the proof of the uniqueness of the Monster (Gorenstein, 1982, 1).

At least, that was claimed at the time. In the late 1980s it was discovered that a part of the proof, on "quasithin" groups, was not quite as complete as had been thought. One gap proved hard to fill in, but was completed by Aschbacher and others in 2001 (Aschbacher, 2001).

2.5 Probabilistic Relations Between Necessary Truths?

The most natural conceptualization of the non-deductive relations between evidence and conclusion is that of objective Bayesianism. The (objective) Bayesian theory of evidence (also known as the logical theory of probability) aims to explain what the nature of evidence is. It holds that the relation of evidence to conclusion is a matter of strict logic, like the relation of axioms to theorems in mathematics but less conclusive—a kind of partial implication. Given a fixed body of evidence—say in a trial, or in a dispute about a scientific theory—and given a conclusion, there is a fixed degree to which the evidence supports the conclusion. It was defended in Keynes's *Treatise on Probability* (Keynes, 1921) and more recently by E. T. Jaynes (Jaynes, 2003; a slightly less objective version in Williamson, 2010; introductions in Franklin, 2001; Franklin, 2009, Ch. 10). It says, for example, that if we could establish just what the legal standard of "proof beyond reasonable doubt" is, then, in a given trial, it is an objective matter of logical fact whether the evidence presented does or does not meet that standard, and so a jury is either right or wrong in its verdict on the evidence.

It is not essential to the Bayesian perspective that the relation of evidence to conclusion should be given a precise number, nor that it be possible to compute the logical relation between evidence and conclusion in typical cases. It is sufficient for objective Bayesianism that it is sometimes intuitively evident that some hypotheses, on some bodies of evidence, are highly likely, or almost certain, or virtually impossible (Franklin, 2011). Keynes certainly believed that it was not always possible even in principle to compute an exact number expressing the relation between an arbitrary body of evidence and a conclusion. Nevertheless, it is usual as an idealization to suppose that for any body of evidence e and any conclusion h, there is a number $P(h|e)$, between 0 and 1, expressing the degree to which e supports h; and that that number satisfies the usual axioms of conditional probability:

$P(\text{not-}h|e) = 1 - P(h|e)$
$P(h_1 \text{ and } h_2|e) = P(h_1|e) \times P(h_2|h_1 \text{ and } e)$

Pólya's qualitative principles of evidence, such as the confirmation of hypotheses by their non-trivial consequences, are then easy deductions from those axioms.

The logical nature of the relation makes it particularly suitable for application to the necessary subject matter of pure mathematics. Conversely, its intuitive agreement with actual evaluation of conjectures supports it as a possible meaningful interpretation of probability (not necessarily the only valid one, as stochastic outcomes or idealized degrees of belief or idealized relative frequencies may also turn out to satisfy the same axioms.)

There is one point that needs to be made precise especially in applying the theory of logical probability or non-deductive logic in *mathematics*. If evidence *e* entails hypothesis *h*, then P(*h*|*e*) is 1. But in mathematics, the typical case is that *e does* entail *h*, though that is perhaps as yet unknown. If, however, P(*h*|*e*) is really 1, how is it possible in the meantime to discuss the (non-deductive) support that *e* may give to *h*, that is, to treat P(*h*|*e*) as not equal to 1? In other words, if *h* and *e* are necessarily true or false, how can P(*h*|*e*) be other than 0 or 1? How can there be probabilistic relations between necessary truths?

The answer is that, in both deductive and non-deductive logic, there can be *many* logical relations between two propositions. Some may be known and some not. To take an artificially simple example in deductive logic, consider the argument

If all men are mortal, then this man is mortal
All men are mortal
_____
Therefore, this man is mortal

The premises entail the conclusion, certainly, but there is more to it than that. They entail the conclusion in two ways: firstly, by *modus ponens*, and secondly by instantiation from the second premise alone. That is, there are two logical paths from the premises to the conclusion.

More complicated and realistic cases are common in the mathematical literature. Feit and Thompson's proof that all finite simple groups have even order, occupying 255 pages, was simplified by Bender (1970). That means that Bender found a different and shorter logical route from the definition of "finite simple group" to the proposition, "All finite simple groups have even order" than the one known to Feit and Thompson.

Now just as there can be two deductive paths between premises and conclusion, so there can be a deductive and non-deductive path, and it may be that only the latter is known. Before the Greeks' development of deductive geometry, it was possible to argue

All equilateral (plane) triangles so far measured have been found to be equiangular
This triangle is equilateral
_____
Therefore, this triangle is equiangular

There is a non-deductive logical relation between the premises and the conclusion; the premises inductively support the conclusion. But when deductive geometry appeared, it was found that there was also a deductive relation, since the second premise alone entails the conclusion. This discovery in no way vitiates the correctness of the previous non-deductive reasoning or casts doubt on the existence of the non-deductive relation. That relation cannot be affected by discoveries about any other relation.

So the answer to the question, "How can there be probabilistic relations between necessary truths?" is simply that those relations are additional to any deductive relations (and may be known independently of them).

2.6 The Problem of Induction in Mathematics

That non-deductive logic is used in mathematics is important first of all to mathematics. But there is also some wider significance for philosophy, in relation to the problem of induction, or inference from the observed to the unobserved.

It is common to discuss induction using only examples from the natural world, such as, "All observed flames have been hot, so the next flame observed will be hot" and "All observed ravens have been black, so all ravens are black". That has encouraged the view that the problem of induction should be solved in terms of natural laws (or causes, or dispositions, or the regularity of nature, or some other contingent principle) which provide a kind of "cement of the universe" to bind the observed to the unobserved.

The difficulty for such a view is that it does not apply to mathematics, which deals in necessary matter. Yet induction works just as well in mathematics as in natural science.

Examples were given above in the second section in connection with the calculation of roots for the Riemann Hypothesis, but let us take a particularly straightforward case:

The first million digits of π are random

Therefore, the second million digits of π are random

("Random" here means "without pattern", "passes statistical tests for randomness", not "probabilistically generated", "stochastic": Ruhkin, 2001; Franklin, 2009, 162–3.) The number π has the decimal expansion

$$3.14159265358979323846264338327950288419716939937\ldots$$

There is no apparent pattern in these numbers. The first million digits have long been calculated (calculations have reached beyond one trillion). Inspection of these digits reveals no pattern, and computer calculations can confirm this impression. It can then be argued inductively that the second million digits will likewise exhibit no pattern. This induction is a good one (indeed, everyone believes that the digits of π continue to be random indefinitely, though there is no proof, Marsaglia, 2005).

It is true, as argued by Baker (2007), that there is a special problem with inductive arguments in mathematics in that all the observed cases are of small numbers. Any number that can be calculated with is very small, compared to numbers in general. That bias in the evidence could raise a question as to whether any induction of the form "All observed numbers have property X, therefore all numbers have property X" could have high probability. That does not imply, however, that inductive arguments in mathematics are generally poor. Firstly, a bias in the evidence towards small numbers does not affect inductive arguments with more modest conclusions, such as "All observed numbers have property X, so the next number calculated will have property X." (The argument above about the randomness of the digits of π only extrapolated a finite distance, keeping to small numbers.) Secondly, many other inductive arguments have a bias in the evidence, without thereby becoming worthless (though they may become less secure). For example, extrapolative inductive inference like "All observed European swans are white, therefore all swans are white" is a worthwhile inductive argument, although the extrapolation beyond the observed range weakens it.

Now there seems to be no reason to distinguish the logic involved in such mathematical arguments from that used in inductions about flames or ravens. But the digits of π are the same in all possible worlds, whatever natural laws may hold in them or fail to. Any reasoning about π is also rational or otherwise, regardless of any empirical facts about natural laws. Therefore, induction can be rational independently of whether there are natural laws (or any other such contingent principle).

This argument does not show that natural laws have no place in discussing induction. It may be that mathematical examples of induction are rational because there are *mathematical* laws or regularities, and that the aim in natural science is to find some substitute, such as natural laws, which will take the place of mathematical laws in accounting for the continuance of regularity. But if this line of reasoning is pursued, it is clear that simply making the supposition, "There are laws", is of little help in making inductive inferences. No doubt mathematics is completely lawlike, but that does not help at all in deciding whether the digits of π continue to be random. In the absence of any proofs, induction is needed to support the law (if it is a law), "The digits of π are random", rather than the law being able to give support to the induction. Either "The digits of π are random" or "The digits of π are not random" is a law, but in the absence of knowledge as to which, we are left only with the confirmation that the evidence gives to the first of these hypotheses. Thus consideration of a mathematical example reveals what can be lost sight of in the search for laws: laws or no laws, non-deductive logic is needed to make inductive inferences.

It is worth noting that there are also mathematical analogues of Goodman's "grue" paradox. Let a number be called "prue" if its decimal expansion is random for the first million digits and 6 thereafter. The predicate "prue" is like "grue" in not being projectible. "π is random for the first million digits" is logically equivalent to "π is prue for the first million digits", but this proposition supports "π is random always", not "π is prue". Any solutions to the "grue" paradox must allow projectible or "natural" properties to be found not only in nature but also in mathematics.

These examples illustrate Pólya's remark that non-deductive logic is better appreciated in mathematics than in the natural sciences (Pólya, 1954, II, 24). In mathematics there can be no confusion over natural laws, the regularity of nature, approximations, propensities, the theory-ladenness of observation, pragmatics, scientific revolutions, the social relations of science or any other red herrings. There are only the hypothesis, the evidence and the logical relations between them.

References

Aschbacher, M. (1980). *The finite simple groups and their classification*. New Haven, CT: Yale University Press.

Aschbacher, M. (2001). The status of the classification of the finite simple groups. *Notices of the American Mathematical Society, 51,* 736–740.

Baker, A. (2007). Is there a problem of induction for mathematics? In M. Leng, A. Paseau, & M. Potter (Eds.), *Mathematical knowledge* (pp. 59–73). Oxford: Oxford University Press.

Baker, A. (2008). Experimental mathematics. *Erkenntnis, 68,* 331–344.

Baker, A. (2009). Non-deductive methods in mathematics. *Stanford Encyclopedia of Philosophy.* Accessed May 2013. http://plato.stanford.edu/entries/mathematics-nondeductive/

Bender, H. (1970). On the uniqueness theorem. *Illinois Journal of Mathematics, 14,* 376–384.

Borwein, J. M., & Bailey, D. H. (2004). *Mathematics by experiment: Plausible reasoning in the 21st century*. Natick, MA: A. K. Peters.

Borwein, J. M., Bailey, D. H., & Girgensohn, R. (2004). *Experimentation in mathematics: Computational paths to discovery*. Natick, MA: A. K. Peters.

Brent, R., van de Lune, J., te Riele, H., & Winter, D. (1982). On the zeros of the Riemann Zeta Function in the critical strip. II. *Mathematics of Computation, 39,* 681–688.

Brown, J. R. (1999). *Philosophy of mathematics: An introduction to the world of proofs and pictures*. London: Routledge.

Chandler, C. (1999). *Hello, I must be going: Groucho Marx and his friends*. Garden City, NY: Doubleday.

Collins, M. J. (1980). *Finite simple groups, II*. London: Academic.

Corfield, D. (2003). *Towards a philosophy of real mathematics*. Cambridge: Cambridge University Press.

Derbyshire, J. (2003). *Prime obsession: Bernhard Riemann and the greatest unsolved problem in mathematics*. Washington, DC: Joseph Henry Press.

Dove, I. J. (2009). Towards a theory of mathematical argument. *Foundations of Science, 14*(1–2), 137–152.

du Sautoy, M. (2003). *The music of the primes: Searching to solve the greatest mystery in mathematics*. New York: Harper Collins.

Echeverría, J. (1996). Empirical methods in mathematics. In G. Munévar (Ed.), *Spanish studies in the philosophy of science, volume 86 of Boston studies in the philosophy of science* (pp. 19–55). Dordrecht: Kluwer.

Edwards, H. M. (1974). *Riemann's zeta function*. New York: Academic.

Epstein, D., Levy, S., & de la Llave, R. (1992). About this journal. *Experimental Mathematics, 1,* 1–13.

Fadiman, C. (1955). *The American treasury*. New York: Harper.

Fallis, D. (1997). The epistemic status of probabilistic proof. *Journal of Philosophy, 94,* 165–186.

Fallis, D. (2000). The reliability of randomized algorithms. *British Journal for the Philosophy of Science, 51,* 255–271.

Feit, W., & Thompson, J. G. (1963). Solvability of groups of odd order. *Pacific Journal of Mathematics, 13,* 775–1029.

Franklin, J. (1987). Non-deductive logic in mathematics. *British Journal for the Philosophy of Science, 38*(1), 1–18.

Franklin, J. (2001). Resurrecting logical probability. *Erkenntnis, 55,* 277–305.

Franklin, J. (2009). *What science knows and how it knows it*. New York: Encounter Books.

Franklin, J. (2011). The objective Bayesian conceptualisation of proof and reference class problems. *Sydney Law Review, 33,* 545–561.

Gorenstein, D. (1979). The classification of finite simple groups (I). *Bulletin of the American Mathematical Society, 1,* 43–199 (New Series).

Gorenstein, D. (1980). An outline of the classification of finite simple groups. In B. Cooperstein
& G. Mason (Eds.), *The Santa Cruz conference on finite groups, volume 37 of proceedings of
symposia in pure mathematics* (pp. 3–28). Providence, RI: American Mathematical Society.

Gorenstein, D. (1982). *Finite simple groups*. New York: Plenum.

Gorenstein, D., Lyons, R., & Solomon, R. (1994–2005). *The classification of the finite simple
groups* (6 Vols.). Providence, RI: American Mathematical Society.

Gourdon, X. (2004). The 10^{13} first zeros of the Riemann Zeta Function, and zeros computation
at very large height. Accessed May 2013. http://numbers.computation.free.fr/Constants/
Miscellaneous/zetazeros1e13-1e24.pdf

Hunt, D. (1980). A computer-based atlas of finite simple groups. In B. Cooperstein & G. Mason
(Eds.), *The Santa Cruz conference on finite groups, volume 37 of proceedings of symposia in
pure mathematics* (pp. 507–510). Providence, RI: American Mathematical Society.

Ivić, A. (2003). On some reasons for doubting the Riemann Hypothesis. Accessed May 2013.
http://arxiv.org/abs/math/0311162

Janko, Z. (1966). A new finite simple group with abelian 2-Sylow subgroups and its characterization. *Journal of Algebra, 3,* 147–186.

Jaynes, E. T. (2003). *Probability theory: The logic of science.* Cambridge: Cambridge University
Press.

Keynes, J. M. (1921). *A treatise on probability.* London: Macmillan.

Kolata, G. B. (1976). Mathematical proofs: The genesis of reasonable doubt. *Science, 192,*
989–990.

Lehrer Dive, L. (2003). *An epistemic structuralist account of mathematical knowledge.* PhD thesis,
University of Sydney.

Linnik, Y. V. (1952). Some conditional theorems concerning the binary Goldbach problem.
Izvestiya Akademii Nauk SSSR, 16, 503–520.

Lyons, R. (1972). Evidence for a new finite simple group. *Journal of Algebra, 20,* 540–569.

Marsaglia, G. (2005). On the randomness of pi and other decimal expansions. Accessed May 2013.
http://www.yaroslavvb.com/papers/marsaglia-on.pdf

Mason, G. (1980). Preface. In B. Cooperstein & G. Mason (Eds.), *The Santa Cruz conference on
finite groups, volume 37 of proceedings of symposia in pure mathematics* (p. xiii). Providence,
RI: American Mathematical Society.

Müller, J., & Neunhöffer, M. (1987). Some computations regarding Foulkes' conjecture.
Experimental Mathematics, 14, 277–283.

O'Nan, M. (1976). Some evidence for the existence of a new finite simple group. *Proceedings of
the London Mathematical Society, 32,* 421–479.

Pólya, G. (1954). *Mathematics and plausible reasoning* (2 Vols.). Princeton, NJ: Princeton
University Press.

Renyi, A. (1962). On the representation of an even number as the sum of a prime and an almost
prime. *American Mathematical Society Translations, 2nd series, 19,* 299–321.

Riemann, B. (1859 [1974]). On the number of primes less than a given magnitude. In H. Edwards
(Ed.), *Riemann's zeta function* (pp. 299–305). New York: Academic.

Ruhkin, A. (2001). Testing randomness: A suite of statistical procedures. *Theory of Probability
and its Applications, 45,* 111–132.

Sabbagh, K. (2002). *Dr Riemann's zeros.* London: Atlantic Books.

Solomon, R. (2001). A brief history of the classification of the finite simple groups. *Bulletin of the
American Mathematical Society, 38,* 315–352.

Tenenbaum, G. (1995). *Introduction to analytic and probabilistic number theory.* Cambridge:
Cambridge University Press.

Tits, J. (1971). Groupes finis simples sporadiques. In A. Dold & B. Eckmann (Eds.), *Séminaire
Bourbaki, volume 180 of Springer Lecture Notes in Mathematics* (pp. 187–211). New York:
Springer.

Van Kerkhove, B., & Van Bendegem, J. P. (2008). Pi on earth, or mathematics in the real world.
Erkenntnis, 68, 421–435.

Wang, Y. (Ed.). (2002). *Goldbach conjecture*. River Edge, NJ: World Scientific.
Weil, A. (1948). *Variétés abéliennes et courbes algébriques*. Paris: Hermann.
Williamson, J. (2010). *In defence of objective Bayesianism*. Oxford: Oxford University Press.
Zeilberger, D. (1993). Theorems for a price: Tomorrow's semi-rigorous mathematical culture. *Notices of the American Mathematical Society, 46*, 978–981.

Chapter 3
Arguments, Proofs, and Dialogues

Erik C.W. Krabbe

To what extent do proofs fall within the scope of a theory of argumentation? In this chapter I shall try to provide an answer. To this end, several types of proof need to be distinguished. Proofs of most types will be seen to be arguments, and therefore amenable to analysis from the point of view of argumentation studies. The last section presents a dialectical view of proof as an argument in dialogue that meets certain supplementary conditions. These conditions can, however, be formulated in dialectical terms.

3.1 Proof and Argument

What is a proof? A set of solid reasons and lucid inferences that clinch the argument? Or do we need more, before we are prepared to accept the credentials of a supposed proof? Are proofs within the range of legitimate subjects for a theory of argumentation? Or must the territory be left to formal logicians?

The Latin word *argumentum* has 'proof' as one of its meanings; *argumentatio* means 'argumentation' or '(the furnishing of) proof;' the correlative verbs are *argumentor* and *arguo*: I argue, I prove. Does this mean that, fundamentally, arguing and proving are one and the same? If so, one may be puzzled by the existence of a branch of logic called 'proof theory' (or metamathematics). Obviously, proof theory and theory of argumentation are quite different disciplines.[1]

---

[1] To see what theory of argumentation is about, one may consult Barth and Martens (1982), and Van Eemeren et al. (1987, 1996); for proof theory see Prawitz (1981). The word 'argument,' in this

E.C.W. Krabbe (✉)
Faculteit Wijsbegeerte, University of Groningen, Oude Boteringestraat 52, 9712 GL Groningen,
The Netherlands
e-mail: e.c.w.krabbe@rug.nl

A. Aberdein and I.J. Dove (eds.), *The Argument of Mathematics*, Logic, Epistemology,
and the Unity of Science 30, DOI 10.1007/978-94-007-6534-4_3,
© Springer Science+Business Media Dordrecht 2013

Webster's Ninth New Collegiate Dictionary renders the relevant meaning of 'to prove' as follows:

3 a: to establish the existence, truth, or validity of (as by evidence or logic)

This covers proofs in mathematics as well as proofs in science and proofs in court. There is, indeed, no reason to postulate three radically different meanings for these cases. On the relevant meanings of 'arguing' the dictionary instructs us as follows:

vi [...] **1**: to give reasons for or against something: REASON [...] *vt* [...] **2**: to consider the pros and cons of: DISCUSS **3**: to prove or try to prove by giving reasons: MAINTAIN **4**: to persuade by giving reasons: INDUCE

Thus the domains of application of the terms 'to prove' and 'to argue' overlap: whenever someone tries, by giving reasons, to establish the existence, truth, or validity of something and *succeeds* in doing so, both terms apply: what she has been doing was not just arguing a case, but proving her conclusion as well. But if she only *tried* to establish the existence, truth, or validity, but without success, she has been arguing all the same, but she has not been proving anything. Hence it is quite possible to argue without proving. Can one also prove without arguing?

A proof is to establish the correctness of a proposition, to justify a point of view or claim. But we are not told that argument is the only way to achieve this end. Suppose I claim to be capable of singing a song. You want a proof. I could start an argument about my reputation as a singer, but the easiest way to justify my claim would be just to burst into song. Similarly, a claim to the effect that it is snowing could be established by drawing back the curtains. If you utter 'I can pronounce an English sentence' you have laid down your claim and handed us a proof of it at one and the same time. In all these cases one justifies a claim without offering an argument. One is proving something, but not arguing for it. Such proofs are intuitively clear, and they are so immediately. We shall use the term *Immediate and Intuitive Proof* for this type of case.

The term 'proof' displays a process-product ambiguity (as does the term 'argument'). Sometimes the *process* of establishing the correctness of a proposition is meant, but at other times it is the *product* (often a text) that is meant. The same holds for 'immediate and intuitive proof,' though, in some cases, it may sound a bit peculiar when the term 'proof' is applied to the product, which could be a song, or a figure, or a gesture, or whatever.

3.2   Mathematical Proof

Among mathematical proofs the *Immediate and Intuitive* ones are at one extreme, whereas the *Formal Proofs* are at the other. In between one finds informal proofs

---

paper, is not used in the technical, logical, sense of a premises-conclusion constellation, but refers to verbal and social means (especially the presentation of reasoning) to convince an addressee that a certain claim is justified. Cf. Walton (1990, esp. 411).

Fig. 3.1 The altitudes of
a triangle *ABC* intersect in
a single point *O*

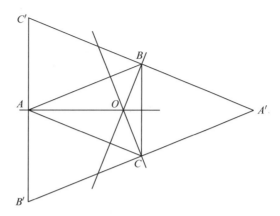

in which arguments are brought into play, possibly within the context of an axiom system. The question is whether these informal proofs would be best regarded as special cases of arguing and argument, or as specimens of a completely different type of process (product). The following case study may clarify this issue:

Theorem 1. *The altitudes of a triangle intersect in a single point.*

As you may remember, each triangle has three altitudes. If *A*, *B*, and *C* are the vertices of a triangle, then the altitude from *A* is the line through *A* that is perpendicular to the side *BC*, etc. We take it to be self-evident that any *two* altitudes (e.g., the altitude from *A* and that from *B*) intersect in a single point (say *O*). The problem is to prove that the third altitude (that from *C*) passes through *O* as well.

A well-known proof of this theorem presents (or has the student draw) a figure (Fig. 3.1).

For some this figure, by itself, may constitute an immediate and intuitive proof. There is no fundamental difference, as far as the method of proof is concerned, between showing some figure (to prove a geometrical theorem) and drawing back the curtains in order to prove that it is snowing outside. But many will not be satisfied by Fig. 3.1 and demand an accompanying argument. Here is one (it is called a 'proof'):

Proof. Consider an arbitrary triangle *ABC*. Draw a line through *A* parallel to *BC*, one through *B* parallel to *AC*, and one through *C* parallel to *AB*. We have constructed a circumscribing triangle *A'B'C'*. It is not hard to see that the altitudes of the inner triangle *ABC* happen to be the perpendicular bisectors of the sides of the outer triangle *A'B'C'* (this is left as an exercise). But, as we all know, the perpendicular bisectors of the sides of any triangle will intersect in a single point which is the center of the circumscribed circle of the triangle. Therefore, the altitudes of *ABC* will intersect in a single point, which is the center of the circumscribed circle of *A'B'C'*. Since *ABC* was chosen arbitrarily, the same result holds for any triangle: its altitudes will intersect in a single point (which may henceforth be called the *orthocenter* of the triangle), QED.

   This illustrates what is meant by *Informal Proof*.[2] Among the characteristics of informal proof are the following: (i) it is an attempt to establish a conclusion by argument, (ii) which is addressed to an addressee who does not yet subscribe to the theorem, but who is willing to let himself be convinced of its truth, (iii) this addressee is assumed to be acquainted with a certain body of knowledge, reference to which is made in the proof (for instance, in the proof given above, the addressee was assumed to be familiar with the fact that the perpendicular bisectors of the sides of a triangle intersect in a single point), (iv) some parts of the proof are left as an exercise to the addressee, (v) some propositions are supposed to be immediately and intuitively obvious (for instance that there is a line through A parallel to BC), and (vi) some procedures of proof are presupposed as well (for instance, how to handle 'arbitrary objects' in order to obtain general conclusions). An informal proof, therefore, is just an argument of sorts, and hence an appropriate object of study for the theory of argumentation. If we take a dialectical view of argument in general, that is, if we look upon all argument as explicit or implicit critical discussion, then informal proofs are no exceptions: a critical discussion is (usually implicitly) contained in them.

   An argument may or may not hold water. Even if it does not, it is an argument all the same. But 'proof' has another type of grammar. Ordinary usage suggests that shaky or fallacious arguments do not count as proofs. It is probably wise to follow ordinary usage in this respect as we are about to recommend some philosophical terminology. This means that we must differ from other, perhaps marginal, phenomena of usage, such as talk of proofs being 'wrong' or 'right.' On closer analysis, a 'wrong proof,' i.e., a 'proof' containing a fallacy or error, is no proof at all, just as a forged Vermeer is not a specific kind of Vermeer.

   This does not at all make it easier to apply the term '(informal) proof' correctly. Suppose we are given an alleged proof. It is an argument for sure, but we cannot tell whether the argument is a proof without knowing whether the argument holds water. However, with arguments, this cannot be seen from the text: we have to study the context. Informal mathematical proof is no exception. What counts as a proof for one person may not count as proof for someone else. The argument given above for the existence of an orthocenter with each triangle can only take effect as a proof for you if you are familiar with certain presupposed facts and methods: you have to be acquainted with perpendicular bisectors and with the theorem that they intersect in a single point. Moreover the homework exercises must be within your reach. And you must be able to understand the use of such phrases as 'consider an arbitrary triangle ABC,' etc.[3]

---

[2] The term *Informal Proof* is here used in a narrow sense, excluding those informal proofs (in a broad sense) that can be classified either as immediate and intuitive or as informal axiomatic proofs (see below). A more explicit, but cumbersome, name for this type of proof would be: *Informal Argumentative Nonaxiomatic Proof.*

[3] Cf. Corcoran: 'A linear chain of reasoning that is cogent for one person need not, and normally will not, be cogent for all other persons.' (1989, 34).

Informal proofs are just arguments. Even so, their authors do not usually announce them as 'arguments,' but as 'proofs.' Why? One reason could be that the author is sincerely convinced that his argument is impeccable, that it should satisfy any member of his intended audience, does not admit of rational objections or reservations, and perhaps fulfills a number of supplementary conditions (see Sect. 3.4). By calling his argument a 'proof,' the author underlines these matters and reminds his audience of the supplementary conditions that an informal mathematical proof should satisfy. In itself there is nothing objectionable to this use of the word 'proof.' But, of course, calling an argument 'sound,' 'correct,' or 'conclusive' does not provide the argument with these meritorious qualities. The self-praise implied in the announcement of an argument as a 'proof' is liable to induce one to overstep the boundaries of sober argument into the realm of propaganda and intimidation. Another message is tagged to the announcement: 'Don't try to find any objections to this, for my argument isn't just an argument, but a proof, meaning a clincher, and if you don't agree, that only goes to show that you failed to grasp the whole thing...' The use of the term 'proof' for these ends is fallacious, it is an *argumentum ad verecundiam*.

Van Eemeren and Grootendorst (1987) reserve the term 'fallacy' for speech acts which hinder in any way the process of conflict resolution in a critical discussion (284). A fallacy violates a rule that has to be observed in a critical discussion. Van Eemeren and Grootendorst (1987) present ten such rules and it is not hard to see which rule is violated by an intimidating use of the term 'proof' to describe one's own argument. The rule formulates what may be called *The Principle of Burden of Proof*:

> *Rule II*: Whoever advances a standpoint is obliged to defend it if asked to do so (285).

This principle applies, not only to the theorem to be proved, but also to the data adduced in the proof and to the methods of proof. For all of these, there is a burden of proof as soon as there is a challenge. Primarily, the principle refers to situations that are explicitly dialogical, for instance, situations in which certain specific parts of an alleged proof are explicitly criticized. But the challenge can also remain implicit. In that case there may, nevertheless, be an identifiable burden of proof. Thus, in a situation where it is reasonable to suppose that a number of addressees will be unable to follow a published argument as it stands, there is, one may say, an implicit challenge: the mathematician is asked (implicitly) to back up her argument. Calling her argument a 'proof,' however, she evades the burden. The move owes its efficacy to the diffidence (*verecundia*) felt when confronted with 'proof' (*argumentum ad probandi verecundiam*).

A strong form of the fallacy would even result in a threat, and thus violate the following dialectical rule formulated by Van Eemeren and Grootendorst (to be called *The Principle of Parrhesia*):

> *Rule I*: Parties must not prevent each other from advancing or casting doubt on standpoints (284).

For instance, our mathematician could claim that she has a full proof to fall back on, safely stored in her desk, implying that any doubt would be futile, and that those of her critics that persist in casting doubt on parts of the proof risk a severe loss of prestige in the near future (*argumentum ad baculum*).[4]

The efficacy of the high-sounding word 'proof' in an *argumentum ad verecundiam* is enhanced by the existence of some rather special, but prestigious, meanings of this term. One of these is the concept of *axiomatic proof* in mathematics. Primarily, an axiomatic proof is a proof within an axiom system, such as (a specified axiomatization of) Euclid's geometry. Working within the confines of a specific axiom system, one has no need to prove the axioms. (Nor is one to prove the system's definitions.) Dependent upon one's epistemology, the axioms are viewed as self-evident, as a matter of choice or convention, or as principles that can be justified from outside the system. Starting from the axioms (and perhaps some basic definitions) the mathematician proves, within the system, one theorem after another. All along, new terms are introduced by definition. In this process, the order of proofs and definitions is crucial. Each proof may fall back only upon axioms, on definitions that were introduced before, and on theorems that were proved before. Definitions may depend upon theorems, but these theorems must precede them.

Proofs within an axiom system often strike one as highly technical, for obviously they utilize much symbolism and quite subtle methods of deduction. Nevertheless these proofs remain *informal*, as long as they are expressed in a language that was never formalized. We shall call such proofs *Informal Axiomatic Proofs*. An informal axiomatic proof is an argument directed at an audience that accepts the axioms and has 'gone through' all the proofs of earlier theorems and the definitions used in them. Such proofs are arguments and are therefore suitable objects of study for a theory of argumentation. The special context in which these arguments are proffered, moreover, makes them especially interesting from an argumentative point of view. There are fallacies and brands of criticism that are peculiar to this context, such as certain forms of the *circulus vitiosus in probando/in definiendo*, the criticism of (perhaps benign) loops in proofs or definitions, and the criticism of inelegance.

3.3   Formal Proof

Formal Proofs are quite different from any kind of informal proof, however technical, in that they presuppose a formalized language. Stipulations that define a formalized language must precede formal proofs formulated within that language. A formalization of the theorem on altitudes and its proof requires a previous specification of a formalized language for geometrical thought. If you wonder about

---

[4]Another relevant rule would be *The Principle of Pertinence*: '*Rule IV*: A standpoint may be defended only by advancing argumentation relating to that standpoint' (Van Eemeren and Grootendorst, 1987, 286). But this rule applies to the argumentation stage and therefore assumes that our mathematician has already acknowledged the existence of critical doubt and agreed to accept a burden of proof. She could then use another type of *ad verecundiam* to try to discharge this burden, e.g. by reference to her expertise in proof construction. This involves more than merely a claim to have a proof.

how to conceive of such a language, imagine something similar to a language for predicate logic with identity in which there are some fixed predicate letters assigned to the 'primitive notions of geometry' (Px: x is a point; Lx: x is a line; Ixy: x is on y, etc.). Alphabet and syntax of the language must be precisely specified. The sentences of such a language are often called *formulas*. *Axioms*, too, are formulas. *Formal proofs* are sequences or tree diagrams of formulas constructed according to syntactically specified rules of derivation such as the well-known *Rule of Modus Ponens* (Rule of Detachment or Arrow-Elimination Rule). Axioms and rules of derivation taken together define a *formal system*. A formula is a *theorem* of a formal system if and only if there is a formal proof for it within the system. To check whether an alleged proof really is a proof, one does not need to know anything about geometry. Nor does one have to be schooled in the theory of argumentation. A purely syntactical check suffices.

Why have formal proofs? For what purpose? Gottlob Frege (1879) certainly had a use for them. He needed them in his attempt to establish that all of mathematics would be derivable from logical principles (logicism). For in order to show this, he had to derive the (completely evident) principles of arithmetic from (equally evident) logical principles, without falling back, inadvertently, upon the use of arithmetic itself. Therefore it was not sufficient to go by intuitively evident steps, as one would go about it in ordinary informal proofs. Everything, including the rules of derivation, needed to undergo a complete formalization:

> Damit sich hierbei nicht unbemerkt etwas Anschauliches eindrängen könnte, musste Alles auf die Lückenlosigkeit der Schlusskette ankommen [To see to it that in this process no intuitive content might without being noticed insert itself, everything had to depend upon having a chain of deductions without any gaps (transl. EK)] (x).

The German mathematician David Hilbert wanted to provide finitary foundations for mathematics. He never explained what, precisely, we are to understand by 'finitary,' but clearly principles and types of reasoning belonging to a very elementary part of arithmetic are meant. Traditionally, mathematics surpasses finitary bounds. For instance, any proof that refers to the set of natural numbers as a completed, actually infinite, totality (and not merely to the sequence of natural numbers as potentially infinite, every natural number n being followed by $n + 1$) would count as nonfinitary. The nonfinitary part of mathematics, however, is not isolated from the finitary part, for a nonfinitary proof may very well have a finitary conclusion. The problem is whether such nonfinitary proofs yield reliable results from a finitary point of view. Hilbert wanted to show (by finitary means) that they do, in other words, he wanted to show that the traditional nonfinitary proofs never yield a conclusion that would be incorrect from a finitary point of view.[5] For this end he needed to enter into certain (finitary) mathematical investigations of formal proofs. This started a discipline called *proof theory* or *metamathematics*. Prawitz (1971, 1981) pointed out

---

[5]Given certain assumptions, this formulation of Hilbert's program is equivalent to the better-known version: to prove arithmetic consistent by finitary means. Cf. Van Dalen (1978, 58 f.) and Prawitz (1981, 235 f.).

that proof theory (taken in a broad sense) does not start with Hilbert (or with Frege) but has been part of logic since Aristotle. His term is *general proof theory*, whereas he uses the term *reductive proof theory* to refer to those studies that are connected with programs like Hilbert's. I hope it will be clear from these remarks that though proof theory is concerned, primarily, with formal proofs, it may, by formalization, yield insights into proof and possibilties of proof that are highly relevant to the study of informal (axiomatic or nonaxiomatic) proof. Formal systems of proof can serve as 'models' for certain techniques of reasoning and arguing. In a sense they give us argumentation-theoretical models (models of argument).

Thus formal proofs can be useful in several respects, but they can never replace informal proofs. For instance, metamathematical proofs themselves (i.e., proofs about formal proofs) are usually informal, and if they are formal they impose a need for a metametamathematics, and so on. In the end, one will find, in each actual case, a level of informality.

Now that we have surveyed a number of types of proof (*Immediate and Intuitive Proofs, Informal Proofs, Informal Axiomatic Proofs, Formal Proofs*) and their uses, it may have become clear that we should drop the idea of an 'absolute' notion of proof. What happens to be a proof for one audience or within the confines of one system, need not be one for some other audience or within some other system. What happens to constitute a proof here and now for a particular audience may not maintain this status forever.[6]

With respect to the theory of argumentation, there are two obvious conclusions:

1. With the exception of *Immediate and Intuitive Proofs* and *Formal Proofs*, every mathematical proof is an argument, and therefore a suitable object of study for a theory of argumentation.[7]
2. *Formal Proof* occurs in systems that can be interpreted as models of reasoning or arguing and of which the theory of argumentation may avail itself.[8]

3.4   The Surplus Value of a Proof

Not every argument is a proof. Setting aside intentions to impress or intimidate, whenever the term 'proof' is applied to an argument, it must be understood to refer to one or more supplementary conditions which the argument is supposed to fulfill.

---

[6]Dummett, when discussing the philosophical and semantic aspects of intuitionistic views on implication and proof, points out the possibility that 'mathematics becomes a subject where results are fallible and liable to revision' (1977, 402).

[7]Presumably, similar observations hold for proofs in science and for proofs in court.

[8]If formal systems of proof (e.g., systems for natural deduction) provide models of certain aspects of argument, the same holds *a fortiori* for formal systems of dialogue rules, such as the dialogue games introduced by Lorenzen and Lorenz (1978), in which dialectical interaction is explicitly taken into account. Cf. also Barth and Krabbe (1982), Haas (1984), Krabbe (1985), Stegmüller and Varga von Kibéd (1984).

These conditions may be different in different contexts of use. Consequently, one should take heed if someone utters the word.

Aristotle, for example, requires the premises of a proof (*apodeixis*) to be (i) true, (ii) themselves indemonstrable, (iii) better knowable than the conclusion, and (iv) giving the cause of the conclusion (*Anal. post.*, I.2, 71b17–33; Aristotle, 1976, 31). The conclusion has to be obtained from the premises by deductive argument (syllogism). Consequently, according to Aristotle, arguments that do not comply with any or some of these conditions are not proofs, but those that do comply have several *surplus values*, such as giving the cause of the conclusion, and can thus yield knowledge.

A recent proposal can be found in a paper by John Corcoran (1989). According to Corcoran, the term 'proof' makes 'tacit reference to a participant or to a community of participants' for whom the alleged proof would constitute a proof (22). In other words, what is a proof for one person is not necessarily a proof for some other person.[9] Not every argument is a proof for everyone, or even for anyone. In order to be a proof for someone, an argument should have the right surplus value for that person:

> Critical evaluation of an argumentation to determine whether it is a proof for a given person reduces to two basic issues: are the premises known to be true by the given person? And does the chain of reasoning deduce the conclusion from the premise-set for the given person? (25).

For special purposes, very special conditions may come into play. We already mentioned Frege's need for proofs starting from purely logical principles that lead, without any gaps, to arithmetical conclusions. Dummett (1977) distinguishes 'mere demonstration' (mathematical proof in a broad sense) from 'canonical proof' (mathematical proof in a narrow sense).[10] Informal proofs are mere demonstrations, but in order to define the concept of an informal proof one needs to refer to the concept of a canonical proof. For canonical proofs, there are supplementary conditions. For instance, no canonical proof is to contain a statement that is more complex than the proof's conclusion (Dummett, 1977, 395). The task of an informal proof (demonstration) is to show that a canonical proof exists for its conclusion (392). Hence an informal proof has some surplus value over mere arguments, and a canonical proof has a surplus value again over a mere demonstration.

According to the philosophy of mathematics in Lakatos (1976), proofs are placed near the beginning, rather than at the end, of the 'method of proofs and refutations,' a heuristic pattern of mathematical discovery. As a stage in this pattern, proof is preceded only by the formulation of a problem and a conjecture (Lakatos, 1976, 127). Once a proof has been obtained, the process of discovery proceeds by criticism and analysis of the proof, by correction of the conjecture, and so on. Hence, according to Lakatos, proof is not primarily a matter of argument: a proof serves to open possibilities for criticism and thus to advance the inquiry, but not to convince

---

[9]Cf. Note 3.

[10]The distinction is important for an intuitionistic explication of the meaning of logical constants.

someone else of the correctness of a theorem. From this point of view, proof may have a surplus value over argument, but so has argument over proof.

These examples may suffice to show that the surplus value of a proof over mere argument can be specified in radically different ways and that each account of this surplus value is closely linked to further philosophical positions taken by its author. Any theorist of argumentation who discusses proof should, therefore, take care to make clear which concept of proof is intended, i.e., what supplementary conditions there are for an argument to count as proof.

3.5   Proof and Implicit Dialogue

In a dialectical theory of argumentation, a monological argument is viewed as an implicit discussion aiming at the resolution of a conflict (explicit or implicit) concerning the acceptability of a point of view.[11] The author defending a point of view in a letter to the editor, for instance, knows or at least assumes that his point of view is not automatically shared by all readers. So he assumes that there is or may be a conflict of opinion about this point of view. To resolve this conflict, the author needs to have a critical discussion with his opponents. But as long as the critics do not actually participate in writing up the letter in a dialogical format, this critical discussion has to remain implicit, whereas the explicit format will be that of monologue. Underlying this monologue, however, there is an implicit critical discussion to which the argumentation analyst refers.

If a proof is to be an argument with a certain surplus value, then obviously one should turn to the underlying discussion to find a good candidate for this surplus value. If the underlying discussion fulfills certain supplementary conditions (over and above compliance with rules that hold for any critical discussion), then the (monological) argument is to be called a (monological) proof, but otherwise it is not. What supplementary conditions would be appropriate?

Along these lines, one may consider the following definition:

Definition 1. A monological argument is a *monological proof* for X if (i) the underlying discussion complies with the rules of a dialectical system that has been accepted by X whereas (ii) X is commited in the strongest sense to the initial concessions in that discussion, and (iii) all possible chains of criticism (i.e., chains of arguments that the Opponent may select) are followed through, and finally (iv) the discussion is won by the Proponent.[12]

---

[11]Cf. Van Eemeren et al. (1983, 9).

[12]Cf. Barth and Krabbe (1982, Ch. III, esp. III.6, III.8, and III.13) for the idea of a discussion as consisting of several *chains of arguments* and for the concepts of *winning or losing* a chain of arguments or the discussion as a whole. The concepts of a *chain of arguments* and of a *chain of criticism* (Dutch: kritieklijn) were introduced by E. M. Barth.

In this definition, 'X' may stand either for a person or for a company or community. The first condition merely states that an argument, in order to count as a proof, should be dialectically correct, i.e., not fallacious. It is a necessary, but not a sufficient condition, which implies that X is committed to the premises of the argument (the initial concessions of the underlying discussion), at least to the extent that there is no reason for X to object to them. The second condition adds that this commitment is to be stronger than a commitment to mere concessions would require: the premises of the argument are to count as *assertions* of X's, implying that X carries a burden of proof for them, if they happen to be challenged.[13] The third condition stipulates that the argument, in order to count as a proof, must deal thoroughly with all relevant objections, possible cases, potential counterexamples, etc. The last condition, together with the third, implies that the underlying discussion shows how a Proponent of the conclussion of the argument can always win *vis-à-vis* an Opponent that grants the premises, no matter how this Opponent selects her moves in the discussion. In other words: there is a *winning strategy* for the Proponent, and this strategy is reconstructible from the explicit (monological) argument.[14]

Notice that this definition is thoroughly nonpsychological and nonepistemical. There is no implication that the rules of the dialectical system in question are rationally or epistemically justified, or that X knows the premises to be true, or that X recognizes a proof for X as such. It is possible for another person, Y, to know that something actually is a proof for X, even though X in all sincerity, denies the fact.

3.6 Proof and Explicit Dialogue

The term 'argument' also refers to the argumentation presented by the Proponent (Protagonist) in the argumentation stage of an explicit dialogue.[15] In this case, as well, we may hunt around for some supplementary conditions that would justify us in saying that a certain critical discussion provides a proof for the Opponent (Antagonist). This would give us a concept of proof as a kind of dialogue: a concept of *dialectical proof*. The simplest choice of conditions would be to copy them from Definition 1, with the Opponent as X:

Definition 2. The Proponent in a critical discussion provides a *dialectical proof* of his point of view for the Opponent, if (i) the critical discussion complies with the rules of a dialectical system that has been accepted by the Opponent whereas (ii) the Opponent is committed in the strongest sense to the initial concessions in

---

[13] For an exposition of different types of commitment in dialogue, see Walton and Krabbe (1995, esp. Sect. 5.4).

[14] Winning strategies are studied in Barth and Krabbe (1982, Ch. V).

[15] For the stages of a discussion, see Van Eemeren et al. (1983, Hstk. 2), Van Eemeren and Grootendorst (1984, 85ff), and Van Eemeren and Grootendorst (1992, 35).

that discussion, and (iii) all possible chains of criticism (i.e., chains of arguments that the Opponent may select) are followed through, and finally (iv) the discussion is won by the Proponent.

The two kinds of proof, monological and dialectical, are straightforwardly linked: a monological argument is a monological proof for X, if and only if the discussion underlying the argument provides a dialectical proof for X, with X in the role of Opponent.

In practice, it will often be very hard to establish that a certain monological argument constitutes proof for a given person X. For one thing, one is to provide a reconstruction of the underlying dialogue. Then, one is to be sure about X's commitments, both with respect to the initial concessions and to the rules of dialogue on which the underlying discussion is based. If too much uncertainty remains on either of these accounts, it will remain unsettled whether the purported proof is a proof for X or not.

In the argumentation stage of some kinds of critical discussion, the Proponent is allowed to present, not only such argumentation as is directly relevant to the Opponent's preceding challenge, but also longer stretches of reasoning that in fact constitute monological arguments, even though they are put forward in a dialogical context. Such monological arguments, presented within an explicit discussion, can themselves be analysed by reference to (another) implicit discussion. Suppose that this implicit discussion happens to provide a dialectical proof for its Opponent. Then the monological argument will be a monological proof for the Opponent of the explicit discussion. This may sound complicated, but it really only goes to show that monological proofs may function also *within* critical discussions. This is yet another example of the ways in which proof ties up with dialogue.

3.7   The Genesis of Proof

If there were no proofs, one would have to invent them. Suppose a certain company of discussants has agreed upon a specific dialectical system S_0 for the regulation of critical discussions among them. Many fruitful and provoking discussions have taken place among the members of this company, all of them regulated by the rules of S_0. Suppose that, after acquiring a certain body of experience with S_0, it has become clear and obvious for the members of this company that whosoever concedes both *if A then B* and *if B then C* can be forced, in a discussion conforming to S_0, to refrain from opposition to the proposition *if A then C*. The company has discovered an important law of dialogical logic that was already implicit in the rules of their agreed dialectical system S_0: whenever the Opponent concedes *if A then B* and *if B then C*, there is a winning strategy for the Proponent of *if A then C*! This company would then be well-advised to skip this type of dialogical fragment and in its stead to adopt a rule of inference:

$$\textit{if A then B, if B then C} \Rightarrow \textit{if A then C}.$$

(The rule is applied by substituting sentences for the variables A, B, and C. To the left of the arrow one then finds the premises of the application, and to the right of the arrow its conclusion.) Adoption of this new rule amounts to a change of dialectical system. S_0 is traded for S_1. Within S_1 it is no longer possible to criticize the steps that conform to the rule of inference just stated. (That is to say, one is not allowed to criticize the relative conclusiveness of the premises for the conclusion in this step, whereas it is of course still possible to criticize the tenability of the premises themselves.) The company has discovered and adopted the rule of the Hypothetical Syllogism!

In the same way this company may go on, discovering ever more laws of dialogical logic relating to the existence of winning strategies in certain situations, and extending its dialogical system, by adding ever more rules of inference so as to obtain the systems $S_2, S_3, \ldots$ Finally, the company winds up with a dialectical system in which a whole system of rules of inference has been incorporated.

Given such a system one may introduce a somewhat special notion of proof:

Definition 3. Let D be a discussion, according to the rules of S between two parties X and Y. Let A be an argument, advanced by Y in the course of this discussion D, in order to convince X of a conclusion C. At moment t, argument A counts as a proof of C in D if and only if the following two conditions hold:

1. Each ultimate premise of A was asserted by X, before t, and not withdrawn by X before t.[16]
2. Each separate argumentative step of A is explicitly sanctioned by one of the inference rules incorporated in S.

Let us suppose that S allows of more ways of arguing than those sanctioned by the incorporated body of rules of inference. And let us further assume that, although every assertion in dialogue counts as a concession, not every commitment to a concession carries a burden of proof which would make it equal to an assertion.[17] Then Definition 3 shows how a meaningful distinction can be introduced between, on the one hand, a merely successful and nonfallacious argument in dialogue, and, on the other hand, a proof in dialogue. Proof is seen to have a surplus value, both with respect to its ultimate premises and with respect to the separate steps of which it consists. Further, it is seen how proof ties up with rules of inference. But the roots of these rules are again dialectical. This concept of proof is, moreover, fully externalized and independent of the concepts of 'knowledge' and 'truth'.[18]

Acknowledgements Previously published in M. Astroh, D. Gerhardus, and G. Heinzmann, editors, *Dialogisches Handeln: Eine Festschrift für Kuno Lorenz*, Spektrum Akademischer Verlag, Heidelberg, 1997, 63–75. An earlier version was published in Dutch: Krabbe (1991).

---

[16] As was pointed out in Section 3.5, this implies that X has a burden of proof for these premises (if challenged, and if X has not discharged this burden before).

[17] See Note 13.

[18] On 'externalization', see Barth and Krabbe (1982, 32–33, 60).

References

Aristotle. (1976). *Posterior analytics. Topica.* (H. Tredennick & E. S. Forster, Trans.). Cambridge: Loeb Classical Library, Harvard University Press. (Original work published 1960).

Barth, E. M., & Krabbe, E. C. W. (1982). *From axiom to dialogue. A philosophical study of logics and argumentation.* Berlin: Walter de Gruyter.

Barth, E. M., & Martens, J. L. (Eds.). (1982). *Argumentation: Approaches to theory formation. Containing the contributions to the Groningen conference on the theory of argumentation, October 1978.* Amsterdam: John Benjamins.

Corcoran, J. (1989). Argumentations and logic. *Argumentation, 3,* 17–43.

Dummett, M. A. E. (1977). *Elements of intuitionism.* Oxford: Clarendon.

Frege, G. (1879). *Begriffsschrift, eine der arithmetischen nachgebildete Formelsprache des reinen Denkens.* Halle: Louis Nebert. (Reprinted from Frege, G., *Begriffsschrift und andere Aufsätze,* Vol. 2, by I. Angelelli, Ed., Hildesheim: Georg Olms).

Haas, G. (1984). *Konstruktive Einführung in die formale Logik.* Mannheim: Bibliographisches Institut.

Krabbe, E. C. W. (1985). Formal systems of dialogue rules. *Synthese, 63,* 295–328.

Krabbe, E. C. W. (1991). Quod erat demonstrandum: Wat kan en mag een argumentatietheorie zeggen over bewijzen? [QED: What can and may a theory of argumentation say about proofs?]. In M. M. H. Bax & W. Vuijk (Eds.), *Thema's in de Taalbeheersing: Lezingen van het VIOT-taalbeheersingscongres gehouden op 19, 20 en 21 december 1990 aan de Rijksuniversiteit Groningen* (pp. 8–16). Dordrecht: ICG.

Lakatos, I. (1976). *Proofs and refutations: The logic of mathematical discovery* (edited by J. Worrall & E. Zahar). Cambridge: Cambridge University Press.

Lorenzen, P., & Lorenz, K. (1978). *Dialogische Logik.* Darmstadt: Wissenschaftliche Buchgesellschaft.

Prawitz, D. (1971). Ideas and results in proof theory. In J. E. Fenstad (Ed.), *Proceedings of the second Scandanavian logic symposium* (pp. 235–307). Amsterdam: North-Holland.

Prawitz, D. (1981). Philosophical aspects of proof theory. In G. Fløistad & G. H. von Wright (Eds.), *Contemporary philosophy. A new survey I: Philosophy of language/philosophical logic* (pp. 235–277). The Hague: Martinus Nijhoff.

Stegmüller, W., & Varga von Kibéd, M. (1984). *Probleme und Resultate der Wissenschaftstheorie und Analytischen Philosophie III: Strukturtypen der Logik.* Berlin: Springer.

Van Dalen, D. (1978). *Filosofische grondslagen van de wiskunde* [Philosophical foundations of mathematics]. Assen: Van Gorcum.

Van Eemeren, F. H., & Grootendorst, R. (1984). *Speech acts in argumentative discussions. A theoretical model for the analysis of discussions directed towards solving conflicts of opinion.* Dordrecht: Foris.

Van Eemeren, F. H., & Grootendorst, R. (1987). Fallacies in pragma-dialectical perspective. *Argumentation, 1,* 283–301.

Van Eemeren, F. H., & Grootendorst, R. (1992). *Argumentation, communication, and fallacies: A pragma-dialectical perspective.* Mahwah, NJ: Lawrence Erlbaum Associates.

Van Eemeren, F. H., Grootendorst, R., & Kruiger, T. (1983). *Argumentatieleer 1: Het analyseren van een betoog* [Theory of argumentation 1: The analysis of arguments]. Groningen: Wolters-Noordhoff.

Van Eemeren, F. H., Grootendorst, R., & Kruiger, T. (1987). *Handbook of argumentation theory: A critical survey of classical backgrounds and modern studies.* Dordrecht: Foris.

Van Eemeren, F. H., Grootendorst, R., Snoeck Henkemans, A. F., Blair, J. A., Johnson, R. H., Krabbe, E. C. W., Plantin, C., Walton, D. N., Willard, C. A., Woods, J., & Zarefsky, D. (1996). *Fundamentals of argumentation theory: A handbook of historical backgrounds and contemporary developments.* Mahwah, NJ: Lawrence Erlbaum.

Walton, D. N. (1990). What is reasoning? What is argument? *Journal of Philosophy, 87*, 399–419.
Walton, D. N. & Krabbe, E. C. W. (1995).   *Commitment in dialogue: Basic concepts of interpersonal reasoning.* Albany, NY: State University of New York Press.
Webster. (1985). *Webster's ninth new collegiate dictionary.* Springfield, MA: Merriam-Webster.

Chapter 4
Argumentation in Mathematics

Jesús Alcolea Banegas

In *The Uses of Argument*, Stephen Toulmin (1958) introduced a model of argumentation, in which what may be called the 'layout of arguments' is represented. This model has become a classic in argumentation theory and has been used in the analysis, assessment and construction of arguments. Toulmin's main thesis is that, in principle, one can make a claim of rationality for any type of argument, and that the criterion of validity depends on the nature of the problem in question. He rejects the idea of universal norms for assessment of argumentation and that formal logic provides these norms. The scope and function of logic is too limited for this purpose. There is an essential difference between the norms which are relevant to the assessment of everyday argumentation or the argumentation of diverse disciplines, on the one hand, and the criterion of formal validity used by formal logic, on the other. Toulmin is convinced that formal criteria are irrelevant for the assessment of the arguments that arise in real life. Logical proof is one thing and the establishment of conclusions in everyday life is another different matter.

In this chapter, I will begin by reviewing Toulmin's model of argumentation, focusing on what is described as the elements of argument; secondly, I will attempt to apply this model to argumentation in mathematics, and finally, I will pursue the consequences that connect it to recent ideas in the philosophy of mathematics.

4.1 The Concept of Argumentation

Arguing is a specific act of social interaction, which may take place in any kind of discussion. If one or more participants in a discussion enunciate something, what they are in fact doing is declaring the validity of that statement. By their proposal,

J. Alcolea Banegas (✉)
Departament de Lògica i Filosofia de la Ciència, Facultat de Filosofia i CC. de l'Educació,
Universitat de València, Avinguda Blasco Ibañez, 30, E46010 València, Spain
e-mail: Jesus.Alcolea@uv.es

A. Aberdein and I.J. Dove (eds.), *The Argument of Mathematics*, Logic, Epistemology, and the Unity of Science 30, DOI 10.1007/978-94-007-6534-4_4,
© Springer Science+Business Media Dordrecht 2013

they show their disposition to act in a rational way and to establish their statement in more detail, if needed. We will describe as *argumentation* the techniques or methods used to establish a statement. In the process, a questioned statement leads to an agreed statement, accepted by all the participants in the discussion. It is not necessary for the development of an argument to proceed in a harmonious manner. Discussions may arise at any stage of the argument, leading to corrections and changes. In this way, the set of those statements of the argument that are agreed by consensus will take shape step by step, as the controversy is overcome. The result of this process can be reconstructed, and this we will call *argument*. By this reconstruction it is possible to draw attention to the functional aspects of the argumentation, showing that the statements in which the argumentation consists have different roles in the internal structure of an argument and in its establishment.

In general, argumentation, is to be understood as a metacommunicative activity, which results when the presumed validity of an everyday action is in doubt. For our purposes, it seems more correct to use the concept of argumentation when describing specific aspects of everyday activities. In the case of mathematics, the process of solving a problem, and following a comprehensible method of reasoning to achieve that solution, already has argumentative traits. Sometimes, the solution of the problem is in itself an argument and there is no need of an additional and separate metacommunicational procedure. Such is the case, for example, in a calculation. In this case, the process followed to obtain a conclusion and that followed to obtain an argument to support it will usually coincide. However, when discussing the concept of argumentation in mathematics, there is a tendency to identify it with the concept of proof. There is an erroneous belief that the analysis of argument is always related to proof. It should be remarked that there is no need for the concept of argument or that of argumentation to be exclusively linked with formal logic, as it is presented in some mathematical proofs. There are other human activities, even in mathematics, which have an argumentative character, but not in a strictly logical sense. Hence, mathematical certainty does not come from the formal verification of formal arguments, but, as William Thurston says, from "mathematicians thinking carefully and critically about mathematical ideas" and "prov[ing] things in a social context and address[ing] them to a particular audience" (Thurston, 1994, 170, 175). As Toulmin points out, if logicoformal conclusions were the only legitimate procedure for argumentation, then the range of rational communication would be extremely restricted and argumentation, as a possible form of communication, would be irrelevant. Every logically valid deduction, for example, contain nothing in its conclusion, which is not already a potential part of the premises. It explains aspects of the meaning of the premises through deduction. For Toulmin, this is an *analytic* argument. It is contrasted with the so-called *substantial* argument, which broadens the meaning of the premises by coherently linking them with a specific case by means of update, modifications or application. In this way, substantial argumentation is informative in the sense that the meaning of the premises increases or changes with the application of new cases, while analytic argumentation is tautological, this is to say, a latent aspect of the premises is elaborated in a visible form. Cases of substantial argumentation do not usually have the logical rigidity of formal deductions, which must not be considered a sign of weakness, but rather a sign that fields of problems exist which are inaccessible to formal logic. For Toulmin, "the only arguments we can fairly judge by 'deductive' standards are those held out as and intended to be analytic, necessary and formally valid" (Toulmin, 1958, 154). As he emphasizes, substantial argumentation must not be subordinated or related to analytic argumentation, as though the latter were the ideal type of argumentation. Substantial argumentation has its own role. It gives graduated support to a statement or a decision. This support is not directed towards a logically necessary, formal conclusion, nor in arbitrariness, but originates in the need to achieve a credible presentation of contexts, relationships, explanations, justification, etc. From this perspective, there is a way to understand mathematical proof as substantial argumentation. The idea rests on the mostly philosophical work of the mathematician Reuben Hersh, for whom a proof is an explanation, where the aim is mainly to pass on comprehension, and for whom the practical meaning of proving is convincing. A mathematician achieves a proof by an argument he considers to be convincing, involving other factors that are not usually included in the published proof. For this reason, in practice, any proof is a convincing piece of argumentation, aimed at the competent expert. Kurt Gödel expressed it in this manner: in its material meaning, "'Proof', does not mean a sequence of expressions satisfying certain formal conditions, but a sequence of thoughts (or rather forms of thought) creating conviction in a sound mind" (Gödel, 1995, 197; see Hersh, 1997).

Consequently, the distinction established above is useful for clarifying the conceptual framework we have selected to analyse mathematical argumentation and we believe that substantial argumentation is the more suitable for reconstruction, without implying that this argumentation is poor or weak. At this point, at least two reasons may be given in support of this decision: (1) Generally, the mathematician does not work on a formal system. According to recent philosophy of mathematics, mathematical knowledge has an empirical and theoretical status, and mathematical statements include the meaning of real experiments on mathematical objects, and (2) Mathematicians do not always reach conclusions of the analytic type. Heuristic, plausible and probable argumentation also count as mathematical activities, as is emphasized in the work of, for example, George Pólya.

4.2 Constituent Elements of Argumentation

We have said that argumentation is a technique used to establish a statement: a technique of the discourse, not a trait of the reasoning subject. It can be characterized as an attempt to transform something open to question into something mutually accepted, in principle, by the participants in a debate. Therefore we must attend to the kinds of statements and to their functionality in order to capture the argumentation. Hence Toulmin proposes what he calls the 'layout of an argument', this is to say, a diagram representing the ideal model of substantial argumentation. This layout may be said to (1) describe a base common to all types of argumentation; (2) serve to reconstruct the informal logic of any argumentation and all forms of explanation convenient for resolution of a debate, and (3) highlight the different roles statements have in any interaction in which they may provoke a substantial argument. Globally, argumentation works in such a way that the questionable statement is finally sustained by being introduced as a conclusion from unquestionable facts. This implies an inferential step from the facts to the conclusion, which the participants must accept as indubitable.

The basic idea behind the functionality of argumentation consists precisely in the support given to a questionable statement when it is deduced from another statement. The argumentative strength, coming from the latter, lies in the level of acceptance of this inference. If we call the *conclusion*, the statement or thesis we are trying to establish C and the statements supporting it D (for *data*), then we can represent this part of the argument as 'D, therefore C'. The data are the facts or reasons we use to ground the statement C. All argumentation requires this starting point upon which the conclusion can be based. Toulmin says that "[d]ata of some kind must be produced, if there is to be an argument there at all: a bare conclusion, without any data produced in its support, is no argument" (Toulmin, 1958, 106). Depending on the sort of thesis in discussion, that data can include experimental observations, statistical data, personal statements, previously established theses, or some other factual data. The soundness and reliability of any thesis is dependent on the data, but no thesis can be stronger than the data. It is obvious that the data cannot be questioned, and if there were no agreement on their validity, further argumentation to provide acceptable evidence would be necessary, involving recursive application of the general layout of argumentation. It might even be that one accepted the data, without realising that it supported the conclusion. The acceptance of the data implies a commitment, by a specific step, to the conclusion. To show that this step is adequate and reliable, it is necessary to produce statements of a functionally different nature to that of the data, which may act as bridges linking the particular inferential steps, since they function as inferential licenses. Toulmin calls them 'warrants' (W) and states that, in a sense, they have an explanatory character, their "task being simply to register explicitly the legitimacy of the step involved and to refer it back to the larger class of steps whose legitimacy is being presupposed" (Toulmin, 1958, 100). For this reason, conclusions may have different degrees of strength. However, it should be stressed that for Toulmin "provided that the correct warrant is employed, any argument can be expressed in the form 'Data; warrant; so, conclusion' and so become formally valid" (Toulmin, 1958, 119). It is clear that the steps from the data to the conclusion are warranted in a different way in each field of reasoning. The data depend on the choice of warrant. For this reason, it must be remembered which data is supposed to be reinforcing the grounds on which the argument has been constructed. By contrast, 'warrants are general, certifying the soundness of *all* arguments of the appropriate type, and have accordingly to be established in quite a different way from the facts we produce as data' (Toulmin, 1958, 100). Consequently, it is not at all unusual for the warrants to take the form of natural laws, formulae, norms, principles, etc.

The soundness of the step from data to conclusion can be ratified by warrants. However, someone could still question the acceptance of a warrant as though it had authority in itself. For Toulmin, behind our warrants there are normally "other assurances, without which the warrants themselves would possess neither authority nor currency—these other things we may refer to as the *backing* (*B*) of the warrants" (Toulmin, 1958, 103). The idea of backing invokes convictions and strategies that can be expressed as categorical statements, linking the data, the warrants and the conclusion with fundamental suppositions usually accepted by the community. Furthermore, warrants may require different sorts of backings, so it is necessary to establish the general informational context in which the backing is situated. We may then observe that the different backings that are offered for each warrant depend upon their relevance to the point of view in question. It may also be that the information given by the conclusion has already appeared in the backing: this is in Toulmin's opinion an analytic argument: "an argument from *D* to *C* will be called analytic if and only if the backing for the warrant authorizing it includes, explicitly or implicitly, the information conveyed in the conclusion itself" (Toulmin, 1958, 125). It is obvious that not all the arguments support their theses or conclusions with the same degree of certainty. Some lead to probable conclusions, other lead to apparent conclusions, etc. Consequently, we require an accurate examination of the different types of *modal qualifiers* (*M*), characteristic of the different types of practical argument, and it must be kept in mind that, because of the difficulty of preventing rebuttals, all arguments that are not necessary are open to refutation (*R*).

We have already said that this model of argumentation is designed to focus on the aspects of argumentation and reasoning that can be found in all types of rational debate and to be illustrated by any field of practical reasoning. The system of analysis involves the elements we have just indicated, and may be summarized as follows: (1) The thesis or conclusion *C* proposed and criticized in a specific context; (2) the data or reasons *D* supporting the thesis; (3) the warrant *W* connecting data and thesis, legitimating the inference applied from *D* to *C*; (4) the backing *B*, available to establish sound and acceptable warrants; and (5) the modal qualifier *M* indicating the strength or the conditions of refutation *R* of the proposed thesis. The general scheme of argumentation is shown in Fig. 4.1, which should be read as follows: "Given reasons, *D*, we may use the warrant, *W*, which has backing, *B*, to justify the thesis *C* or the presumption, *M* that *C*, in the absence of any specific refutation or impediment, *R*". The diagram can change in respect to the specific exchange and allows for the reconstruction of emergent rationality, this is to say, the informal logic of the debate in question.

What Toulmin describes as the 'field dependency' of arguments is a very important issue in his work. A field may be identified with a social institution, a law court, a scientific society, etc, and each one has its own specific canons. The distinctions Toulmin make between fields "are very broad ones, and a closer examination could certainly bring to light further more detailed distinctions, which would improve our understanding of the ways in which arguments in different fields are related" (Toulmin, 1958, 42). Toulmin and his collaborators undertaken this closer examination in another work, (Toulmin et al., 1979), in which he deals with

Fig. 4.1 Toulmin's general
scheme of argumentation

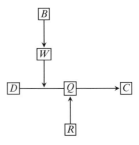

special fields of argumentation, science among them, without dealing directly with
the field of mathematics. Since, as I pointed out at the beginning, I want to apply
Toulmin's model of argumentation to this field, I will try to pose my problem in the
right perspective, and it is with this aim that I will make use of some of his ideas.

4.3   The Nature of Mathematical Activity

Quite independently of content, there are three basic general traits that the mathematics of any age exhibits, which to a considerable degree determine the scope of
argumentation and rational criticism:

(1) It has to deal with specific issues about mathematical objects and phenomena,
    in the hope that it will explain their nature. What types of object are there?,
    what are their characteristic properties, functions and relationships?, etc. These
    are some of the sort of questions mathematicians try to answer when devising
    their theories.
(2) It must provide a set of ideas leading to the explanation of the source of
    mathematical facts, together with acceptable procedures for criticizing and
    improving these explanations. Mathematicians develop systematic procedures
    to represent the mathematical world and its structure, including theories,
    conjectures, axioms, inferential methods, techniques of calculation, computer
    software, etc. It can be said that the system of representation accepted at any
    specific moment defines the content of the mathematical tradition, this is to say,
    the best attempt up until that moment to generate realist conceptions about the
    mathematical world.
(3) There must be groups of people in the society undertaking the responsibility
    for preserving and transmitting this tradition of criticism. When a mathematical
    community opens its ideas to critical debate, the relationship between argu-
    mentation and the aims of mathematics becomes clear. Criticism is an effective
    way of improving established beliefs about mathematical objects, and, at the
    same time, permits the development of new ideas and implications, which will
    in turn be opened to assessment, so that the valuable ones may be added to
    the established mathematical tradition. Toulmin and his collaborators point

out that "[t]hose ideas that survive this critical evaluation will be 'good' as scientific ideas. If sufficient grounds and solid enough arguments are produced to demonstrate their merits clearly, that will mean that their scientific basis is also 'sound.' Where critical evaluation shows both requirements fulfilled, we can be satisfied that practical argumentation has demonstrated the 'rational' basis of those novel ideas" (Toulmin et al., 1979, 232).

The general purpose of mathematics guarantees that the problems, reasoning and argumentation tend to consensus or rational agreement between the parties involved. For example, when a mathematician proposes a new idea, a conjecture or an axiom, there will always be differing opinions among the mathematicians involved with the topic. They will argue, in groups or individually, about its acceptance or rebuttal, or whether it should somehow be modified. Although conflicts are usually temporary and mathematicians have objectives and interests in common, the acceptance of Cantor's ideas or the explicit acceptance of the axiom of choice are examples of the form in which some questions have been received in the field of mathematics.

It is clear that there are many types of mathematical explanations and arguments. So much so that indicating the general character of the elements involved in argumentation would appear to be a complicated task. If one assumes, besides, that the general aim of mathematical activity is to improve the correspondence between mathematical ideas and theories and experience of the mathematical world, then suitable questions for research and discussion will be apparent, since it will be possible to identify gaps in the mathematical concepts, to be filled by the appropriate activity. By way of example, we may mention two big questions mathematicians have set out in the last century: the first relates to the possibility of expanding a mathematical theory to explain hitherto unexplained mathematical facts and the second concerns the possibility of restructuring a set of mathematical theories in order to obtain a tidier and better organized overall explanation of the mathematical world. The expansion of set theory by the addition of new axioms, in answer to the first question, and the introduction and systematisation of set theory or the creation of category theory, in answer to the second, are examples of the mathematical solutions offered by mathematicians which have provided abundant material for a specific type of mathematical theses, in the sense we may discuss in the framework of Toulmin's model.

However, we must not overlook an important distinction, especially relevant in the case of mathematics. We refer to the distinction between those arguments, including proofs, that mathematicians make within a specific theory, and those arguments used by mathematicians to challenge current ideas and attempt to give alternatives or improvements in their place. There are plenty of examples of this sort at the heart of mathematics. There is, for example, the geometrical work from Gauss to Riemann that led to non-Euclidean geometry, or the no less provocative arguments of Brouwer which led to intuitionist mathematics. Using Toulmin's terminology, we will describe the first sort of argument as *regular* and the second sort as *critical*, and we will take both to be substantial arguments in the sense previously discussed, although most mathematicians, motivated by formal logic, would consider regular arguments to be generally analytic. Regular mathematical arguments take it for granted that the ideas within the framework of argumentation are coherent relevant, and applicable to the facts under consideration. For this reason, these elements conform to currently accepted theories and rules, for which they do not represent any type of challenge. Critical mathematical arguments are created precisely by challenging those current ideas, which thereby lose their reliability, relevance or applicability and are opened to critical discussion, rebuttal or readjustment. The arguments that mathematicians have to give in support of as yet unproven statements (axioms, postulates, problem hypotheses and conjectures) are also critical, since either they are grounded in intuition, or the mathematician must give reasons for or against them. This argument need not necessarily be limited to a proof that would count as a justification of the truth of the proved statement. It is obvious that there will be important systematic differences between both types of mathematical arguments. It can be pointed out, however, that regular arguments can be interesting and new enough to provoke a critical discussion when they are presented in public. The proof of the four colour theorem or the recent proof of Fermat's last theorem are examples of regular arguments provoking critical arguments. The first case, for example, initiated a debate as to whether or not it was legitimate, from the mathematical point of view, to use computers to obtain proofs that could not have been obtained otherwise and, in general, whether computers could be used as tools for mathematical exploration: a debate made even more interesting since it propelled mathematicians and philosophers to discuss the concept of mathematical proof. In this context, I would like only to point out that, proof has an important role as part of the critical mathematical interpretation in any case, because it is a way to produce ideas and relationships between ideas, and also a way to communicate and critically contrast them.[1] It is also necessary to point out that critical argumentation itself provokes the creation of regular mathematical arguments. For example, the criticism of the introduction of the axiom of choice into set theory motivated the proofs of the independence and consistency of the axiom in relation to the rest of the axioms of the theory. Although it appears as if the critical arguments of mathematics were 'less' mathematical and more discursive than the regular arguments, whose acceptability they try to adjust, it is clear that critical and regular argumentation are mutually complementary, and it is not sustainable to suppose that mathematicians' work is limited exclusively to elaborate regular arguments. The exclusive emphasis on this activity of mathematicians has created the belief that mathematics is an a priori science, which has been sustained by foundational currents in the philosophy of mathematics. In considering the capacity mathematicians have for critical argument, we are closer to understanding the similarity between mathematics and science in general, and consequently, to considering it as a quasi-empirical science. The history of mathematics favours this quasi-empirical conception, for which it provides decisive evidence.

---

[1] I have developed these ideas in more detail in (Alcolea Banegas, 1997).

4.4 Regular and Critical Mathematical Arguments

Next we will show the adequacy of Toulmin's model of argumentation to mathematics. In regular arguments, the aim is to establish a conclusion by appealing to generally accepted mathematical ideas. These arguments give support to more or less simple conclusions, backing them with other reasons or mathematical facts. Let us consider the four colour conjecture (see Appel and Haken, 1977) as proved with the help of a computer, stating:

(C) Four colours suffice to colour any planar map.

As the necessary reasons or data we can provide theoretical results or factual reports giving evidence in support of the thesis. So:

(D_1) Any planar map can be coloured with five colours.
(D_2) There are some maps for which three colours are insufficient.

These mathematical reasons or data are theorems of graph theory, which have been justified from the axioms, definitions and other theorems of this theory. Consequently, they are the conclusions of other mathematical arguments. We also have available a factual report:

(D_3) A computer has analysed every type of planar map and verified that each of them is 4-colourable.

How will a mathematician justify the step from this data or other reasons (D) to the original thesis or conclusion (C)? He needs a warrant for this step. How can one be sure, for example, that the computer has done the expected work? The warrant authorising us to give a rational support to our thesis would be:

(W) The computer has been properly programmed and its hardware has no defects.

The argumentation proceeds with some confidence because the accumulated previous experience supports the safety of the given warrant. This is to say, the warrant is reliable and can be used because it rests on the proper backing:

(B) Technology and computer programming are sufficiently reliable.

It can be remarked at that point that the task of showing in what way particular warrants are properly backed by theories, observations or scientific experiments is a scientific rather than a logical task. In our case, the representation of our argument does not completely warrant the relevance of B for W, but highlights the specific requirements that our theories must incorporate to make certain that their warrants are well-based. One of these requirements, for example, is the verification of the computer software. Even then, the given argument supporting the four colour conjecture is not the only one, in the sense that, as indeed was the case, other computers or software can be used to verify the conjecture, obviously giving other arguments. This reinforces the *presumptive* character of the conclusion of our argument. Furthermore, since no strictly mathematical warrant is available, the door remains open for a specific counterexample, that is to say, a particular map that cannot be coloured with four colours might still exist. The importance of this regular argument lies less in the conclusion it establishes than in the fact that it highlights, as we have already mentioned, the issue of the computer-assisted proof of mathematical theorems. Mathematicians find themselves in the uncomfortable situation of having to accept both the fallibility and the indispensability of the computer. Meanwhile, some of them are still searching for satisfactory reasons that would validate the conjecture.

Mathematicians should not only use arguments in this regular manner, but must also criticize and improve them. For this reason, there is a need for critical arguments. They represent the way mathematicians argue about the proper form of their own regular arguments. That is to say, when we move to the level of criticism the rational procedures of explanation, justification, *etc.* used in regular mathematical arguments fall under scrutiny. Consequently, the subject matter of the critical arguments is constituted not only by objects and facts of the mathematical world, but also by mathematical theories about these objects and facts. The immediate questions in judging these arguments have nothing to do with either the mathematical universe or with the suitability and aptitude of mathematicians' present ideas about this universe.

As an example of critical mathematical argument, we will try to reconstruct, this time in more detail, what we may describe as the argument for the indispensability of the axiom of choice. In simple terms, Zermelo's work on the axiomatisation of Cantor's set theory showed that the latter could be placed in a framework free from the difficulties posed by the paradoxes. To give some sense to the fundamental mathematical concepts, the original idea of 'set' (which was too wide) had to be replaced by the Zermelian idea of 'set' in axiomatic terms, with all that this implies. Having reviewed the issue from the point of view adopted in this work, the most important problem Zermelo confronted was the explicit introduction within his theoretical framework of the axiom of choice. Since the concept of well-ordered set was essential to Cantor's theory of sets of points and he thought that any well-defined set should be considered to be well-ordered, he soon realized the need to demonstrate this fact. In 1904, Zermelo demonstrated what would come to be known as the well-ordering theorem. His argument was based on a new supposition, that is to say, the axiom of choice. It is not an easy axiom for non-mathematicians to understand, but Zermelo thought that it could not be reduced to a simpler principle and that it was self-evident. Furthermore, he pointed out in his own defence, that it was necessary as an axiom of mathematics, in order to explain and establish many elementary and fundamental facts, and that it had already been used in an implicit way to this purpose. However, the reception of the axiom by the mathematical community was not wholly positive. It is not our purpose here to enter "the great debate resulting from Zermelo's note (1904)", in the words of one of the critics (René Baire) at the time of its publication. We will say only that, in 1927, Nicolai Luzin was still expressing the belief that "all the arguments which can be invented to support this axiom are psychological in nature...", maintaining a psychological perspective adopted, since the axiom's introduction, by such mathematicians as the aforementioned Baire, Emile Borel and Henri Lebesgue.[2]

The problem is one of a class of general scientific questions, that is to say, a set of mathematical phenomena or problems which cannot be explained, or whose explanation is very difficult without the axiom, and which could be fitted perfectly into the greater fabric of mathematics, as a result of the conceptual changes proposed by Zermelo in his axiomatisation. For this reason, Zermelo's *thesis* will take the following form

(C) The axiom of choice is indispensable (Zermelo, 1908b, 187–188).

What kind of *reasons* are needed to support this thesis? If we propose to mathematicians that they change their regular form of argument, that is to say, use the axiom of choice explicitly, they will ask the reason for them to do so, that is to say, what advantages will the adoption of the proposed changes bring, etc. The kind of reasons more directly relevant to this sort of critical mathematical thesis include proofs that Zermelo's axiomatic theory can be used to deal with phenomena and problems which cannot be explained without the presence of the thesis. The best evidence Zermelo could offer for the virtues of his new concepts and of his new axiomatic presentation of set theory was, for example, to show how the use of these concepts could, in a much simpler fashion, account for phenomena which Cantor's explanation had left unfinished or surrounded by mystery, as is the case with the theory of transfinite cardinals, the theory of real functions or projective sets, or even Dedekind's theory of finite sets. Thus we may codify these reasons:

(D) There are many problems and phenomena that can be treated with the axiom of choice (Zermelo, 1904, 141; Zermelo, 1908b, 187–188).

Hence these critical mathematical arguments have a *pragmatic* character. Mathematics has a task and a mission to accomplish, and the proposed changes in the mathematical procedures have to be justified by demonstrating the way they contribute to this mission. In accordance with this, critical mathematical arguments have *warrants*. In our case the warrant could be stated as follows:

(W) Despite the presence of unproved principles, an axiomatic set theory that makes sense of mathematical phenomena, deserves to be accepted in view of the facts (Zermelo, 1908b, 187).

In each of our possible examples, the best reasons for modifying our ideas about the specific issue are based on demonstrating the real results of the proposed changes, that is to say, demonstrating what else can be explained, or providing a more elegant and comprehensible explanation in terms of the modified ideas. In this sense, the introduction of the axiom makes for a better and more effective theory.

---

[2]Baire's words are quoted, and the Luzin citation is discussed, in (Moore, 1982, 313, 288, respectively).

Although we will now proceed to search for a *backing* for the introduced warrant, it is not very clear that its reliability or coherence needs a backing. In introducing the reasons these critical arguments require it is enough to show that the suggested theoretical changes would make a genuine contribution to mathematics. This does not imply that such a warrant lacks the backing that could correctly be demanded of it. In fact, in Zermelo's work we can find more than one backing. For example, he says that the theory of sets "is an indispensable component of mathematical science" (Zermelo, 1908a, 200) and that "principles must be judged from the point of view of science" (Zermelo, 1908b, 189). Perhaps the most suitable backing, the one most in agreement with Zermelo's work, and the one we will use here, could be:

> (*B*) Our mathematical theories must explain the biggest possible area of the mathematical universe.

If the basic mission of mathematics convinces us, in any way, then the warrants backing Zermelo's arguments will also be convincing. Beyond this there is nothing more to say about the backing of this critical mathematical argument. On the other hand, the strength of the *modal* qualifiers will weaken the conclusion by indicating the possibility of exceptions or refutations. For our argument, the most suitable modal qualifier is, without any doubt,

> (*M*) In the light of the facts.

But, despite the successes of Zermelo's axiomatic theory, with the later additions and modifications made by other mathematicians, nobody has proved its consistency, nor the existence of a contradiction. For this reason, in giving a complete analysis of our example, we have to anticipate all possible and conceivable refutations. So it is necessary, as Zermelo realized, to state:

> (*R*) Unless a contradiction is found as a consequence of the axiom (Zermelo, 1908b, 189).

Consequently, with all the mathematical requisites in place, we have completed the pattern of argument by which conviction may be leant to Zermelo's thesis. The relevant diagram given in Fig. 4.2 should be read as follows: Since there are numerous problems and phenomena that can be dealt with by the axiom of choice, and because axiomatic set theory can make sense of mathematical phenomena, since our mathematical theories explain the largest area of the mathematical universe, in the light of these facts, the axiom of choice is indispensable, unless a contradiction is found as a consequence of the axiom.

4.5 Conclusions

Mathematical practice has been the starting and the end point for our application of Toulmin's argumentative model. We have attempted to show how it is possible to set out and analyse the structure of arguments without using the traditional notion of logical form. Mathematicians are, like all human beings, generally committed

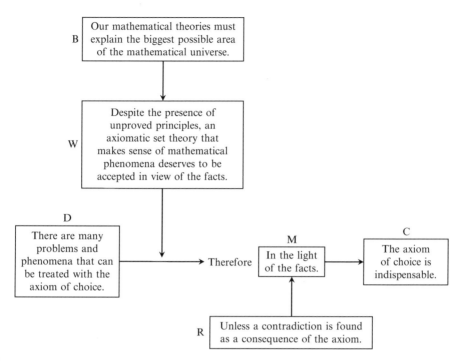

Fig. 4.2 The pattern of argument by which conviction may be leant to Zermelo's thesis

to their ideas and opinions, but they share a common interest in establishing coherent and sound mathematical results, not dependent on the peculiarities of the mathematicians who suggested them. For this reason, the debate over these ideas and opinions within the mathematical community is a way to highlight its rational and scientific character. As long as the content of a mathematical argument is explicit and exposed to criticism, it is not necessary for the argumentative procedures of mathematical debate to be rigid and formal. We are not talking about the formal validity of arguments, but of their relevance, sufficiency and acceptability, which depend on the strength the warrant has from the data to the thesis that is to be established, and on their ability to convince those participating in the debate. In this way, new light is shed on real mathematical argumentation. It is possible that the interpretation of proofs as argumentation will not serve to generate new mathematical knowledge, but may instead permit progress in the comprehension of specific aspects of mathematics and create an alternative route to the knowledge of already established theses. Furthermore, the explanatory capacity of the proposed model of argumentation allows those not familiar with mathematics to acquire mathematical knowledge, where otherwise they would be left in the dark.

Finally, the application of Toulmin's model to mathematics highlights once more the parallelism between mathematics and natural science. In the light of the regular argument that established the four colour conjecture, in which we we saw the experimental character of some of the data and of the warrant, or considering that there is no strong difference between the way Zermelo justifies his axiom of choice and the way other scientific principles are introduced, it is difficult to maintain that mathematics is totally different from the rest of the sciences. Mathematical arguments, especially critical ones, clearly support the quasi-empirical nature of mathematics. Mathematicians have to search for the reasons supporting a thesis, reasons that are only clear when examining those reasons' potential for developing a theory. In this way, it is possible to imagine scientists and mathematicians fighting to achieve theories of great explicative and predictive power, which are at the same time simple and fruitful, and this image can be applied to the scientists and mathematicians of any period of history. While searching for simplicity, efficiency and suitability of explanation and prediction, we can detect a rational development of the ideas of explanation and prediction through critical argumentation. These methodological ideas are always connected to more general scientific objectives and to theories yet to be presented, and they arise from reflection upon the most successful methods of mathematical research, in an effort to pinpoint the secret of success, and to use it as a pattern for any mathematical activity. It is clear that in both cases argumentation exercises a fundamental role.

Acknowledgements Originally published in Catalan as 'L'argumentació en matemàtiques', in E. Casaban i Moya, editor, *XIIè Congrés Valencià de Filosofia*, València, 1998, 135–147. Translated into English by Miguel Gimenez and Andrew Aberdein.

References

Alcolea Banegas, J. (1997). Demostración como communicación. In A. Estany & D. Quesada (Eds.), *Actas del II Congreso de la Sociedad de Lógica, Metodología y Filosofía de la Ciencia en España* (pp. 73–77). Barcelona: Servicio de Publicaciones de la Universidad Autónoma Barcelona.

Appel, K., & Haken, W. (1977). The solution of the four-color-map problem. *Scientific American, 237*(4), 108–121.

Gödel, K. (1995). *Unpublished philosophical essays*. Basel: Birkhäuser.

Hersh, R. (1997). Prove—Once more and again. *Philosophia Mathematica, 5*, 153–165.

Moore, G. H. (1982). *Zermelo's axiom of choice*. Berlin: Springer.

Thurston, W. P. (1994). On proof and progress in mathematics. *Bulletin of the American Mathematical Society, 30*, 161–171.

Toulmin, S. (1958). *The uses of argument*. Cambridge: Cambridge University Press.

Toulmin, S., Rieke, R., & Janik, A. (1979). *An introduction to reasoning*. London: Macmillan.

Zermelo, E. (1967 [1904]). Proof that every set can be well-ordered. In J. van Heijenoort (Ed.), *From Frege to Gödel: A source book in mathematical logic, 1879–1931* (pp. 139–141). Cambridge, MA: Harvard University Press.

Zermelo, E. (1967 [1908a]). Investigations in the foundations of set theory I. In J. van Heijenoort (Ed.), *From Frege to Gödel: A source book in mathematical logic, 1879–1931* (pp. 199–215). Cambridge, MA: Harvard University Press.

Zermelo, E. (1967 [1908b]). A new proof for the possibility of a well-ordering. In J. van Heijenoort (Ed.), *From Frege to Gödel: A source book in mathematical logic, 1879–1931* (pp. 183–198). Cambridge, MA: Harvard University Press.

Chapter 5
Arguing Around Mathematical Proofs

Michel Dufour

> In the history of Mathematics, you can notice that during some
> epochs no damage is done to the truth of particular
> propositions, but their systematic linking has changed because
> of the rapprochements allowed by new discoveries.
>
> *(Lacroix, 1797, Preface)*

Any theory of argumentation will certainly have to pronounce on the status of mathematical proofs. Formally, a typical proof is a string of regular inferential steps between statements. So, it can be seen as a string of arguments since each statement, except the first one, is supported by the reasons offered by previous statements. But is this enough to claim that argument practically matters to proof or even that proof is a kind of argument? Some authors have argued that, in spite of this strong family resemblance, mathematical proofs are not arguments: hence the temptation—and its methodological consequences for a theory of argumentation—to keep the study of mathematical proofs away from the study of arguments. But the exclusion of mathematical proof from the field of argument theory has also been disputed by philosophers who argued, mostly on practical grounds or by stressing that the border of mathematics has changed over time and place, that the use of such a proof is not incompatible with an argumentative practice (Corfield, 2002; Finocchiaro, 2003; Dove, 2007; Aberdein, 2011). This paper aims at supporting this view, and claims that a theory of argumentation should encompass at least the proof process and would benefit from looking beyond the idealized situation of the presentation of the proof to an audience of expert peers.

We shall begin with a discussion of the views of two authors—Perelman and Johnson—who claim, on different grounds, that proofs are not arguments. The

M. Dufour (✉)
Department of Communication, Sorbonne Nouvelle, 13 rue Santeuil,
75231 Paris Cedex 05, France
e-mail: mdufour@univ-paris3.fr

A. Aberdein and I.J. Dove (eds.), *The Argument of Mathematics*, Logic, Epistemology, and the Unity of Science 30, DOI 10.1007/978-94-007-6534-4_5,
© Springer Science+Business Media Dordrecht 2013

second part of the study asserts that, even when you focus only on proofs, the practice of mathematics is full of arguments, different in styles and goals but amenable to a classification based on their temporal relation to the publication of the proof.

5.1   A Proof Is Not an Argument

Perelman is well known in the folklore of argumentation studies as one of the two founding fathers of the mid-twentieth century renewal of academic reflection on argumentation, Toulmin being the other. His very project of a "new rhetoric" (now 50 years old) is based on a radical distinction between proof and argument that will be discussed first. Then we shall turn to Ralph Johnson, a leading figure in the informal logic movement. Johnson's position is also based on a radical distinction between proof and argument. But his view is less systematic than Perelman's, for (mathematical) proof and everyday argument would belong to what Johnson calls the "spectrum" of applications of the term "argument".

5.1.1   Argument and Proof, the Two Poles of Perelman's System

Perelman's starting point was his dissatisfaction with the principle held by some philosophers—especially in the context of logical positivism[1]—that (formal) logic could provide a general theory of human reasoning and a suitable tool to analyse human inferences. He did share the positivist idea that logic is convenient for science, but denied that beyond the area of logic, mathematics and empirical sciences, human thinking is fuzzy and even irrational since it does not lend itself to logical analysis. As a jurist, Perelman could not discard value or moral judgments as irrational since they are an essential part of legal argumentation.

A consequence of this criticism is that he and Lucie Olbrechts-Tyteca, in their famous *New Rhetoric: A Treatise on Argumentation* (1958), defined and investigated a whole domain spreading beyond the field of hard sciences but precluding logic, mathematics and reasoning used in the empirical sciences. The way of reasoning specific to that domain is even characterized in contradistinction to the style that Perelman, like the positivists, held to be typical of science. The *Treatise*, like many of Perelman's other writings, states this point explicitly, its opening pages being devoted to this principle which justifies the whole enterprise:

> Formal logic constituted itself as the study of the means of demonstration used in mathematics. But a consequence is that its scope is limited, for all that is ignored by the mathematicians is foreign to formal logic. Logicians have to add a theory of argumentation to their theory of demonstration (Perelman and Olbrechts-Tyteca, 1958, 13).

---

[1]The term "positivism" is used loosely. Deeper philosophical subtleties are not essential for the point made here.

It is worth noticing the shift made here from logic to mathematics. The authors seem to think that mathematicians (rather than logicians?) frame the scope and program of formal logic. This tendency to take both sciences as very close to each other and, sometimes, to identify them was not unusual among some philosophers of Perelman's era.

But it is in Aristotle that the authors of the *Treatise* found two important ideas for their program. First, they explicitly borrowed a distinction between analytical, dialectical and rhetorical arguments that seemed to confirm their idea of a sharp contrast between logic and argumentation. They also granted that different domains of activity require different kinds of reasoning. As the master said in his *Nicomachean Ethics* (Aristotle, 2009, I, 1094b 25), do not expect a probable reasoning from a mathematician and a proof from a rhetorician.

This fundamental dichotomy between logico-mathematical reasoning and unscientific arguments that the *Treatise* qualified as rhetorical, after hesitating over dialectical, is supported by a whole set of other distinctions systematically used in an exclusive way. Let us mention, for instance, demonstrative/non conclusive, analytic/dialectical, true/persuasive, certain/probable, or rational/reasonable. Although Perelman does not explicitly use such an expression, his view of logic and scientific reasoning is "purely semantic", in the sense that it lacks any pragmatic dimension brought about by human interactions. Hence, in science, the only concern would be the truth of propositions. On the contrary, argumentation "never happens in a void. For it presupposes a contact between the minds of the orator and of his audience: a speech must be heard, a book read, otherwise they would not act" (Perelman, 1977, 28).

For sure, the *Treatise* and the *Realm of Rhetoric* are logic free. The most logical arguments discussed therein are the ones dubbed "quasi-logical". What makes them so? They appeal either to "logical structures" (contradiction, total or partial identity, transitivity) or to "mathematical relations" (whole/part, smaller/bigger, frequency ratios). Hence, although they are driven by logic or mathematics they keep being arguments since they are inconclusive, a distinctive feature of perelmanian arguments (Perelman and Olbrechts-Tyteca, 1958, 261). They look formal but are not and should not be taken as such. Their misplaced prestige comes from the prestige of logic or mathematics. The *Treatise* adds that "explicitly based on mathematical structures" they were cogent "in the old days, and especially among the Ancients", but now "just like formal logic allowed separating demonstration and argumentation, the development of the sciences certainly helped to limit their use to the field of calculation and measurement".

Besides providing an example of the Perelmanian confusion between logic and mathematics, this quotation confirms that the distinction between demonstration and argumentation overlaps precisely with those between conclusive and inconclusive and between scientific and practical reasoning. Although the introduction of the *Treatise* announces that examples of arguments will be borrowed from the "human sciences", a careful reading shows that no argument or, to be more careful, almost no example of argument is borrowed from any science.

Does that mean that the authors held "scientific practical reasoning" or "the logic of an unscientific discourse" to be inconsistent expressions? It is not that simple. Although they see argumentation as essentially pragmatic (interactive) and science as purely semantic since "demonstrative and impersonal" (Perelman, 1977, 28), they make some general comments about scientific meetings between peers. It is worth noting that these remarks are closely connected with their discussion of the "universal audience", a notion that blurs their numerous dichotomies for "ultimately, the rhetoric efficient for a universal audience would be the one handling only the logical proof" (1958, 42).

Perelman's notion of a universal audience is a bit fuzzy. It reflects his methodological hesitation between an empirical approach and a normative one, more or less inspired by Kant. The universal audience is neither the whole of humanity nor "any rational being": in the *Treatise* it is "at least, adults and normal men" (1958, 39) and in the *Realm of Rhetoric* "at least, its competent and reasonable members" (1977, 32). So, let us ask two sets of questions. First, to the *Treatise*: Why adults? Isn't "normal" enough? Wouldn't normal teenagers or even children make the cut? Then, to the *Realm*: Doesn't competence imply it is used reasonably in the field concerned? In other words, if someone becomes unreasonable (in her usual field of competence), isn't it a good reason to think that her competence has gone astray or has been cancelled for a moment or a joke?

The fact that the very notion of competence normally implies its reasonable use certainly explains why competence alone appears in Perelman's discussion of the cogency of arguments. According to him it depends only on two factors, usually held to be independent: the size of the audience and its "quality", a notion he takes to be closely related (if not identified) with competence.

Who is entitled to evaluate the cogency of an argument? For Perelman, as far as argumentation is concerned, it is up to the audience to decide (1958, 32). This is why an evaluation made by a (the?) universal audience would have the last word since it is the widest competent audience. But this does not disentangle the problem of the balance between number and competence, a problem especially important in science where audiences are often small and competent.

According to Perelman, competence comes first for two kinds of audiences.

One is the elite audience. It is small because it can listen to arguments that do not convince many people. Hence its tendency to disqualify opponents, claimed to be stupid or not normal, when it sees itself "endowed with exceptional and infallible means to get knowledge" (1958, 44). Sometimes, an elite audience is widely praised by outsiders. Other times it is taken as ridiculous when compared with "the number and the intellectual value" of the people who have been rejected.

The second is the specialised audience. A paradigmatic example is a peer audience listening to a scientific talk. It is often assimilated to the universal audience by the speaker "supposing that all men with the same training, the same competence and the same information would grant the same conclusions". But this is just the opinion of the speaker and the authors of the *Treatise* do not say if this view is shared by the members of the audience or what happens if the audience, or part of it, disagrees with the orator. Here again, Perelman seems to believe that

scientific proofs cannot be controversial since they are "logical". Appealing here
to the universal audience is a comfortable way to escape the pragmatic problem
raised by the opposition of competent members of the audience. Finally, even when
Perelman stresses the variety of actual audiences and the fragility of the link between
the orator and her audience, his conception of scientific reasoning precludes the
possibility of an argument among scientists, especially in mathematics.

5.1.2 Johnson: The Autonomy of Experts

Ralph Johnson's conception of argument is not based on an opposition between
logic and dialectic or rhetoric, nor associated with different fields of knowledge or
specific epistemic attitudes as in Perelman's system. Rather, in *Manifest Rationality*
logic and dialectic are federated into a single entity, for Johnson contends that an
argument has two faces. One of them is what he calls the illative core, namely a
discursive structure where reasons support a thesis. But he claims that this logical
aspect cannot account for the pragmatic dimension of an argument. Therefore, it
has to be supplemented with a second face named the dialectical tier. It must be
added that this does not mean that an argument is always put forth in a context
of divergence of opinions. It suffices that "the conclusion is at least *potentially*
controversial" (Johnson, 2000, 206).

   The notion of dialectical tier is closely connected with the practical behaviour
of the arguer. For it is the way she discharges what Johnson calls her "dialectical
obligations", bound to the illative core of the argument, that makes her rationality
manifest. So, the utterance of an illative core is not enough: it must be accompanied
by a convenient dialectical behaviour to be evidence of rationality (2000, 164).
These dialectical obligations—which have been widely discussed—are not limited
to the critical attitude of examining and anticipating objections and considerations
running against the conclusion of the argument, but also take into account mis-
guided or irrelevant criticisms, because "to ignore such criticisms compromises the
appearance of rationality" (2000, 270).

   The main point for us is that the dialectical tier is the extra part which makes
the main difference between a mathematical proof and an argument, for "no
mathematical proof has or needs to have a dialectical tier" (2000, 232).

   Is that true? To illustrate the potential of controversy of an argument Johnson
says: "There are those who take a different view; there are adversary views; there are
typically well-known objections. [...] An argument that does not take into account
these dialectical realities is in some important sense incomplete" (2000, 206). I shall
argue below that unless you presume that mathematical proofs are complete they
often have to deal with these kinds of dialectical realities. Some proofs are more
than "potentially controversial" and even if Johnson is right that, sometimes, they
do not need a dialectical tier, they often do urgently need an explicit one. To be fair,
this reading of Johnson should not be exaggerated for in other places he supports
a spectrum theory claiming that the word argument can be applied to scientific

theories and proofs (2000, 168). But he insists that the core notion, the prototype of the concept of argument, is outside of the scientific field. And he has a social epistemic comment about the distance between proof and argument: "The proof that there is no greatest prime number is conclusive, meaning that *anyone who knows anything about such matters* sees that the conclusion must be true for the reasons given" (2000, 232, my emphasis).

Let us take this remark about mathematics and competence as an opportunity to make a comparison with Perelman's position, which relies on a sharp contrast between his semantic view of science—in short, science is impersonal, essentially related to the world and only concerned with truth—and a conception of argumentation that is essentially audience relative, hence pragmatic and mostly concerned with agreement. We know this contrast is crucial for the intellectual and social independence of Perelman's vivid rhetorical realm spreading beyond the stark world of logic and calculus. And we remember that it is only in the context of an exchange between peers that Perelman and Olbrechts-Tyteca made a timid step toward the notion of scientific argument. But they finally canceled the possibility of a scientific argumentative idiosyncrasy by immediately calling upon the normative principle of the expertise of scientists which led them to assert that such an audience "is generally considered by the scientist, not as a particular audience, but as the true universal audience" (Perelman and Olbrechts-Tyteca, 1958, 45).

Johnson does not share Perelman's principles and introduces a concept of argument which is less exclusive. But when he is concerned by an audience consisting of "*anyone who knows anything about such matters*", his view becomes somewhat similar to Perelman's. First, a necessary condition to make an argument a candidate to become a proof is to get rid of the idiomatic and epistemic differences giving rise to the dialectical tier. Johnson's "anyone knowing anything about such matters" makes a quite acceptable equivalent to Perelman's "competent audience". Then, Johnson's quasi-pragmatist statement that a proof is conclusive when "everyone recognizes the proof as a proof and as a result, there is no longer any debate about whether the conclusion is true" (Johnson, 2000, 232) reminds us of Perelman's statement that "ultimately, the rhetoric efficient for a universal audience would be the one handling only the logical proof". A vexing question is: could there be an ultimate agreement over a proof that would not amount to a logical proof? Perelman answers no and Johnson answers that a proof is conclusive when any expert "sees that the conclusion must be true for the reasons given" (2000, 232), a claim reminiscent of Aristotle's thesis that scientific syllogisms are not open to discussion.

5.2 A Mathematical World Besides Proofs

So far, our discussion about the possibility of mathematical arguments has focused on the status of mathematical proofs and their dialectical potential. We have just seen that most of Perelman and Johnson's objections are based on an antagonism between their concepts of argument and features, for instance necessity, that they think typical of mathematical proof. Perelman's position goes even further since he uses this opposition as grounds for excluding the whole domain of the logico-mathematical sciences from the realm of argumentation.

This extreme position, although reliant on considerations about proof, forgets that proof is not all there is in mathematics. According to ethnomathematics, for instance, other cultures, especially traditional cultures having no writing, developed mathematical ideas and skills which are out of touch with the activity of proof and the focus put on it by Western professional mathematicians. In this case, mathematics is embedded in activities or general conceptions which are not traditionaly identified as mathematical, such as religious or metaphysical ideas, games, administrative tasks, economical activities, kinship relations (Ascher, 1991; D'Ambrosio, 2001). Accordingly, although proof is important in mathematics, at least for Western mathematics, it may not be a necessary condition for defining an activity as mathematical. As we shall see, loose notions of abstraction and systematicity seem to have been more important as organizing concepts for Ancient Greek mathematics than necessity and deduction. So, the extension of the field in which mathematical argumentation could be sought is far from well-defined, especially when the distinction between pure and applied mathematics is blurred (as it was for Western mathematics until the mid-nineteenth century). This matters for the very notion of a mathematical argument.

However, my focus will stay on mathematical proof since it is the core of Perelman's and Johnson's objections to the notion of mathematical argument. And, as I subscribe to their main insight that argumentation is essentially pragmatic, I suggest that it is not by looking at proof itself but at its use that its argumentative dimension can be illuminated. This is why, rather than wondering where arguments can be found in a mathematical proof, I shall address the question "When are mathematicians arguing?". The leading idea is that a proof looks like a totem-pole and that arguments can be found if you look at people bustling around it, especially at three typical moments: before, during or after its construction.

5.2.1 Before the Proof

Any beginner in mathematics knows that the trouble with proofs is that they have to be found out. It suffices that the master says: "Show that p" to infer and believe first that p is true (but beware of devilish teachers who do not hesitate to ask for the proof of a false p), second that there is a path to the solution (although there may be none), third that it can be discovered (although it may be impossible for a human mind).

Expert mathematicians usually have no omniscient master anymore except those who, like Plato, Leibniz or Cantor, live in the shadow of a mathematician God. When they try to show that p, they (usually) try to answer an open question which, sometimes, depends on definitions and methods that they have to frame or reframe. The search for a proof is an important part of the activity of professional mathematicians, maybe the essential part.

Yehuda Rav expressed a similar view by saying that there is a way to escape foundational problems, including the priority given to axioms. It suffices to "realise that *proofs rather than the statement-form of theorems are the bearers of mathematical knowledge.* Theorems are, in a sense, just tags, labels for proofs, summaries of information, headlines of news, editorial devices" (Rav, 1999). But the same move can be applied to proofs. Even if mathematical truths and proofs exist somewhere in a Platonist heaven, to contemplate them may not be the highest good. You may feel the compulsive demand to lay your hand on a methodological scale and, finally, think that the supreme mathematician is the one who finds the scale to the proof. Mathematical know-how is certainly praised by mathematicians as much as the theoretical knowledge of theorems and proofs.

Time spent hunting for a proof is a good time for arguments. First, you have to be convinced that trying to show that p is worthwhile. This granted, methodological questions matter and arguments for or against such and such an approach come to the forefront of the mathematician's activity. In the last century, philosophy of science labeled as "context of discovery" a notion reminiscent of what ancient philosophy and rhetoric called "analysis", understood as the art of invention (*ars inveniendi*), that is of discovering convincing arguments or clever starting points for a proof. This analytical stage offers many opportunities to argue. For instance, in March 1847, two rival French mathematicians, Cauchy and Lamé, claimed to have found a proof of Fermat's famous Last Theorem (Singh, 1997). Each of them published independently part of his proof after having left a complete proof in a sealed envelope at the French Academy of Science. Two months later, after having read their partial proofs, Kummer, a German mathematician, criticized their approaches. He thought they were not necessarily false but likely on a wrong path. Lamé granted Kummer's objection but Cauchy argued that his approach was less exposed to Kummer's criticism which he thought was not clearly demonstrated. Cauchy resisted for a few weeks but finally made up his mind and stopped publishing on this topic.

Lakatos's famous book (1976) about Euler's conjecture provides many other examples of arguments produced during the gestation of a proof, especially in Chap. 5 concerning hidden lemmas working as hidden premises and the fallacious appeal to what "everybody knows" or "any expert in the field knows". The history of enduring problems—and this is not unique to mathematics—is often full of arguments about methods, definitions or provisional solutions. Some of them even aim to renew the very problem at stake. Famous examples are also provided by the cantorian "continuum hypothesis" (Hallett, 1984).

Descartes's confessions about his intellectual formation and the way he came to look for a *mathesis universalis* shed some more light on the arguing process prior to the proof and the mythical status of the analysis, at least since Ramus's time (Timmermans, 1995). After stating his fourth "rule for the direction of the mind" he explains that when he began to study mathematics he read "almost everything of what is usually taught by the authors dealing with it" (Descartes, 1908). But he felt unsatisfied because "they did not show clearly enough to the mind why it [the proof] is done that way and how its invention was made". Descartes was certainly not very interested in the mere checking of the validity of a proof. He even claimed that after having "tasted" mathematics, most "men of talent and knowledge" are usually not interested anymore by such a "childish and trivial" activity. Men of talent— like him—look for something more exciting, the path to a discovery. This tends to confirm Rav's idea that theorems are not very important and are just tags, but also that the search for proofs is more exciting for mathematicians than the proof itself. Complete and stabilized proofs are themselves milestones, totem-poles, as I said, which have become "trivial" except when they stimulate new challenges and new arguments.

Descartes shared with many of his contemporaries a dream about the future of mathematical arguments: he believed that the legendary analysis of the Ancients opened the path to what he called a "true mathematics". This famous *mathesis universalis,* whose obviousness and simplicity should convince any rational mind, would avoid the stumbling process of critical dialectic by opening a direct path to sound demonstrations.

A new Perelman, while still borrowing from Aristotle's classification of reasonings, could object that in spite of a partial homophony, analysis, understood as the search for a proof, should not be confused with the analytical proof that may emerge afterwards. He could claim that arguments preliminary to the proof pertain to metamathematics, for they are not so much the proof itself as about it. Hence they should be left in the surroundings of mathematical knowledge rather than incorporated into it. But aside from the slightly paradoxical fact that Descartes's "true mathematics" would then also belong to these surroundings, a reply to our new Perelman is that a sharp distinction between language and metalanguage or theory and meta-theory presupposes a clear-cut, if not formalized, language, theory or discipline. But this is precisely the stance that opponents to the perelmanian dichotomies between argument and proof or scientific and non scientific reasoning do not want to take.

5.2.2 During the Proof

When a candidate to the status of proof has been established, the time has come for its public and critical evaluation by competent experts who may not have participated in its discovery.

If we assume that a proof is the linguistic expression of a reasoning which is the achieved and flawless production of an expert mind, as expert as any other expert mind, the distinction between producer and receiver of the proof is only a contingent matter. This presumption of equality or equivalence of expertise seems to lie behind the opinions of Johnson and Perelman when they claim that proofs do not need dialectical (Johnson) or rhetorical (Perelman) support. An expert needs nobody to grasp a proof, otherwise she is not an expert.

This view leads to the idea that anything added to a minimal version of a proof is at most mere decorum, a rhetorical ornament, a "colour" to borrow Frege's word (see Dubucs and Dubucs, 1994). This adjuvant would bring about only side effects since the proof is supposed to be self-sufficient to be convincing. If the very notion of argument implies a human interaction—let us say a dialectical tier—a perfect proof (i.e. a proof raising no critical comment, no request for explanation and so forth) certainly differs from an argument. This is probably the idea at the core of Johnson's view.

But from a practical point of view, if you drop the normative hypothesis of equal and perfect expertise such a situation seems to be the exception rather than the rule. It is likely that perfect proofs and equal expertise exist mostly in "the world of novels" as people said in Descartes's time.[2] So, let us have a look at the conditions of our starting assumption.

How do you know that a reasoning is flawless? The correction of a mistake of reasoning is likely to come from someone other than the proponent of the proof. This offers the opportunity for a refutation. Moreover, the conversion, not to say the translation, of the reasoning into its linguistic expression may produce specific defects like confusion or equivocation. Like anybody, a mathematician may fail to express clearly a correct reasoning and this may be another occasion for an objection.

The works of Cauchy provide several examples of the case at hand (Belhoste, 1985). Besides his own innovative works, Cauchy criticised and renewed several aspects of the works of his masters. In his *Cours d'Analyse* he explains that he tried to give to his new methods "all the rigour that one can demand in mathematics so that I never rely on the reasons based on the generality of algebra". The trouble that he had with these "algebraic" reasons was that they lacked rigour: "although commonly granted ... they can only be considered as inductions which can sometimes make you feel the truth, but fit badly with the celebrated exactness of the mathematical sciences" (Cauchy, 1821, Introduction). Among these "new methods", there is the introduction of a concept of continuity based on the concept of limit and breaking with the previous notions, especially Euler's. Cauchy's definition states that "$f(x)$ will stay continuous for x taken between two given limits, if, between them, an infinitely small increase of the variable always produces an infinitely small increase of the function itself" (1821, 43).

Two years later, some of his critics are more explicit in the *Résumé* of the lectures on the calculus that he gave at the Ecole Royale Polytechnique. He argued against the systematic use of Taylor's formula that was common among mathematicians because "although the famous author of the *Mécanique analytique* [Lagrange] based his theory of derivative functions on this formula ... most geometers now grant the uncertainty of the results that can be obtained when using diverging series" (Cauchy, 1823, Avertissement). Cauchy's point was that Lagrange's approach was limited, owing to his ignorance of the notion of interval of convergence.

---

[2]"Mostly", for a friend told me that a celebrated mathematician visiting his university wrote his lines of proofs on the board without uttering a single word.

The new concepts introduced by Cauchy, and the demonstrations that they allowed, have framed part of the mathematical doxa since the beginning of the nineteenth century and they still inspire the corresponding notions and demonstrations taught in our schools. They are real mathematical milestones.

But unfortunately, in spite of its alleged "rigour", Cauchy's concept of continuity was not rigorous enough. It blurred the distinction between continuity and uniform continuity that our contemporary definitions, using quantifiers, make clear. And this confusion led Cauchy to a confused demonstration of the mean value theorem which required an argument to be rectified.

Up to now, we have supposed that the mathematical audience is composed of experts and that arguments occur because of mistakes, confusions or wrong moves made by the proponent. But some of them can come from the audience. A sound and correctly expressed proof can be misunderstood by some members of the audience, unless you again presume an idealized situation of communication with an audience as perfectly competent and rational as Perelman's universal audience. For instance, a member of the audience may object that the proof has not been achieved because a formula of the demonstration does not follow from the previous formulae. This is not an uncommon criticism and Cauchy's reply to Kummer can be seen as an example.

Standard logical derivations progress by making moves as short as possible, that is, in accordance with only one acknowledged rule. But many mathematical proofs do not comply with these canonical requirements of logic even if most of them can be rewritten in a format that makes their checking purely mechanical. Unfortunately, this option was not open for past mathematics and is still often practically impossible. Accordingly, a gap between two formulas may look hardly intelligible for the audience although it seems clear for the proponent. According to Johnson, the proponent will manifest her rationality by supplementing her proof with answers and replies to questions and objections from failing experts.

On this account, the outstanding romantic mathematician Evariste Galois seems to have had a hard time making his rationality manifest. But the problem may have come from his audiences. We know that the reason of his first failure when he tried to be admitted to the French Ecole Polytechnique was that his answers were too abrupt and unclear. But things got worse, for a few years later, in 1830, he had the same problem when, to win the Grand Prix of Mathematics of the Science Academy, he submitted an already improved version of his seminal *Mémoire sur les conditions de résolubilité des équations par radicaux*. In the preface to Galois's works that he published in 1846 after having deciphered his last writings, Joseph Liouville wrote that: "In their report, the Commissioners reproached the young analyst for the obscurity of his paper, and indeed, this reproach, already made to his previous communications (as we learnt from Galois himself) was justified. An exaggerated desire of brevity was the cause of this imperfection that you should avoid, especially when you deal with the abstract and mysterious topics of pure Algebra." Liouville alludes here to Poisson's comment that Galois's paper was unclear and insufficiently developed to be accepted. But more than a century later, in a new preface, another celebrated mathematician, Jean Dieudonné, expressed his admiration for Galois's conciseness that he praised much more than the "laborious presentations that his immediate followers felt obliged to make" (Galois, 1997).

5.2.3 After the Proof

Let us suppose now that the proof is professionally settled, that is accepted by acknowledged competent experts. This is not the end of the story, for mathematics is not the exclusive property of a handful of industrious shareholders but the common good of an open society.

For centuries, mathematics has been familiar to pupils from the elementary to, sometimes, the academic level. From a less institutional point of view, the didactical use of a proof can be tentatively defined as any situation in which the proponent has to convince people who stand on a lower epistemic footing, if we grant that the notion of "lower epistemic footing" is clear. A didactical situation is then similar to the previous case of an epistemic asymmetry due to failing expertise. In a didactical context, it would be harsh to qualify a very common inability or failure to take up the inferential proposal made by a proof as a defect or lack of rationality of the tutee. Nor is it always the sign of a lack of propositional knowledge, for he may know that p and grant q but not "see" that q follows from p. Even limited to a field, logical omniscience is rare and Perelman's requirement of a universal *adult* and/or *expert* audience is a convenient way to escape any didactical situation coming from epistemic or rational discrepancies.

Although they are commonly associated with pupils or students, didactical situations are not uncommon among scientists, especially since last centuries' history has taught us not to take scientific axioms or principles as obvious. This marks a shift from scientific to dialectical and even didactical argument, at least if we go back to Aristotle's classification of arguments.

In the *Sophistical Refutations* (1955, II, 165a–b) Aristotle claims there are four kinds of διαλέγεσθαι λόγων, an expression generally translated as "argument (or reasoning) involved in a discussion". But it can also be translated as "dialogue" or "dialectic", in the broad sense of "talking together". Although Aristotle here uses neither the word "syllogism", nor "enthymeme", "argument" seems a good translation since the Philosopher stresses that these discourses have premises. And these premises make the main difference between the four kinds of argument. Dialectical arguments are rooted in an *endoxa*, a common opinion. Critical arguments start from premises accepted by the answerer but also granted by the arguer whose discourse aims at "showing that he knows". Eristic arguments reason from premises that appear to be generally accepted but are not so. And finally, didactical arguments do not reason from the opinions of the answerer but from "principles appropriate to each μαθήματος". This word, μάθημα, is usually translated by "branch of knowledge" or "discipline" but it also means "lecture" or "lesson", two notions often related to an educational context. It is also close to μαθηματιχος which means "someone who studies" or "relative to a field of knowledge" and, of course, to μαθηματιχά, usually translated by "mathematics".

For Aristotle, what makes something "mathematical" is the way you consider it. It depends on the properties dropped in the process of abstraction and the principles finally taken into account, some of them being proper and some others not proper to the said science (Aristotle, 1976, I, 10, 76a, 35–40). So, a science is mathematical in the broad sense of "systematic". And a didactical argument is a deductive argument based on the principles of a field of knowledge, of a discipline. If we follow the idea that this kind of argument is based on the principles of the discipline and not on the opinions of the audience, its premises may not have the typical property of Aristotelian scientific arguments announced in the *Posterior analytics* (1976, I, 2, 71b, 20) namely "true, primary, immediate, better known than, prior to, and causative of the conclusion". To put it shortly (perhaps too shortly) the premises of a didactical argument only have to be granted, they are not necessarily "believed" (Dufour, 2011).

Granting premises is certainly the most accommodating way to share them, but it is sometimes a submissive one. Moreover, a pedagogical tragedy is likely to happen if the tutee is accommodating enough to grant a logical step because of the authority of the field instead of acknowledging its necessity or its strength. This is the case with students who learn a proof by heart to pretend they know and understand it, a situation not uncommon but notoriously fallacious.

Avoiding such cases is probably one of the reasons why there is often a long distance between the original proof and textbook or oral classroom versions, full of hints aiming at making it accessible to a wider audience. Although there may be several demonstrations of a theorem, we usually do not hesitate to qualify seemingly different proofs as versions of the same one. This allows to modify or rephrase the initial version to make it more explicit or to make its necessity more salient for people who are not experts. In my opinion, these strategic rewordings belong to the field of mathematical argumentation. Some people would certainly prefer the word "rhetoric", even if these manoeuvres aim at enhancing the process of rational persuasion, for instance by paraphrases which are not mere rhetorical embellishments of the initial proof. Some of these manoeuvres may fall under Johnson's notion of dialectical tier but only if you grant, against him, that mathematical proofs may need one.

The fate of a proof resembles the pragmatist conception of truth for although a proof is sometimes provisionally satisfactory it will attain perfection only at the end of the enquiry that is when nobody or, more faithfully to Peirce (1871), when no member of the community of inquirers will complain about it. Are didactical efforts to make it intelligible to less expert people a step in this direction? A pragmatic pragmatist could answer no, because the more you open the community the more remote perfection becomes. Perelman's small elite communities seem to be forerunners of Peirce's paradise, but unfortunately they are mostly examples of wishful thinking. A less eager pragmatist would answer that a larger paradise is a better paradise; therefore, the teaching of mathematics widens the mathematical community, the possibility of criticism and the expected paradise. But the history of mathematics also shows that conceptual distinctions and the proofs that follow are not always more intelligible for non-experts and so, are open to argument. This seems to have happened when Cauchy had some trouble with his students at the Ecole Polytechnique, although political considerations also mattered in this particular case.

The logicist episode also provides good examples. Frege, like Peirce, complained that natural languages lacked the rigour and precision required to make them decent places to harbour arithmetic. Hence, Frege's investigation in the *Begriffsschrift* (1879) of a new and purely logical language which would allow the rewriting of arithmetical proofs with no logical gaps and no appeal to the benevolence of intuition. This is the core idea that Frege and his followers perceived as improving mathematical proofs. But it required a reformulation, if not a revision, of previous demonstrations.

Unfortunately for the logicist program, not everybody agreed. Numerous arguments were raised, which were not isolated conflicts but a battle in the wider debate about the foundations of mathematics. This affair was neither purely mathematical, i.e. involving only professional mathematicians and field-dependent concepts, nor "above" mathematics, that is free from mathematical technicalities. Mathematics, logic, philosophy and their practitioners were all concerned by the logicist controversy (Heinzmann, 1986). Henri Poincaré was a famous opponent to the logicist program. Anticipating and denouncing Perelman's confusion of logic and mathematics, he contended that mathematics harbours intuitive principles— notoriously the principle of complete induction—irreducible to the principles of the new logic. For him, logicism was viciously circular. Either it stealthily called to mathematical principles to show their reducibility to logical considerations or it inflated the meaning of the word "logic" by directly incorporating mathematical principles or knowledge (Poincaré, 1905a, 808). The most interesting point for us is that one of his criticisms of logicism and formalism was a didactical objection, for he argued that their claim to improve mathematical proofs failed to make them more accessible. Poincaré ironically wrote that Peano's formal definition of "one" is "eminently apt to give an idea of it to people who would have heard nothing about it" (Poincaré, 1905b, 823), that he "would not advise it [Hilbert's mechanical formalism] to a high school student" who would very soon drop it (Poincaré, 1908, 68) and that logicists had a talent to "define what is clear by what is obscure". However, he did not disagree with any program intended to improve mathematical proofs for he thought that proofs lacking rigour have to be reworked. For him, a most significant improvement in mathematics, is an inductive move towards a greater abstraction and a greater generality. And in this case higher accessibility to beginners is not required. According to him, such an improvement had already begun around the mid-nineteenth century and he stressed that this move towards a kind of formalization was not foreign to mathematics since it was not "logicized" but "arithmeticized" (Poincaré, 1908, 69; 1905a, 32). Right or wrong, this example shows that there is room for argumentation in mathematics not only before and during the proof but also after, at least as long as it can be criticized.

References

Aberdein, A. (2011). The dialectical tier of mathematical proof. In F. Zenker (Ed.), *Argumentation: Cognition and community. proceedings of the 9th international conference of the Ontario Society for the Study of Argumentation (OSSA)*. Windsor, ON: OSSA.

Aristotle (1955). *On sophistical refutations. On coming-to-be and passing-away. On the cosmos.* (E. S. Forster, Trans.). Cambridge, MA: Loeb Classical Library, Harvard University Press.

Aristotle (1976). *Posterior analytics. Topica.* (H. Tredennick, Trans.). Cambridge: Loeb Classical Library, Harvard University Press. (Original work published in 1960)

Aristotle (2009). *The Nicomachean ethics.* (D. Ross, Trans.). Oxford: Oxford University Press.

Ascher, M. (1991). *Ethnomathematics: A multicultural view of mathematical ideas.* Belmont: Wadsworth.

Belhoste, B. (1985). *Cauchy: Un Mathématicien Légitimiste au XIXe Siècle.* Paris: Belin.

Cauchy, A. L. (1821). *Cours d'Analyse de l'École Royale Polytechnique.* Paris: de Bure. (Reprinted from *Œuvres Complétes*, p. XV, 1882–1974, Paris: Gauthier-Villars).

Cauchy, A. L. (1823). *Résumé des Leçons données à l'École Royale Polytechnique sur le Calcul Infinitésimal.* Paris: de Bure. (Reprinted from *Œuvres Complétes*, p. XVI, 1882–1974, Paris: Gauthier-Villars).

Corfield, D. (2002). Argumentation and the mathematical process. In G. Kampis, L. Kvasz, & M. Stöltzner (Eds.), *Appraising Lakatos: Mathematics, methodology and the man* (pp. 115–138). Dordrecht: Kluwer Academic Publishers.

D'Ambrosio, U. (2001). *Ethnomathematics: Link between tradition and modernity.* Rotterdam: Sense Publishers.

Descartes, R. (1908). Regulae ad directionem ingenii. In C. Adam & P. Tannery (Eds.), *Oeuvres de Descartes* (Vol. X, pp. 349–469). Paris: Leopold Cerf.

Dove, I. J. (2007). On mathematical proofs and arguments: Johnson and Lakatos. In F. H. Van Eemeren & B. Garssen (Eds.), *Proceedings of the sixth conference of the international society for the study of argumentation* (Vol. 1, pp. 346–351). Amsterdam: Sic Sat.

Dubucs, J., & Dubucs, M. (1994). Mathématiques: La couleur des preuves. In V. De Coorebyter (Ed.), *Rhétoriques de la Science* (pp. 231–249). Paris: Presses Universitaires de France.

Dufour, M. (2011). Didactical arguments and mathematical proofs. In F. H. van Eemeren, B. Garssen, D. Godden, & G. Mitchell (Eds.), *Proceedings of the 7th conference of the international society for the study of argumentation* (pp. 390–397). Amsterdam: Rozenberg/Sic Sat.

Finocchiaro, M. A. (2003). Physical-mathematical reasoning: Galileo on the extruding power of terrestrial rotation. *Synthese, 134,* 217–244.

Frege, G. (1879). *Begriffsschrift, a formula language, modeled upon that of arithmetic, for pure thought.* J. Van Heijenoort (Ed. & Trans.). Cambridge, MA: Harvard University Press.

Galois, E. (1962 [1997]). *Écrits et Mémoires Mathématiques d'Évariste Galois: Édition Critique Intégrale de ses Manuscrits et Publications.* Paris: Jacques Gabay.

Galois, E. ([1989] 1846). *Œuvres Mathématiques: Publiées en 1846 dans "Le Journal de Liouville".* Paris: Jacques Gabay.

Hallett, M. (1984). *Cantorian set theory and limitation of size.* Oxford: Oxford University Press.

Heinzmann, G. (Ed.) (1986). *Poincaré, Russell, Zermelo et Peano.* Paris: Blanchard.

Johnson, R. H. (2000). *Manifest rationality: A pragmatic theory of argument.* Mahwah, NJ: Lawrence Erlbaum Associates.

Lacroix, S. F. (1797). *Traité du Calcul Différentiel et du Calcul Intégral.* Paris: Duprat.

Lakatos, I. (1976). *Proofs and refutations: The logic of mathematical discovery* (edited by J. Worrall & E. Zahar). Cambridge: Cambridge University Press.

Peirce, C. S. (1992 [1871]). Fraser's *The Works of George Berkeley.* In N. Houser & C. Kloesel (Eds.), *The essential Peirce* (Vol. 1, pp. 83–105). Bloomington, IN: Indiana University Press.

Perelman, C. (1977). *L'Empire Rhétorique.* Paris: Vrin.

Perelman, C., & Olbrechts-Tyteca, L. (1958). *La Nouvelle Rhétorique. Traité de l'Argumentation.* Paris: Presses Universitaires de France.

Poincaré, H. (1905a). *La Valeur de la Science.* Paris: Flammarion.

Poincaré, H. (1905b). Les mathématiques et la logique. *Revue de Métaphysique et de Morale, 13*, 815–835. (Reprinted from Heinzman (1986) pp. 11–34 and, modified, in Poincaré (1908), pp. II, chap III).

Poincaré, H. (1908). *Science et Méthode.* Paris: Flammarion.

Rav, Y. (1999). Why do we prove theorems? *Philosophia Mathematica, 7*(3), 5–41.

Singh, S. (1997). *Fermat's last theorem.* London: Fourth Estate.

Timmermans, B. (1995). *La Résolution des Problèmes de Descartes à Kant.* Paris: Presses Universitaires de France.

Part II
Argumentation as a Methodology
for Studying Mathematical Practice

Chapter 6
An Argumentative Approach to Ideal Elements in Mathematics

Paola Cantù

6.1 Introduction

It is quite common in mathematics to speak about ideal, imaginary, or even impossible elements. One might consider these linguistic practices as a mere matter of words, arguing that complex numbers are no less real (or ideal) than real numbers, and that these names are just conventional ways to denote mathematical entities. One might also consider these practices from a historical point of view, and argue that ideal, imaginary, and impossible are just names used for newly introduced entities or theories, revealing the revolutionary impact of recent mathematical developments. Or one might take the linguistic practices at their face value, and explain the partition by means of some kind of distinction: metaphysical, epistemological, psychological, pragmatic, or genetic.

Metaphysically, one might distinguish between two different kinds of existence, or features of existence. Introducing the distinction between actual things and ideas in themselves, Bernard Bolzano (1837) considered all mathematical entities as being in themselves, i.e. non-existent. The Fregean and the Platonic analytic tradition shared a similar conception of arithmetical entities. Geometry, on the contrary, was mainly considered in the nineteenth century as a part of physics, following a tradition that had begun with Carl Friedrich Gauss and that was theorized by Hermann Grassmann (1844), as he distinguished between formal and real sciences. Giuseppe Veronese (1891) considered geometry as a mixed science, partly formal and partly real: for example, he distinguished between a line segment, which is a representation of something real, and an infinite straight line, which has no counterpart in the physical world. In *Formal and factual science* (1935) Rudolf

P. Cantù (✉)
CNRS and CEPERC, Université Aix-Marseille, 29, Avenue Robert Schuman,
13621 Aix-en-Provence Cedex 1, France
e-mail: paola.cantu@univ-amu.fr

A. Aberdein and I.J. Dove (eds.), *The Argument of Mathematics*, Logic, Epistemology, and the Unity of Science 30, DOI 10.1007/978-94-007-6534-4_6,
© Springer Science+Business Media Dordrecht 2013

Carnap refuted the ontological interpretation of the distinction between ideal and real, claiming that it does not yet legitimate

> the opinion of some philosophers who believe that the 'real' objects of the factual sciences must be contrasted with the 'formal', 'geistig' or 'ideal' objects of the formal sciences. The formal sciences do not have any objects at all; they are systems of auxiliary statements without objects and without content (Carnap, 1935, 128).

From an epistemological perspective the distinction might be traced between two different grades of knowledge: higher or lower certainty, greater or smaller perspicuity, higher or lower difficulty in the calculus, greater or smaller efficiency of proofs. For example, one might distinguish between theories that can be proved to be consistent and theories that cannot, or between finitary and infinitary theories.

From a psychological perspective there is a distinction between elements that can be intuitively conceived, and elements that cannot be intuited or represented. For example, several mathematicians, including Hermann von Helmholtz (1876) and Veronese (1891), and philosophers, such as Bertrand Russell, searched for an axiomatic nucleus common to Euclidean and non-Euclidean geometries that might express the main features of the Kantian *a priori* intuition of space (Torretti, 1978; Cantù, 2009).

Finally, one might trace a pragmatic distinction between two kinds of elements that have different functions in mathematics: e.g. one might assume, like Otto Hölder (1901), that the function of real numbers is measurement, or, like Hans Hahn (1907), that the introduction of imaginary numbers has the instrumental function of allowing a better understanding of the properties of real numbers.

I will not enter into this debate here, but I have highlighted it both for its intrinsic interest, and because the descriptive shortcomings of some of the aforementioned approaches, especially when taken in isolation, urged me to develop a different kind of approach to the topic of ideal elements. The method that will be applied here follows a quite recent but significant thread of investigations in the philosophy of mathematics that concerns the relations between argumentation theory and mathematics (Aberdein and Dove, 2009).[1] In our case, it will consist in the application of technical tools from argumentation theory and informal logic to the study of mathematical discourses about ideal elements.

From the analysis of several passages by mathematicians such as Hilbert, Gauss, Veronese, Cantor and Dedekind[2] I will claim that the use of the terms 'ideal', 'imaginary', and 'impossible' occurs either in the context of a practical argument aimed at justifying a particular kind of modification of a theory under given circumstances, or in the context of other kinds of arguments aimed at legitimating the former. The real-ideal distinction will thus be considered in the light of a dialectic argumentative practice concerning mathematical methodology.

---

[1] The emphasis on mathematics as a rational activity and as a human practice also contributed to the increased interest in investigations of the relations between mathematical reasoning and argumentation (Mancosu, 2008b; Mancosu et al., 2005; Hanna et al., 2009).

[2] Unless otherwise indicated, translations are mine.

Firstly, I will show that some of the aforementioned passages contain an argument that can be reconstructed as an instance of the argumentation scheme[3] that Douglas Walton (2008, 234) calls Value-Based Practical Reasoning:

(1) I have a goal G
(2) G is supported by my set of values V
(3) Bringing about A is a means for me to bring about G
(4) Therefore, I should (practically ought to) bring about A

In our case, A stands for "the method of introduction of ideal elements", and the conclusion of the argument is a prudential statement suggesting that any mathematician who acts like a rational agent, having a specific goal, a set of values that support the goal, and the capability of carrying out actions based on those goals, might be practically committed to introduce ideal elements into a theory, whenever this allows them to achieve the goal, unless certain further conditions occur. As we will see, the main goal and the set of values that support it might vary from author to author, as well as rebuttal conditions.

As for any instance of Value-Based Practical Reasoning, there are some critical questions that might facilitate an evaluation of the acceptability of the previous argument:

1. What other goals do I have that might conflict with goal G?
2. How well is goal G supported by the set of values V?
3. Are there any alternative means to A in order to bring about G?
4. Are there any negative consequences of A that might offset the positive value of the goal G?

Drawing attention to such critical questions and to the answers that one might find in the works by Hilbert, Gauss, Veronese, Cantor and Dedekind, I will claim that some further arguments about ideal elements could be interpreted as meta-arguments. A meta-argument is an argument claiming that some other argument is or is not legitimate: facing one of the critical questions above might produce a meta-argument.[4] What is at stake here is the justification of a heuristic procedure

---

[3]I will not give here an exact definition of argument and argumentation scheme (see other essays in this volume). Anyway, I will illustrate the meaning by an example, and also by analogy with the difference between syllogism and form. The Aristotelian syllogism "All humans are mammals. All mammals are mortal. Therefore, all humans are mortal." differs from its form which is usually referred to as 'Barbara' ("All S are M. All M are P. Therefore, all S are P"). Analogously, the argument "I want to eliminate social iniquities, because I believe that all humans are equal. Redistribution of wealth is a way to eliminate social iniquities. Therefore, I practically ought to bring about a redistribution of wealth" differs from the argumentation scheme mentioned in the text. 'Barbara' and the argumentation scheme express the form of the syllogism and the form of the argument, respectively.

[4]Maurice Finocchiaro (2007) has defined a meta-argument as "an argument about one or more arguments", and applied the notion to several examples from the philosophy of science, especially from authors of the early modern time (Galileo, Newton). I will consider if this notion can be interestingly applied to nineteenth century mathematicians too.

by mathematicians themselves. Attention is focused here on the analysis of goals, values and rebuttal conditions: different levels of freedom are allowed by different authors.

Finally, I will suggest some reasons of interest and possible developments of this argumentative approach to the method of introduction of ideal elements, including new insights on the epistemology of mathematicians such as Hilbert, Gauss, Veronese, Cantor, and Dedekind, and new questions concerning the nature of mathematical reasoning. The significance of an argumentative investigation of the method of introduction of ideal elements concerns the distinction between arguments and meta-arguments, and thus between instances of practical reasoning and the discussion about the legitimacy of practical reasoning in mathematics, as well as between informal arguments and generalization procedures used in mathematics and in other scientific disciplines.

6.2 A Value-Based Pragmatic Argumentation Scheme on Ideal Elements

There are several ways to define and to understand what a mathematical practice is, and what kind of theoretical investigations could be more appropriate to understanding it (see e.g. Mancosu, 2008b; van Kerkove et al., 2010). I will not enter into this debate here, but I would like to mention some reasons why the following approach could contribute to the investigations on mathematical practice. My approach is centered on a linguistic practice and takes as primary objects of investigation the texts of mathematicians themselves: not the formal presentation of their system, or some standard proofs, but rather the arguments they formulate in order to defend the methodology they follow and appreciate. I will thus consider what mathematicians explicitly said about the reasons for the introduction of ideal elements in mathematics, and why and in which cases they considered this move as legitimate.

In this section, I analyze and compare several passages by Dedekind and Hilbert concerning arithmetic and geometry showing that their discourses can be reconstructed as a practical argument based on three premises and a prudential conclusion, like the following:

• Any mathematician who acts like a rational agent, having the goal of eliminating exceptions from mathematical theories, some values such as generality and simplicity to support this goal, and some method of introduction of ideal elements that allows the mathematician to achieve that goal, is practically committed to introducing ideal elements.

Problematic issues will concern the goal, the values, and the understanding of what the method of introduction of ideal elements amounts to:

1. Why should one aim at eliminating exceptions? What kind of exceptions?
2. What is meant here by generality?

3. What does the method of introduction of ideal elements consist in?

A passage by Gauss will facilitate the understanding of possible hierarchies of values, while further passages by Gödel and Veronese will be used to analyze what is meant by the adjective 'ideal'.

6.2.1 Premises (1) and (2): Goal and Values

In the 1877 article *On the number of ideal-classes in different orders of a finite field* Dedekind writes:

> The main and most fruitful step was made by Kummer in 1847 as he introduced ideal numbers. Even if his researches also concerned at first only the partition of the circle and some related domains, *the ideas they are based on have a much more general meaning*. The extraordinary success obtained by Kummer had in 1856 already induced me to devote my main efforts to this topic, and I finally managed to build *a theory* of algebraic integer numbers *that is general [allgemein] and without exceptions [ausnahmslos]* (Dedekind, 1877/1930, 110).

Dedekind claims that a theory, such as Kummer's theory of ideals, is general if it can be applied without exceptions and has a general meaning. He implicitly assumes that a general theory is better than a restricted theory that has exceptions. Dedekind's goal could thus be formulated as follows:

(G) Goal G. Exceptions should be removed as far as possible.

The goal is legitimized by the fact that generality is considered as a positive value in mathematics. So, one could formulate Dedekind's value as follows:

(V) Value V. The generality of a theory, i.e. its being without exceptions, is a desirable value in mathematics.

In the case of Kummer's ideals Dedekind had the goal G and supported it by the value V. Not only did Hilbert agree with the goal and the value accepted by Dedekind, but while discussing the question of the complex numbers, he added an argument to support the preference for the value V. In the 1919 lecture *On the role of ideal entities* Hilbert remarked:

> A good example is here offered by the question of the existence of the algebraic root. If one takes here into consideration only real numbers as solutions, then one should say that some equations have no roots. One could also leave it at that, but then the general theory of equalities would be too difficult. So, one extends the system of real numbers by introduction of complex numbers, so as to obtain without exception the existence of the root (Hilbert, 1919, 90).

Hilbert claims that whenever there are exceptions (for example some equations have roots while other equations do not), the theory might become too difficult, but difficult theories are not preferable to easy and simple theories. So, to support the value V of generality, Hilbert claims that it is a condition for the simplicity of a theory, which is, according to him, an even more fundamental value in mathematics.

(V′) Value V′ as a warrant for value V. Generality is desirable because it increases simplicity.

But is simplicity a value in itself or is it in turn necessary to grant some other properties of mathematical theories? In the second treatise on the *Theory of biquadratic residues* (1831/1863) Gauss connected simplicity to the role of being a true foundation for other theories:

> This amounts to extending the field of higher arithmetic, which is usually extended only to real integer numbers, to imaginary numbers too, as a true foundation of the theory of biquadratic residues. To the imaginary one should concede the same right of citizenship that one concedes to the field of the real numbers. As soon as one understands this, the theory appears in a completely new light, and its results gain a most extraordinary simplicity. But before the theory of biquadratic residues can itself be developed in this broader domain of numbers, one should also extend to this domain the preceding doctrines of higher mathematics. Concerning these present investigations I can only mention here something in brief (Gauss, 1831/1863, 171).

So, Gauss presents an argument to support the fact that simplicity is a value as he claims that a simple theory can serve as a true foundation of some other theory. But he also acknowledges a possible disadvantage: a change in a new theory involves a change in several preceding theories. Since Gauss is aware that the generalization introduced to simplify the theory of biquadratic residues will induce changes in all domains of higher mathematics, he apparently considers the value of invariability of mathematical theories as less important than simplicity, because the latter can grant a true foundation. So, imaginary entities have the same right of citizenship as real ones.

In the formulation given above, propositions (G) and (V) do not contain any reference to the examples explicitly mentioned by Dedekind, Hilbert and Gauss. Dedekind had the theory of Kummer's ideals in mind, while Hilbert and Gauss referred to the algebraic theory of numbers. One might ask if the same kind of generality is at stake in the two examples, and what kind of exceptions are eliminated in each case.

Depending on how one interprets what Dedekind and Hilbert meant by an '*ausnahmslose*' theory, one could accept or refute the mentioned premises. Are exceptions generally bad in a theory, or did the authors have some special kind of exceptions in mind?

I suggest, as a tentative interpretation of passages that are quite vague and unclear, that exceptions might concern the definition of some inverse operation: for example, the operation of extracting the root, which is the inverse of the raising to a power, or the operation of division, which is the inverse of the operation of multiplication. It is true that in the case of Kummer's ideals the attention was actually focused on the elimination of another kind of exception: the theorem of the unique factorization of numbers, which could be demonstrated for real numbers but not for complex numbers in general. Anyway, unique factorization is a divisibility property, and thus a relevant feature of the inverse operation of division.

(G′) Goal G′. Exceptions should be removed, allowing direct and inverse operations to satisfy closure properties.

In the 1852 inaugural dissertation *On the elements of a theory of Euler's integration* Dedekind underlines the problematical and at the same time heuristically advantageous nature of the asymmetry between direct and inverse functions:[5]

> It is well known that the execution of indirect operations in analysis often encounters more important difficulties than the execution of direct operations. But exactly this apparent unfortunate circumstance has always exerted the most advantageous influence on mathematics. On the one hand, the defeat of these difficulties, whenever possible, always has a peculiar attractiveness for mathematicians. On the other hand, the cases in which it was not possible to defeat the difficulties by means of the previously introduced concepts and auxiliary tools always opened up entirely new fields to the further development of mathematics. For example, the operations of subtraction, division and root extraction have led to the concept of negative, fractional and imaginary numbers, by means of which the domain of mathematics was extraordinarily generalized [erweitert] (Dedekind, 1852/1930, 1).

Proposition (G′) might seem obvious to the contemporary reader, but it is not obvious at all. In the nineteenth century, for example, Hermann Grassmann, who is widely acknowledged as the founder of vector theory and as one of the main contributors to what will afterwards be called universal algebra, did not consider operations as necessarily defined on a fixed domain, and therefore did not consider the closure of a domain with respect to a given operation as an increase in generality. On the contrary, the increase in generality could rather arise when the same structural properties of an operation (for example commutativity and associativity with respect to some other operation) could be ascertained in different mathematical domains: arithmetic, vector theory, physics, and so on (Cantù, 2010).

Up until now, all the examples considered have been taken from arithmetic, but Hilbert also considered points and lines at infinity as examples of ideal elements. Are they introduced to eliminate the same kind of exceptions? In the 1926 paper *On the infinite* Hilbert writes:

> But it is known that through introduction of ideal elements, in particular of points and a line at infinity, it can be obtained that the proposition that two straight lines always intersect in one and only one point becomes generally valid (Hilbert, 1926, 166, Engl. transl. by S. Bauer-Mengelberg).

In the case of projective geometry, the closure of incidence between lines allows the construction of a dual transformation between the model interpreting incidence as two points lying on a line and the model interpreting incidence as the intersection

---

[5]It should be remarked that Dedekind refers to 'the' domain of mathematics being generalized, as if the generalization consisted in the transformation of the universe of the model. This is just one possible way to conceive the introduction of ideal elements.

of two lines in a point. In order to include the case of projective geometry, premise (G′) could be modified into:

(G″) Goal G″. Exceptions should be removed, allowing direct and inverse operations to satisfy closure properties, and dual transformations between models to be introduced, whenever possible.[6]

Hilbert's reason for allowing duality would then be similar to Dedekind's reason for introducing the closure of the inverse operation: it offers an increase in simplicity.

To summarize, mathematicians who dealt with arithmetic and geometry had the goal of eliminating certain exceptions, e.g. guaranteeing whenever possible the closure of direct and inverse operations and the formulation of duality transformations, because they held generality to be an important value in mathematics, given that it increases simplicity, which in turn is especially desirable in the foundation of mathematics. From the aforementioned passages by Gauss, Dedekind and Hilbert we have thus reconstructed a goal and a value that might occur in the first two premises of the argumentation scheme mentioned in the introduction:

Premise (1)   I, as a mathematician, have the goal (G″)
Premise (2)   The goal (G″) is supported by the set of values (V), and (V′).

6.2.2   Premise (3): A Way to Bring About the Goal

The method of introduction of ideal elements is presented as a way to bring about the goal (G″), i.e. the closure of direct and inverse operations and the formulation of duality transformations in some specific cases. This goal is achieved by some kind of 'Erweiterung', or generalization.[7]

In the aforementioned lecture on ideal entities, Hilbert asserts that ideal entities occur in mathematics

> when a system is generalized [erweitert] by the introduction of new elements that then have the character of ideal elements with respect to the original system. [...] one speaks about existence only with reference to a determinate given system, which might vary according to the theory under consideration (Hilbert, 1919, 90–1).

---

[6]In category theory the two conditions might be unified, and one might formulate the goal as follows: (G*) Exceptions should be removed, allowing for any category to construct the opposite or dual category, whenever possible.

[7]I translate the term 'Erweiterung' as 'generalization' not only to follow a linguistic choice adopted in the English translation of the *Foundations of geometry* by Townsend (Hilbert, 1902), but also to make it clear that the expression is non-technical. Any understanding of it as a particular kind of transformation is a matter of interpretation, and given the problematic and complex nature of this task, I will not attempt it here.

The method of introduction of ideal elements is presented as some kind of procedure that concerns a generalization [Erweiterung] of a system into another system, which could be interpreted as the transformation of a theory into another theory by the modification of some axiom, or as the transformation of a model into another model by modification of the universe and of the relations and functions defined on it.[8]

> This procedure of introduction of ideal elements is one of the most important mathematical methods being applied again and again up to the higher parts of mathematics. More precisely, the procedure consists in the fact that from an original system that is complicated with respect to some unsolved question one goes over [übergehen] to a new system in which those relations are granted more easily, and which in addition has the property of containing a subsystem that is isomorphic to the original system (Hilbert, 1919, 90–1).

The expression 'ideal' is again used in the case of the generalization of geometry by the addition of Hilbert's axiom of completeness in the first edition of the *Foundations of Geometry*.

> However, it is always possible to generalize [erweitern] the original system of points, straight lines, and planes by the addition of 'ideal' or 'irrational' elements, so that, upon any straight line of the corresponding geometry, a point corresponds *without exception* to every system of three real numbers. By the adoption of suitable conventions, it may also be seen that, in this generalized [erweitert] geometry, all of the axioms I–V are valid. This geometry, generalized [erweitert] by the addition of irrational elements, is nothing else than the ordinary analytic geometry of space (Hilbert, 1902, 35–6, Engl. transl. by E. J. Townsend).

But in the 1926 paper on the infinite and in the 1919 lecture, Hilbert does not mention irrational numbers as a typical example of ideal elements.[9]

What is meant by the adjective ideal? In his reconstruction of Hilbert's thought, Michael Detlefsen (1993; 1986) suggested a Kantian interpretation of ideal elements as statements concerning symbols that do not constitute genuine contentful statements. Other mathematicians have used the expression 'ideal' or 'fictive' to denote entities that cannot be perceived or that have no possible counterpart in the empirical world. Gödel, for example, wrote in the 1944 paper on *Russell's Mathematical Logic*:

> When [Russell] started on a concrete problem, the objects to be analyzed (e.g., the classes or propositions) soon for the most part turned into 'logical fictions'. Though perhaps this need not necessarily mean (according to the sense in which Russell uses this term) that these things do not exist, but only that we have no direct perception of them (Gödel, 1944/1990, 121).

---

[8]This is the topic of a research project that I started in 2008 thanks to a generous grant by the Université de Nancy and the ANR Research Project on Ideals of Proof directed by Michael Detlefsen. The research is still in progress, but partial results have been presented at the Paris-Nancy PhilMath Workshop in 2009 and at the Seventh European Congress of Analytic Philosophy in 2011.

[9]Interestingly enough, the term 'ideal' used in the quoted passage from the first German edition published in 1899 (Hilbert, 1899, 39), was eliminated in following editions of the *Grundlagen*.

Similarly, Veronese called 'ideal' all mathematical entities that lack an external representation:

> If someone wants to call me rationalist or idealist because of the idea discussed here, I will accept this denomination to distinguish myself from authors who unjustifiably want to negate the biggest possible logical freedom to the mathematical and geometrical mind, and who ask, for any new result or hypothesis, whether it has or not a pure external representation. But we accept the denomination only on condition that no properly philosophical meaning be associated to it. (Veronese, 1891, xiv).

According to Hilbert, ideal elements can be introduced axiomatically, or by construction from older elements of the system:

> The way of speaking concerning ideal elements has its justification only from the point of view of the initial system. In the newly obtained system one cannot any more distinguish between real and ideal elements. The transition to the broader [weiterem] system might happen either in a constructive way, if the new elements are derived from the old elements through mathematical construction, or axiomatically, if the new system is characterized by relational properties. In the second case one also needs a proof that the assumption of the required properties does not contain a contradiction in itself (Hilbert, 1919, 90–91).

A possible explanation of the reasons why Hilbert expresses himself as if the elements were the entities belonging to the domain of a model of the theory, could be the fact that he has in mind Dedekind's way of eliminating exceptions by the introduction of closure properties for the inverse operation. The previous passage concerning an isomorphism between the old system and a part of the new system (see above page 87) could also be understood in the same way. Setting aside here the effort to find an adequate interpretation of what Hilbert describes as a 'system generalization', I will only recall that, however exactly it might be spelled out, the method of introduction of ideal elements is some kind of procedure that allows the closure of operations and dual transformations between relevant models. So, I can now instantiate premise (3) of the argumentation scheme mentioned in the introduction:

Premise (3)   The method of introduction of ideal elements is a means for mathematicians to bring about the goal (G'').

6.2.3   Conclusion (4): A Prudential Claim

According to my reconstruction, if one accepts the premises (1–3), then one should also accept a claim that is an instantiation of the conclusion of the argumentation scheme presented in the introduction:

Premise (1)   I, as a mathematician, have the goal (G'').
Premise (2)   The goal (G'') is supported by the set of values (V) and (V').
Premise (3)   The method of introduction of ideal elements is a means for me, as a mathematician, to bring about (G'').
Conclusion (4)   Therefore, I should (practically ought to) introduce ideal elements.

In other words, if I, as a mathematician, have the goal of eliminating certain exceptions in the sense described above, which is a valuable goal in mathematics because it helps generality, simplicity, and better foundation, and if introducing ideal elements is a means to eliminate exceptions, then I should (practically ought to) introduce ideal elements. The argument is based on a means-end relation and has a normative conclusion, suggesting what a rational agent would be expected to do in the given circumstance.

My claim is that Gauss, Dedekind and Hilbert all advanced this kind of argument. I am not claiming that there are no differences concerning (1) the relevant application cases, (2) the role of simplicity, (3) the specific kind of exceptions that need to be eliminated, and (4) rebuttal conditions.

1. Application cases differed. Gauss considered the case of complex numbers extending real numbers, while Dedekind considered both complex numbers and Kummer's ideals as a simplification of the theory of polynomials. Hilbert added points and line at infinity in projective geometry.
2. Concerning the role of simplicity, it is sometimes foundational and sometimes epistemological. Gauss claimed that an increase in generality would allow a true foundation of the theory, Dedekind believed one could reduce the less general theory to the more general one (see Sect. 6.3.2), while Hilbert remarked that ideal elements would increase the perspicuity and the fecundity of the theory.
3. As for the kind of exceptions that need to be eliminated, I already mentioned operations that are not closed under a domain, inverse operations that lack a definition, modifications that allow the construction of a dual relation and, more generally, of interesting conceptual transformations, as in the case of the construction of the opposite or dual of a category in category theory.
4. As in any practical argument, there are relevant rebuttal conditions: one should (practically ought to) introduce ideal elements (a) unless contradictions arise (a rebuttal condition assumed by all authors), (b) unless no new results can be obtained (see Dedekind's, Veronese's and Hilbert's remarks on fruitfulness), (c) unless relevant properties of the original theory get lost (compare Veronese's requirement that projective and non-Archimedean theories conservatively extend a finite part of Euclidean geometry: Cantù, 1999). The discussion of rebuttal conditions often involves a meta-analysis of the validity of the argument (1)–(4), as we will see in Sect. 6.3.

Notwithstanding all these differences, I claim that the arguments by Dedekind, Gauss and Hilbert can be reconstructed as instantiations of the same argumentation scheme, because they are all value-based practical arguments. This kind of argument is usually advanced in a deliberation dialogue, when one of the participants tries to convince the others of the opportunity to take some decision. As such, the argument does not pretend to legitimate the theoretical necessity of the method of introduction of ideal elements, nor to legitimate their existence. It is rather an example of the kind of rhetoric arguments advanced by mathematicians trying to defend and put forward a methodological move.[10] Like any practical argument, it is defeasible and a set of critical questions can be used to test how strong the argument is. For example, one could ask:

1. What other goals do I have that might conflict with the goal of eliminating exceptions?
2. How well is the goal supported by the values of generality, simplicity and foundational role?
3. Are there any alternative means to the introduction of ideal elements in order to eliminate the same exceptions?
4. Are there any negative consequences of the introduction of ideal elements that might offset the positive value of the goal?

If the argument above recommends the use of the method of introduction of ideal elements in certain cases, are ideal elements thereby legitimated? Is the newly obtained theory as legitimate, as consistent, as fruitful, or as intuitive as the former? Concerning consistency, I already mentioned Hilbert's point of view. Concerning legitimacy, we have seen Gauss's claim that imaginary elements have the same right of citizenship as real elements. But the latter belief, which would be sufficient to ground conclusion (4), was not generally shared either by contemporaries or by later mathematicians.[11] This is the reason why Hilbert, Dedekind, Veronese and Cantor advanced further arguments to support the method of introduction of ideal elements.

6.3   Further Arguments on Ideal Elements

In the following, I will consider other arguments concerning ideal elements, and classify them according to the argumentation scheme they instantiate. I will further investigate whether they answer the aforementioned critical questions, and what role they play with respect to the previous argument. Are they independent reasons in support of claim (4) asserting that ideal elements should be introduced in mathematics, or are they meta-arguments, i.e. claims about the legitimacy of ideal elements as a successful heuristic strategy?

---

[10]For this reason, the argument differs radically from the application of the indispensability argument to mathematics itself. Although the argument discussed by Paolo Mancosu (2008a, 139) with reference to Solomon Feferman (1964) is a practical argument too, its conclusion claims that we must not only postulate abstract entities, but also believe in their existence, given that they are needed in certain parts of mathematics itself. Here on the contrary the conclusion (4) is a prudential statement claiming that under certain circumstances I, as a mathematician, should make use of the method of introduction of ideal elements. No commitment to the existence of such elements is required.

[11]Actually Gauss gave other arguments too, for example he showed that complex numbers could be geometrically represented in a plane, but I will not discuss them here.

6.3.1 Answering Critical Question No. 4 by Means of a Meta-Argument

While discussing the method of introduction of ideal elements in the 1919 lecture, Hilbert mentions complex numbers, points and lines at infinity, and Kummer's ideals, but remarks that there are many more applications:

> Of the many diverse applications of the method of ideal elements in mathematics I will discuss two more (Hilbert, 1919, 91).[12]

Notwithstanding the "many diverse applications of the method", in his 1926 paper on the infinite Hilbert cites the same three successful examples mentioned in 1919:

> The ideal elements 'at infinity' have the advantage of making the system of the laws of connection as simple and perspicuous as is at all possible. As is well known, the symmetry between point and straight line then yields the duality principle of geometry, which is so fecund.
>
> The ordinary complex magnitudes of algebra likewise are an instance of the use of ideal elements; they serve to simplify the theorems on the existence and number of the roots of an equation.
>
> Just as in geometry infinitely many straight lines, namely, those that are parallel to one another, are used to define an ideal point, so in higher arithmetic certain systems of infinitely many numbers are combined into a number ideal, and indeed probably no use of the principle of ideal elements is a greater stroke of genius than this. When this procedure has been carried out generally within an algebraic field, we find in it again the simple and well-known laws of divisibility, just as they hold for the ordinary integers 1, 2, 3, 4... Here we have already entered the domain of higher arithmetic (Hilbert, 1926, 373–74, Engl. transl. by S. Bauer-Mengelberg).

Hilbert's statement concerns the legitimacy of the practical argument reconstructed in Sect. 6.2.3: if the heuristic strategy expressed by the argument applies successfully in three cases (complex numbers, points at infinity in geometry, and Kummer's ideals), one might apply it in other cases too. Of course, the success in previous cases does not allow us to expect that the move will always be successful, but it is enough to assume it as a good heuristic strategy. This argument plays the role of a meta-argument. It does not furnish further support to the premises (1–3), and does not therefore make conclusion (4) stronger, but claims that the argument

---

[12]Another example of ideal elements is suggested by Gödel, who, criticizing Russell's 'no-class theory' remarked that Russell's argument could at most prove that the null class and the unit class should be considered as ideal elements as well as points at infinity in geometry: "Russell adduces two reasons against the extensional view of classes, namely, the existence of (1) the null class, which cannot very well be a collection, and (2) the unit classes, which would have to be identical with their single elements. But it seems to me that these arguments could, if anything, at most prove that the null class and the unit classes (as distinct from their only element) are fictions (introduced to simplify the calculus like the points at infinity in geometry), not that all classes are fictions." (Gödel, 1944/1990, 131).

(1–4) is on the whole acceptable in mathematical reasoning. Hilbert also makes a more general claim about the legitimacy of practical strategies in mathematics:

> We shall carefully investigate those ways of forming notions and those modes of inference that are fruitful; we shall nurse them, support them, and make them usable, wherever there is the slightest promise of success. No one shall be able to drive us from the paradise that Cantor created for us (Hilbert, 1926, 375–76, Engl. transl. by S. Bauer-Mengelberg).

Hilbert is aware of possible negative consequences of the strategy (e.g. inconsistent results) and does not ignore the objection suggested by the fourth critical question—there might be undesirable effects of this strategy—but, provided they do not offset the positive value (the fruitfulness) of the move, and provided one applies a rigorous inferential method, there is no reason to abandon the strategy as illegitimate:

> But there is a completely satisfactory way of escaping the paradoxes without committing treason against our science. [...] It is necessary to make inferences everywhere as reliable as they are in ordinary elementary number theory, which no one questions and in which contradictions and paradoxes arise only through our carelessness (Hilbert, 1926, 375–76, Engl. transl. by S. Bauer-Mengelberg).

Giuseppe Veronese was not as confident as Hilbert in the possibility of a general application of the strategy to any case where exceptions need to be eliminated. He remarked that this heuristic strategy might be more effective and fruitful in certain kinds of investigations than others. Yet he did not give a counterexample. On the contrary, he suggested further cases of application of the method of introduction of ideal elements, such as non-Archimedean numbers and hyperspaces, following a line of reasoning that is very similar to the value-based practical argument reconstructed above. On the one hand, the transformation from lower to higher spaces and vice-versa is justified inasmuch as it is useful to prove certain theorems more easily, or to order theorems in a different way, which is also a part of the investigations on foundations. On the other hand, different aims might require different strategies.

> The advantage of this method with respect to the usual space consists exactly in the fact that from general or special things one can derive classes of things in the plane and in the space of three dimensions, or in a space of lower dimension, and vice-versa, and in the fact that new or already known properties or things are thereby proved more easily, and can be ordered according to a general point of view. [...] Besides, the geometry of n dimensions is also in some other respect advantageous: many things about the plane and the usual space can be represented by means of more simple things, for example through points of a higher space, and their properties can thus be investigated more easily. Certainly, it cannot be affirmed that a similar method could be applied to any geometrical investigation. For example, it can be used more successfully in projective rather than in metric geometry. Any method has its advantages and defects. Especially, one should dominate the method completely in order to obtain from it the utility it is in a position to offer. It is already enough if the method is truly fruitful at least in one class of significant investigations (Veronese, 1891, xxxii–iii).

6.3.2   Answering Critical Question No. 2: A Side-Argument by Hierarchical Reduction

The argument by hierarchical reduction, first advanced in an ontological version by Augustine of Hippo in his book on *The Teacher* (1995, Ch. 9.27, 129), claims that anything that is subordinated to some other thing is less worthy than the thing it is subordinated to. Notwithstanding the religious origin, or maybe just for this reason, mathematicians often use the argument in the discussions on the foundations of mathematics. For example, Dedekind claimed that a general law is preferable to a special case of the same law:

> Since here the divisibility of numbers only forms a special case of the divisibility of ideals, one has in the end only to establish the laws of ideals, which are in effect more simple (Dedekind, 1877/1930, 116).

The possibility of expressing the content of the original theory inside the framework of a more general theory not only aims at increasing simplicity, but also at reducing one theory to the other. As in many arguments based on a hierarchy of things, the higher in the hierarchy the better: the higher theory can be used to express and better understand the lower theory in the hierarchy. The difference in generality and the hierarchical relation between the two theories is explicitly acknowledged by Dedekind:

> One obtains this important generalization if one substitutes the field of rational numbers with any field that is contained as a divisor in [the fundamental ideal] Ω [of the field], where besides the usual norms, discriminants, and partial or relative traces, one should also introduce norms, discriminants, etc. that refer to that field, and substitute certain rational numbers with ideals of this field (Dedekind, 1882/1930, 352).

The argument by hierarchical reduction could be interpreted as an answer to the second critical question, because it discusses the relations between the goal of eliminating exceptions and the values that support it: generality is here intended as a reduction to the superior theory that simplifies mathematical treatment.

6.3.3   Veronese and Cantor: A Meta-Argument of Direction

Finally, I will consider several passages by Cantor and Veronese who also made a meta-argument, i.e. an argument about the practical argument reconstructed in Sect. 6.2. Both claim that the heuristic strategy expressed by the argument (1–4) is legitimate not only because of its fruitfulness, but also because of the possibility of its being applicable again and again with a continuous increase in value. Yet Cantor believes that there is just one 'natural' way to introduce ideal elements to generalize the notion of number, whereas Veronese, like Hilbert, remarks that there might be different ways of introducing ideal elements: given a plurality of strategies, the best strategy to follow could be determined only on a case by case basis taking into account the aims of the researcher.

In traditional rhetoric an argument that claims that one can always go further in a determinate direction, without a limit being reached, and with a continuous increase in value is called an argument of direction (Perelman and Olbrechts-Tyteca, 1969, 287 ff.). While defending the right of mathematicians to introduce new hypotheses, Veronese writes:

> When a determinate thing A is given (each thing considered in its concept [...]) and when it cannot be derived that A is the group of all possible things that we want to consider, then we can imagine another thing that is not contained in A and that is independent from A (Veronese, 1891, 16, Sect. 9.37).

According to Cantor, the direction to be followed in the assumption that there is something beyond a set of things that we are considering (for example natural numbers $\mathbb{N}$), is unique:

> The concept of an 'order type' [Ordnungstypus] that we have developed here includes, once extended [übertragen] to multiple ordered sets, not only the 'cardinal number or power' [...], but also anything that is at all conceivable 'as a number'. In this sense it allows no further generalization [...]. It does not contain anything arbitrary, but it is the 'natural' generalization [Erweiterung] of the concept of number. In particular, it should be underlined that the identity criterion follows with absolute necessity from the concept of 'order type', and thus allows no variation. In the misunderstanding of this state of affairs one can see the main reason for the serious mistakes to be found in the work by G. Veronese, *Foundations of Geometry* (Cantor, 1895–97/32, 300).

This is of course also a strategy to refute alternative introductions of transfinite numbers.[13] As a matter of fact, Veronese did consider an alternative generalization, based on the introduction of infinitely great but also infinitely small numbers. Quoting Crelle, he expressed his anti-foundationalism in a clear and at the time not so widely shared form: indeterminacy can be found both in the foundations and in the development of a theory.

> Crelle says that the establishment of principles of geometry offers no less difficulty than the development of the most complicated theories: "The former are turned towards the heights and the latter towards the depths, but both heights and depths are equally unlimited and obscure." It is certain that the difficulties one encounters in the former case ask for much more time and perseverance than the investigations of higher kind, where a new and fruitful idea might quickly lead to very important results. On the contrary, it has little or no value to bring forward ideas on the foundations, if one does not show that they can be effectively realized (Veronese, 1891, 632).

Unlike Cantor's transfinite numbers, which are characterized by the fact that there is a first transfinite ordinal corresponding to the class of finite numbers,

---

[13]I am assuming here that Cantor's transfinite numbers could be considered as an example of ideal elements. This opinion is clearly shared by Veronese, but need not be ascribed to Cantor as such. While the distinction real–ideal that I have suggested concerns a relation to the domain of the original theory, Cantor introduced a quite different distinction between an intrasubjective (immanent) and a transsubjective (transient) notion of reality (Cantor, 1895–97/32, 182, 106).

Veronese's system of numbers is so construed as to have no first infinitely great number and no first infinitely small number.[14]

In particular, Veronese criticized the refusals to take new hypotheses into account, because this would unnecessarily limit the domain of mathematics and the freedom of mathematicians, and also prevent the proof of independent results.

> It is certainly possible to get along without a determinate hypothesis or stipulation, for example without the imaginary numbers. But this would result in an absolutely unjustified limitation of the mathematical domain, and besides, it would not prove, independently from the hypothesis itself, anything against the hypothesis and its consequences in that domain (Veronese, 1891, 647).

The possibility of introducing ideal elements in different ways is also admitted by Hilbert, who mentions the case of plane geometry:

> Plane geometry also offers us a good example of the fact that from the given system taken as starting point there is not just one possibility of generalization [Erweiterung] but rather the way one generalizes the system could vary depending on the aim one has. For example, in the complex function theory we find a different generalization of elementary plane geometry. Here one adds just one ideal point to the points of the plane represented by complex numbers (the number $u + iv$ represents the point with coordinates u,v) (Hilbert, 1919, 94).

If one accepts the possibility of various alternative ways of introducing ideal elements, then the heuristic strategy reconstructed in Sect. 6.2 becomes even more undetermined. Yet, in the case of infinitely great numbers, there is an evident difference between Cantor's transfinite numbers which require a change in the features of the operation (e.g. addition is not commutative) and Veronese's non-Archimedean numbers, which constitute a non-standard model of the theory of real numbers.

6.4 The Interest of an Argumentative Investigation of the Method of Ideal Elements

The analysis of meta-arguments is philosophically interesting because it reveals significant differences in the epistemological approaches of Dedekind, Hilbert, Cantor and Veronese to the method of introduction of ideal elements. Dedekind had a tendency to conceive generality as the achievement of a theory that might include the original theory as a special case, sharing a cumulative conception

---

[14] Veronese himself had rightly acknowledged that his own generalization was different from the one by Cantor: "We do not make any use of G. Cantor's numbers in the *Foundations of geometry*, and we mention them in this Introduction only to compare them with our infinitely great numbers" (Veronese, 1905, 349). Nonetheless, misunderstanding on this point remained, especially because Poincaré wrote that Veronese had developed "a true non-Archimedean arithmetic and a true non-Archimedean geometry where the transfinite numbers of Cantor play a predominant role." (Poincaré, 1904, 14).

of mathematical knowledge and a preference for hierarchically ordered theories. Hilbert aimed at justifying practical arguments in heuristic mathematical reasoning. Cantor considered his generalization of numbers as 'the' natural one, and therefore unique, while Veronese considered that different alternative generalizations were not only possible but also equally legitimate with respect to different aims.

Besides, the above analysis of what mathematicians said about ideal elements has shown that the method of introduction of ideal elements can be interestingly investigated not only with respect to the ontological status of mathematical objects, but also with respect to the normative value of certain heuristic strategies. In particular, the reconstruction of the specific arguments has shown that the discourses on the method of introduction of ideal elements contain distinct lines of reasoning that might be roughly divided in two groups. The arguments discussed in Sect. 6.2 have been reconstructed as instances of a value-based practical argumentation scheme that can be used to convince the mathematical community of the opportunity to choose a given heuristic strategy under certain circumstances. The arguments discussed in Sect. 6.3 have on the contrary the function of answering some possible critical objections to the arguments presented in Sect. 6.2, and therefore play the role of meta-arguments, i.e. arguments concerning the legitimacy of an argument. While the argument reconstructed in Sect. 6.2 has a normative conclusion recommending the introduction of ideal elements to eliminate certain exceptions of a theory, the meta-arguments in Sect. 6.3 aim at evaluating the legitimacy of that heuristic strategy, and thus offer interesting insights on the epistemological conception of their proponents.

The method of introduction of ideal elements is here investigated as a procedure that is typical of mathematics: it focuses our attention not only on the investigation of the mathematical reasoning displayed in the axiomatic formulation of a theory or in the development of a proof, but also on the mathematical reasoning displayed in mathematical argumentation about mathematical reasoning itself. The philosophical interest and the possible fruitful developments of this approach might concern several aspects of the investigation of mathematical reasoning.

Firstly, the method of introduction of ideal elements is investigated here as a complex phenomenon, that includes ontological, epistemological, historical, and especially methodological features that are not always taken into due account in the literature. Attention is not focused on what makes real numbers 'real' (some kind of relation to the world or to material signs), but on the rationality of an agent that takes a theory as a starting point, transforms it in a legitimate way, and obtains an extended theory whose domain can be compared to the domain of the initial theory, thereby gaining some further knowledge. The notion of ideality concerns the use of mathematical theories inside mathematics, and the meta-theoretical relations between such theories. The regulative value of ideal elements can be accounted for as an increase of knowledge on the elements of the first theory obtained by meta-theoretical means.

Secondly, the argumentative approach takes mathematical practice into account, provided of course that under mathematical practice one considers not only the discovery and proof activity of mathematicians, but also the broader sphere of their investigations of the legitimacy of mathematical heuristic strategies and efforts to produce adequate support for normative claims about preferable strategies under given circumstances. Hintikka (2004) remarked that logic (or philosophy of logic) should concern successful strategies rather than only correct strategies of reasoning; I am suggesting that the investigation of ideal elements might shed some light on successful strategies of reasoning about mathematical theories. Moreover, when mathematical discourse is reconstructed in the light of arguments over heuristic moves, attention is focused on another important feature of mathematical practice: the relations between the discovery of a new theory and the presence of specific difficulties or unsolved problems (the exceptions discussed by Dedekind and Hilbert).

Thirdly, a detailed analysis of texts from nineteenth century mathematicians might not only be historically interesting, but also theoretically fruitful, because it helps to discover relevant changes in the general mathematical framework. For example, the innovative role played in the second half of the century by closure properties of operations and dual transformations might go unobserved in contemporary practices, because it is often taken for granted in the symbolism used to formalize the models of a theory.

Fourthly, a precise and detailed reconstruction of the arguments involved helps to point out problems of interpretation, and to underline the vague formulation of certain concepts (e.g. generalization, exceptions, . . .), whose detailed analysis constitutes a preliminary condition for subsequent investigations on the possible strategies to generalize a theory and on the kind of generality to be thus achieved.

From a foundational perspective, the approach turns the attention from the ontological privilege of certain entities to the methodological relevance of heuristic moves, showing the need for a more detailed investigation into the mechanisms of theory-change in mathematics and into the question of cumulative growth of knowledge: what is preserved and what is lost in the transformation of the original theory into a new theory? What is the role of models in theory-change?

Finally, this approach might focus the attention on the question of the specificity of mathematical reasoning with respect to other argumentative and scientific practices. If the discourse on ideal elements can be reconstructed as a value-based practical argument often accompanied by further arguments that aim at justifying the former, would it be possible to reconstruct the discourse on ideal elements in physics in the same way? Hilbert had already raised the question in his 1919 lecture on ideal entities: is the method of reasoning by appeal to ideal elements typical of mathematics or is it common to other disciplines? Hilbert claimed that 'ideal' in physics usually means something different, because it concerns an operation of abstraction from some properties that are considered as irrelevant in the construction of a model, but he also suggested that the notion of a 'juridical person' in legal theory might have interesting analogies with the discourse on ideal elements in mathematics.[15] The answer to this question may be relevant for philosophical

---

[15]A 'juridical person' (*persona ficta*) is a non-human entity that is regarded in law as having the same status (name, rights, responsibilities, . . .) as a human person.

discussion on mathematical reasoning and explanation: are there several field-dependent notions of explanation or rather a unique notion of explanation for all scientific disciplines?

Acknowledgements The author is indebted for useful comments to Mic Detlefsen, Andy Arana, Andrew Aberdein, Ian J. Dove, and to the participants in the MidWest Philosophy of Mathematics Workshop (MWPMW 12), where a first version of the paper was presented in November 2011.

References

Aberdein, A., & Dove, I. J. (Eds.). (2009). *Mathematical argumentation. Foundations of Science, 14*(1–2) (Special issue), 1–8.

Augustine. (1995). *De magistro* [Against the academicians and the teacher] (P. King, Trans.). Indianapolis, IN: Hackett Publishing Company.

Bolzano, B. (1837). Wissenschaftslehre [Theory of science] (Sulzbach, 4 Vols.). Stuttgart: Frommann Holzboog, 1985–2000. (Reprinted in J. Berg (Ed.), Bolzano's Gesamtausgabe (Vols. 11–14)).

Cantor, G. (1895–97/32). Beiträge zur Begründung der transfiniten Mengenlehre (1895–97/32) [Contributions to the foundation of transfinite set-theory]. In E. Zermelo (Ed.), *Gesammelte Abhandlungen mathematischen und philosophischen Inhalts* (pp. 282–356). Berlin: Springer.

Cantù, P. (1999). *Giuseppe Veronese e i fondamenti della geometria* [Giuseppe Veronese and the foundations of geometry]. Milano: Unicopli.

Cantù, P. (2009). Le concept de l'espace chez Veronese. Une comparaison avec la conception de Helmholtz et Poincaré [The concept of space in Veronese. A comparison with the conceptions of Helmholtz and Poincaré]. *Philosophia Scientiae, 13*(2), 129–149.

Cantù, P. (2010). Grassmann's epistemology: Multiplication and constructivism. In H. J. Petsche, A. C. Lewis, J. Liesen, & S. Russ (Eds.), *From past to future: Graßmann's work in context. The Graßmann Bicentennial Conference. September 2009* (pp. 91–100). Basel: Birkhäuser.

Carnap, R. (1935/1953). Formalwissenschaft und Realwissenschaft [Formal and factual science]. In H. Feigl & M. Brodbeck (Eds.), *Readings in the philosophy of science* (pp. 123–128). New York: Appleton-Century-Crofts.

Dedekind, R. (1852/1930). Über die Elemente der Theorie der Eulerschen Integrale [On the elements of a theory of Euler's integration]. In *Gesammelte mathematische Werke* (Vol. 1, pp. 1–26). Braunschweig: Vieweg & Sohn.

Dedekind, R. (1877/1930). Über die Anzahl der Ideal-Klassen in den verschiedenen Ordnungen eines endlichen Körpers [On the number of ideal-classes in different orders of a finite field]. In *Gesammelte mathematische Werke* (Vol. 1, pp. 105–158). Braunschweig: Vieweg & Sohn.

Dedekind, R. (1882/1930). Über die Diskriminanten endlicher Körper [On the discriminants of finite fields]. In *Gesammelte mathematische Werke* (Vol. 1, pp. 351–397). Braunschweig: Vieweg & Sohn.

Detlefsen, M. (1986). *Hilbert's program. An essay on mathematical instrumentalism.* Dordrecht: Reidel.

Detlefsen, M. (1993). The Kantian character of Hilbert's formalism. In J. Czermak (Ed.), *Philosophy of mathematics (Kirchberg am Wechsel, 1992)* (pp. 195–205). Vienna: Hölder-Pichler-Tempsky.

Feferman, S. (1964). Systems of predicative analysis. *Journal of Symbolic Logic, 29*, 1–30.

Finocchiaro, M. A. (2007). Arguments, meta-arguments, and metadialogues: A reconstruction of Krabbe, Govier, and Woods. *Argumentation, 21*(3), 253–268.

Gauss, C. F. (1831/1863). Theoria residuorum biquadraticum. Commentatio secunda [Theory of biquadratic residues. Second treatise]. In *Werke* (Vol. 2, pp. 169–178). Göttingen: Königlichen Gesellschaft der Wissenschaften. (Reprint in 1863, Hildesheim: Olms).

Gödel, K. (1944/1990). Russell's mathematical logic. In S. Feferman, J. W. Dawson, S. C. Kleene, G. H. Moore, R. M. Solovay & J. van Heijenoort (Eds.), *Kurt Gödel: Collected works. Volume 2: Publications 1938–1974* (pp. 119–141). New York: Oxford University Press.

Grassmann, H. G. (1844). Die Wissenschaft der extensiven Grösse oder die Ausdehnungslehre. In F. Engel (Ed.), *Gesammelte mathematische und physikalische Werke* (Vol. I.1). Leipzig: Teubner, 1894. [Trans. by Lloyd C. Kannenberg in A New Branch of Mathematics: The "Ausdehnungslehre" of 1844 and Other Works. Chicago: Open Court, 1995.]

Hahn, H. (1907). Über die nichtarchimedischen Grössensysteme [On the non-Archimedean systems of quantities]. *Sitzungsberichte der Kaiserlichen Wissenschaften, Mathematisch-Naturwissenschaftlichen Klasse, 116*(IIa), 601–655.

Hanna, G., Jahnke H. N., & Pulte, H. (Eds.). (2009). *Explanation and proof in mathematics: Philosophical and educational perspectives*. Dordrecht: Springer.

Hilbert, D. (1899). *Grundlagen der Geometrie*. Leipzig: Teubner.

Hilbert, D. (1902). *Foundations of geometry* (E. J. Townsend, Trans.). Chicago, IL: Open Court.

Hilbert, D. (1919/1992). Die Rolle von idealen Gebilden [The role of ideal elements]. In D. Rowe (Ed.), *Natur und mathematisches Erkennen*. Basel: Birkhäuser.

Hilbert, D. (1926/1967). Über das Unendliche [On the infinite]. In J. van Heijenoort (Ed.), *From Frege to Gödel: A source book in mathematical logic, 1879–1931* (pp. 369–392). Cambridge, MA: Harvard University Press.

Hintikka, J. (2004). A fallacious fallacy? *Synthese, 40*(1–2), 25–35.

Hölder, O. (1901). Die Axiome der Quantität und die Lehre vom Mass. Berichten der mathematisch-physischen Classe der Königl. Sächs. Gesellschaft der Wissenschaften zu Leipzig, *53*(4):1–64. [Trans. in J. Michell and C. Ernst. The axioms of quantity and the theory of measurement. *Journal of Mathematical Psychology, 40–41*(4):235–252; 345–356, 1997.]

Mancosu, P. (2008a). Mathematical explanation: Why it matters. In P. Mancosu (Ed.), *The philosophy of mathematical practice* (pp. 134–150). Oxford: Oxford University Press.

Mancosu, P. (Ed.). (2008b). *The philosophy of mathematical practice*. Oxford: Oxford University Press.

Mancosu, P., Jørgensen, K. F., & Pedersen, S. A. (2005). *Visualization, explanation and reasoning styles in mathematics*. Dordrecht: Springer.

Perelman, C., & Olbrechts-Tyteca, L. (1969). *The new rhetoric: A treatise on argumentation*. Notre Dame, IN: University of Notre Dame Press.

Poincaré, H. (1904). Rapport sur les travaux de M. Hilbert [Report on Hilbert's works]. *Bulletin de la Société Physico-Mathématique de Kazan, 14*, 10–48.

Torretti, R. (1978). *Philosophy of geometry from Riemann to Poincaré*. Dordrecht: Reidel.

van Kerkove, B., De Vuyst, J., & van Bendegem, J. P. (Eds.). (2010). *Philosophical perspectives on mathematical practice*. London: College Publications.

Veronese, G. (1891). I fondamenti della geometria [On the Foundations of Geometry]. Padova: Tipografia del Seminario. [Germ. trans. Grundzüge der Geometrie by A. Schepp, Leipzig: Teubner, 1894.]

Veronese, G. (1905). La geometria non-Archimedea. Una questione di priorità [Non-Archimedean geometry. A matter of priority]. *Rendiconti della Reale Accademia dei Lincei, 14*, 347–351.

von Helmholtz, H. (1876). Über den Ursprung und die Bedeutung der Raumanschauung und der geometrischen Axiome [The origin and meaning of geometrical axioms]. In *Populäre Wissenschaftliche Vorträge* (Vol. 3, pp. 21–54). Braunschweig: Vieweg und Sohn. (Engl. Trans. in W. Ewald (Ed.), From Kant to Hilbert, Vol. 1, pp. 663–685. Oxford: Clarendon Press, 1996).

Walton, D. N., Reed, C., & Macagno, F. (2008). *Argumentation schemes*. Cambridge: Cambridge University Press.

Chapter 7
How Persuaded Are You? A Typology of Responses

Matthew Inglis and Juan Pablo Mejía-Ramos

Several recent studies have suggested that there are two different ways in which a person can proceed when assessing the persuasiveness of a mathematical argument: by evaluating whether it is personally convincing, or by evaluating whether or not it is publicly acceptable. In this chapter, using Toulmin's (1958) argumentation scheme, we produce a more detailed theoretical classification of the ways in which participants can interpret a request to assess the persuasiveness of an argument. We suggest that there are (at least) five ways in which such a question can be interpreted. The classification is illustrated with data from a study that asked undergraduate students and research-active mathematicians to rate how persuasive they found a given argument. We conclude by arguing that researchers interested in mathematical conviction and proof validation need to be aware of the different ways in which participants can interpret questions about the persuasiveness of arguments, and that they must carefully control for these variations during their studies.

The issue of what types of arguments students find persuasive and convincing has been a recurring theme in the mathematics education literature. Mason et al. (1982) placed conviction at the heart of mathematical practice by proposing that learners first need to convince themselves, then a friend, and finally an enemy. Harel and Sowder (1998) analysed the notion of conviction in depth, by publishing a taxonomy of 'proof schemes': the types of arguments that students use both to ascertain for themselves (to remove their own doubts) and to persuade others (to remove others' doubts) about the truth of a statement. Other authors have suggested

M. Inglis (✉)
Mathematics Education Centre, Loughborough University, Loughborough, Leicestershire
LE11 3TU, UK
e-mail: m.j.inglis@lboro.ac.uk

J.P. Mejía-Ramos
Graduate School of Education, Rutgers University, 10 Seminary Place, New Brunswick,
NJ 08901-1183, USA
e-mail: pablo.mejia@gse.rutgers.edu

A. Aberdein and I.J. Dove (eds.), *The Argument of Mathematics*, Logic, Epistemology, and the Unity of Science 30, DOI 10.1007/978-94-007-6534-4_7,
© Springer Science+Business Media Dordrecht 2013

that these two processes may be disassociated: that the types of arguments that students find personally persuasive may not necessarily be the same types that they would use to persuade third parties such as their mathematics teachers or lecturers (Healy and Hoyles, 2000; Mejía-Ramos and Tall, 2005; Segal, 1999; Raman, 2002). In this paper we suggest that the notion of persuasion and conviction is yet more complicated than this two-way categorisation suggests. Here, we argue that there are (at least) five different interpretations that a participant may reasonably make when asked how persuasive they find a mathematical argument.

7.1 Toulmin's Argumentation Scheme

To situate our proposed taxonomy of the ways in which participants may respond to questioning about how 'persuaded' or 'convinced' they are by a mathematical argument, we first introduce Toulmin's (1958) argumentation scheme. Toulmin advocated an approach to analysing arguments that departed dramatically from traditional approaches to formal logic. He was less concerned with the logical validity of an argument, and more interested in the semantic content and structure in which it fits. This manner of analysing argumentation has become known as 'informal logic' in order to emphasise its differences from formal logic.

Toulmin's scheme has six basic types of statement, each of which plays a different role in an argument. The conclusion (C) is the statement of which the arguer wishes to convince their audience. The data (D) are the foundations on which the argument is based, the relevant evidence for the claim. The warrant (W) justifies the connection between data and conclusion by, for example, appealing to a rule, a definition, or by making an analogy.[1] The warrant is supported by the backing (B), which presents further evidence. The modal qualifier (Q, henceforth qualifier) qualifies the conclusion by expressing degrees of confidence; and the rebuttal (R) potentially refutes the conclusion by stating the conditions under which it would not hold. Importantly, in any given argument, not all of these statements will necessarily be explicitly verbalised. These six components of an argument are linked together in the structure shown in Fig. 7.1.

In the field of mathematics education, many researchers have applied Toulmin's scheme to analyse arguments constructed by students. However, it has become commonplace to use a reduced version of the scheme by omitting the qualifier and rebuttal. Krummheuer, for example, adopted this reduced version of the scheme to analyse pupil behaviour throughout his long programme of research on the development of collective argumentation practices in primary school classrooms

---

[1]Toulmin's (1958) use of the word 'warrant' is not identical to the way in which the term has been used by some mathematics education researchers. Rodd (2000), for example, saw a warrant as removing uncertainty, whereas Toulmin was more flexible, accepting that a warrant can be qualified so as to merely reduce uncertainty.

Fig. 7.1 The layout of
Toulmin's (1958)
argumentation scheme,
showing data (D), warrant
(W), backing (B), qualifier
(Q), rebuttal (R) and
conclusion (C)

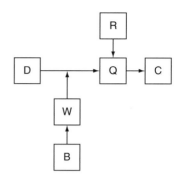

(e.g. Krummheuer 1995). A similar stance has been adopted by researchers studying classroom interaction at the university level (Stephan and Rasmussen, 2002), basic number skills (Evens and Houssart, 2004), logical deduction (Hoyles and Küchemann, 2002; Weber and Alcock, 2005), geometry (Cabassut, 2005; Pedemonte, 2005), and general proof (Yackel, 2001). Inglis et al. (2007) argued that without using Toulmin's full scheme it may be difficult to model accurately the full range of mathematical argumentation. They gave research-active mathematicians a series of conjectures, and asked them to decide whether or not the conjectures were true, and to provide proofs. It was found that these mathematicians regularly constructed arguments with non-deductive warrants in order to reduce rather than remove their doubts about a conjecture's truth value. Inglis et al. pointed out that it would be impossible to model such arguments accurately without incorporating the qualifier component of Toulmin's scheme. They concluded that (i) using the restricted version of Toulmin's scheme in the manner adopted by earlier researchers reduces the range of mathematical arguments that can be successfully modelled; and (ii) rather than concentrating on the appropriateness of the warrants deployed by students, researchers should instead study the appropriateness of the warrant-qualifier pairings constructed in student argumentation.

In this chapter, we use Toulmin's full scheme to derive a classification of the ways in which the question "how persuaded are you by this argument?" can be interpreted. We note that, as in the case of argument construction, a comprehensive study of argument evaluation cannot be conducted using a reduced version of the scheme.

7.2 How Persuaded Are You?

Some previous researchers have studied the types and levels of conviction and persuasion students place in an argument simply by asking them (e.g. Mejía-Ramos and Tall 2005; Segal 1999; Raman 2002). But what do students understand by such a question? We suggest that there are (at least) five distinct and reasonable ways of answering the question how persuaded are you by this argument? To demonstrate our typology in a general context, we introduce the following fictional day-to-day argument about train times:

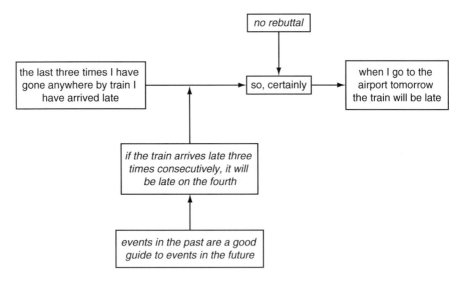

Fig. 7.2 Late-train argument modelled using Toulmin's (1958) scheme, with inferred warrant, backing and rebuttal components (the inferred components are italicised)

> The last three times I have gone anywhere by train, I have arrived several hours late. So, it is certain that, when I go to the airport tomorrow, the train will be late.

Using Toulmin's (1958) layout we can model this argument by identifying its different types of statements (see Fig. 7.2). We claim that a person evaluating how persuaded he or she is by this argument may focus on different parts and aspects of the argument. He or she may focus on: (0) the data of the given argument, and how significant/trustworthy it is; (1) the likelihood of its conclusion; (2) the strength of the warrant (and its associated backing); (3) the given qualifier (and its associated rebuttal), and the extent to which this qualifier is appropriate considering the rest of the argument; and (4) the particular context in which the given argument may take place. We now focus on each of these evaluation types in turn.

7.2.1 Type 0

One possible evaluation occurs when the participant focuses on the data of the given argument and evaluates the whole argument in terms of how strongly he or she trusts these data.

One possibility is that the participant distrusts the data and projects these doubts onto his/her evaluation of the whole argument. For instance, in the train example someone may suspect that the arguer is lying about the irregularities of the three train trips mentioned. For that person, this argument could be rated as unpersuasive mainly because he or she considers its data unreliable.

On the other hand, it is possible that someone's feeling of affinity for the data would be so strong that he or she would feel persuaded by the argument without taking into account how these data fit in the whole argument. For example, someone who has herself arrived several hours late in her last few train trips may find the argument persuasive mainly because of her empathy for its data.[2] Such an evaluation may involve what is known in the psychology literature as myside bias, a tendency to evaluate data from one's own perspective, having difficulty in decoupling one's prior beliefs and opinions from the evaluation of evidence and arguments (Stanovich and West, 2006). In this particular case, the evaluator could focus on the data and evaluate them in a manner biased towards his or her own opinions. These opinions and beliefs could then be projected on to the evaluation of the whole argument.

7.2.2 Type 1

Another way of evaluating this argument is by focusing on the likelihood of its conclusion alone. In the train example, someone focusing on the argument's conclusion (i.e. "the train to the airport will be late tomorrow") may report being highly persuaded by the argument, since he or she knows that scheduled track repairs in the vicinity of the airport will indeed delay trains that day. Alternatively, some people may not feel persuaded by the argument as, knowing that the train to the airport is the most reliable journey in the local train network, they expect this train to be on time.

In these two cases, evaluators would have reported the qualifier component of an entirely separate, and self-constructed, argument to that which they were asked to evaluate; the only similarities with the original argument being that they shared conclusions . In other words, in these cases the evaluators have their own evidence and reasons for trusting/distrusting the conclusion, and this information is projected onto the evaluation of the whole argument. However, as in Type 0 evaluations, it may also be the case that evaluators' uninformed intuition regarding the argument's conclusion, and not explicit information external to the argument, influence their reported level of persuasion.

7.2.3 Type 2

Another possible evaluation occurs when participants focus their attention on the warrant of the argument (with its associated backing). Unlike the data and the conclusion, the warrant of the argument is inextricably linked to other parts of the

---

[2]Elena Nardi has pointed out to us that this may be the type of persuasion enlisted by propagandists who use sensationalist data in the hope that the affinity of the audience for these data will sway them to support whatever conclusion he or she wishes to draw.

argument: it is a statement linking the data and the conclusion. Furthermore, any question regarding the trustworthiness of the warrant would lead to an evaluator querying its (explicit or implicit) backing and to consider possible rebuttals : if the warrant is appropriately backed and accepts no rebuttals (or only extraordinary ones), then one would say that the argument strongly supports the conclusion; but if the warrant is not satisfactorily backed and one can think of critical rebuttals, one would say that the warrant weakly supports the conclusion. Therefore, focusing on the warrant of an argument, a person may evaluate the strength with which it links the data with the conclusion, taking into account its backing and possible rebuttals. In this case, a participant's evaluation of the whole argument essentially consists of completing this core part of the given argument (data-warrant-conclusion) with what he or she believes is the appropriate qualifier for the argument, and then reporting this qualifier as his or her level of persuasion in the whole argument.

In the train example, a person might reply to the request of stating his or her level of persuasion in the whole argument by saying that (given the data, implicit warrant and the possible rebuttals associated with it) it is reasonable to reach the conclusion with a plausible qualifier. It is important to note that in this case the person is paying little or no attention to the absolute qualifier that was actually given in the train argument; he or she would be reporting what they believed to be the appropriate qualifier. It is also important to note that this way of evaluating the argument differs from Type 1 evaluations: a Type 2 evaluation focuses on the given warrant and takes into account certain information from the argument that is associated with that warrant, whereas a Type 1 evaluation focuses on its conclusion and may involve the (possibly implicit) construction of an entirely new argument.

7.2.4 Type 3

In contrast to the previous types of evaluation, a participant's attention may be drawn to the qualifier given in the argument (and its associated rebuttal). A Type 3 evaluation occurs when the evaluator decides to what extent he or she believes that the given qualifier is appropriate, considering the rest of the argument.

In the train example, someone may state that he or she is not at all persuaded by the argument as, although it might be reasonable to be worried about the possible lateness of the train based on prior experience, it is completely inappropriate to pair such a warrant with an absolute qualifier as the arguer appears to have done. Unlike a Type 2 evaluation, where the evaluator decides what type of qualifier would be appropriate given the rest of the argument, in a Type 3 evaluation the issue is whether the given qualifier is appropriate on account of the rest of the argument. It is clear that one could simultaneously consider an argument to be Type 2 persuasive but Type 3 unpersuasive. Indeed, believing (based on the prior experience cited in the argument) that the train would plausibly—but not certainly—be late could lead someone to such a judgement: they would assess the appropriate qualifier to be relatively high (i.e. Type 2 persuasive), but the qualifier as given in the argument to be inappropriate (i.e. Type 3 unpersuasive).

7.2.5 Type 4

Finally, instead of focusing on a particular part of the argument, the participant may attend to the context in which the argument is situated, and the kinds of arguments that are admissible in such contexts. In this case, when asked how persuaded they are by a given argument, participants may answer by considering how acceptable the argument would be in a particular context. It is well known in the context of jurisprudence that some arguments, no matter how persuasive, are not admissible in court. In England and Wales, for example, a prosecuting lawyer may not refer to a defendant's criminal record during the case. An argument based on such data may well carry an extremely high qualifier, but in the given context it is inadmissible. Naturally what constitutes an admissible argument will depend on the particular context: what is admissible in a criminal court is different from what is admissible in a civil court which, in turn, is different from what is admissible during an argument in a pub.

The example of the train argument may well be admissible when talking informally, but if one were attempting to convince one's departmental finance officer to issue an advance to pay for a taxi fare to the airport, it could be considered inadmissible. Such matters are governed by a set of rules which state what kinds of data, warrants, backings, qualifiers and rebuttals can be used in an admissible argument; and a hunch about the possible lateness of the train is unlikely to meet these rules.

The five types of persuasion we have discussed are summarised in Fig. 7.3.

7.3 Illustrating the Typology in Mathematics

Our primary aim in the second half of the paper is to illustrate the applicability of this typology to the evaluation of mathematical arguments. A second aim is to provide 'existence proofs' of each of the types: to show that the different types of evaluations can be, and are, made by mathematicians and students when evaluating mathematical arguments (or, at least, that mathematicians and students claim to be making evaluations of each type). To accomplish these aims, we draw on evaluations of a heuristic argument collected as part of a study on the role of authority in mathematical argumentation (Inglis and Mejia-Ramos, 2009). The argument used in the study was given by Gowers (2006) and supports the conjecture that there are one million consecutive sevens somewhere in the decimal expansion of π:

> All the evidence is that there is nothing very systematic about the sequence of digits of π. Indeed, they seem to behave much as they would if you just chose a sequence of random digits between 0 and 9. This hunch sounds vague, but it can be made precise as follows: there are various tests that statisticians perform on sequences to see whether they are likely to have been generated randomly, and it looks very much as though the sequences of digits of π would pass these tests. Certainly the first few million do. One obvious test is to see whether any short sequence of digits, such as 137, occurs with about the right frequency in

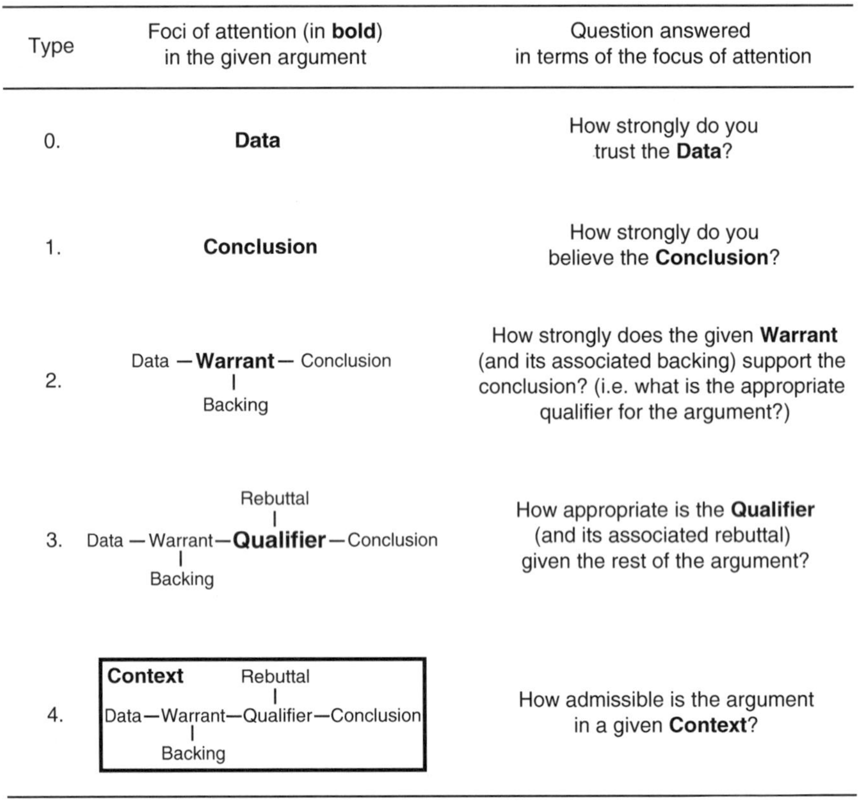

Type	Foci of attention (in **bold**) in the given argument	Question answered in terms of the focus of attention
0.	**Data**	How strongly do you trust the **Data**?
1.	**Conclusion**	How strongly do you believe the **Conclusion**?
2.	Data —**Warrant**— Conclusion | Backing	How strongly does the given **Warrant** (and its associated backing) support the conclusion? (i.e. what is the appropriate qualifier for the argument?)
3.	Rebuttal | Data — Warrant—**Qualifier**—Conclusion | Backing	How appropriate is the **Qualifier** (and its associated rebuttal) given the rest of the argument?
4.	**Context** Rebuttal | Data—Warrant—Qualifier—Conclusion | Backing	How admissible is the argument in a given **Context**?

Fig. 7.3 A summary of the types of persuasiveness identified in this paper, expressed using Toulmin's (1958) scheme

the long term. In the case of the string 137 one would expect it to crop up about 1/1000th of the time in the decimal expansion of π.

Experience strongly suggests that short sequences in the decimal expansion of the irrational numbers that crop up in nature, such as π, e or $\sqrt{2}$, do occur with the correct frequencies. And if that is so, then we would expect a million sevens in the decimal expansion of π about $10^{-1000000}$ of the time—and it is of course, no surprise, that we will not actually be able to check that directly. And yet, the argument that it does eventually occur, while not a proof, is pretty convincing (p. 194).

The two stages of this argument are shown graphically, using Toulmin's (1958) scheme, in Fig. 7.4.

Our sample consisted of two groups, undergraduate students and research-active mathematicians. Participants completed the task online (for a discussion on the reliability of internet studies see, for example, Krantz and Dalal 2000). The undergraduate students were studying at one of four highly-regarded UK universities, and were asked to participate by means of an email from their departmental secretary. The email explained the task and asked them to click through to the experimental

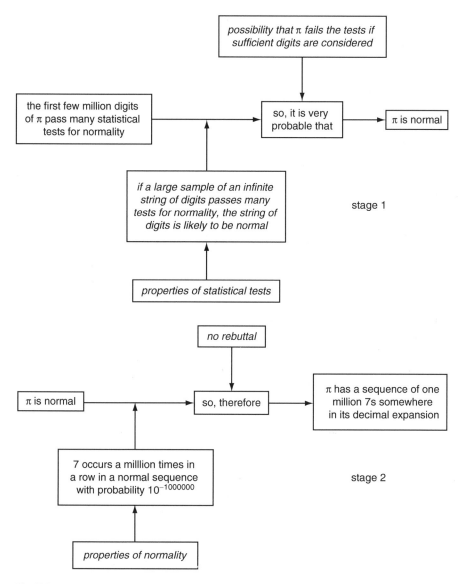

Fig. 7.4 Two stages of Gowers's (2006) argument modelled using Toulmin's (1958) scheme, with inferred components italicised (A number is said to be *normal* if its digits show a random distribution)

website should they wish to participate. The research-active mathematicians were recruited in two different ways. Some were recruited in a similar manner as the undergraduates, via emails from their departmental secretaries; others were recruited through an advertisement posted on a mathematics research newsgroup. In the study, participants were presented with Gowers's (2006) argument, and were asked to state to what extent they were persuaded by it, using either a five point Likert scale or a continuous 0–100 scale (depending on whether participants took part in the pilot or main study). In addition, participants were invited to leave explanatory comments on their reported level of persuasion. It is these comments that we use in the following sections to illustrate our theoretical classification.

Our focus in the analysis of the extracts is on the aspects of the argument which participants focus upon when explaining their evaluations. While we accept that the full complexity of participants' judgements may not be fully reflected by such short explanations (especially given factors such as myside bias, Stanovich and West 2006), we nevertheless believe that focusing on these reported comments will allow us to illustrate the utility of the typology for researchers interested in how mathematicians and students become persuaded by mathematical arguments.

We should emphasise that the classification we introduce in this paper is derived from a theoretical analysis of Toulmin's (1958) argumentation scheme; an analysis that considers the type of statement upon which participants may focus their attention when asked to evaluate their degree of persuasion in an argument. The extracts reported in the following sections, therefore, should be viewed as existence proofs of each of the categories, not as a data set from which we are attempting to generalise.

7.3.1 Type 0

When asked to evaluate and explain their level of persuasion in Gowers's argument regarding the decimal expansion of π, one research-active mathematician wrote:

> This is not an argument. To be more precise, in the statement no concrete evidence is presented. He only explains how the statistical evidence could look like, but does not specify the empirical results of the tests (Researcher).

In this case, the evaluator's explanation of his or her rating clearly focused on the data of the first stage of the argument. This researcher's comment concentrated on the "statistical evidence" of the argument, reporting dissatisfaction with the lack of concreteness of its presentation. Although this factor may be related to concerns that the researcher may have had with the warrant and backing of the argument, it is clear that their focus was on the data: to persuade this participant, at a minimum the data of the argument would need to be presented in a considerably more formal fashion.

7.3.2 Type 1

Type 1 evaluations were not uncommon among participants' comments. For example, one research-active mathematician wrote:

> Normalcy of (the digits) of pi is not unreasonable given almost all reals are normal (Researcher).

This researcher's comment focused on the conclusion of the first stage of the argument (i.e. "π is normal"), and reported the qualifier component (i.e. "not unreasonable") of an entirely separate, and self-constructed, argument to that which participants were asked to evaluate. He or she did not mention the data, warrant, backing or rebuttal given in Gowers's argument, and instead appears to have constructed a separate argument that merely shares a conclusion with the given argument. The data ("almost all reals are normal"), implicit warrant, and implicit backing in this new argument are entirely distinct, and the evaluator reported the new argument's qualifier by stating that the conclusion is "not unreasonable".

Another example of a Type 1 evaluation came from the following student:

> I am mainly not persuaded because I have seen a formula which can calculate the n-th digit of pi, suggesting that is not a random series of numbers (Undergraduate Student).

Again, the student's comment focused on the conclusion of the first stage of Gowers's argument, and evaluated the qualifier of an entirely new argument; the data, warrant and backing of the given argument are not taken into account. Of course, the construction of an entirely new argument—necessary for a Type 1 evaluation—would only be possible if the participant had a strong background knowledge of the domain in which the argument is situated.

7.3.3 Type 2

The following response typifies a Type 2 evaluation of Gowers's argument:

> The evidence lends decent weight to the conjecture; but naturally as proof is impossible it is unrealistic to assume certainty (Undergraduate Student).

Here the student seemed to be suggesting that the evidence presented—the data, warrant and backing—indicates that the conclusion may be true, but that any stronger qualifier would be inappropriate (possibly considering the existence of rebuttals). This student has considered the warrant of the argument (with its associated data, backing, and possible rebuttals) and has decided that he or she would be willing to pair a qualifier with it that "lends decent weight" to the conclusion. Characteristically for Type 2 evaluators, this student does not seem to be addressing the argument's given qualifier; this comment only refers to the new qualifiers that he or she considers appropriate given the rest of the argument.

7.3.4 Type 3

When participants were asked to evaluate Gowers's argument there were many examples of Type 3 evaluations:

> Despite the statistical evidence that pi is a 'normal number' (only testing short sequences) there could still be some subtle numerical invariant that prevents this particular very long sequence from occurring (Researcher).

In this comment, the researcher first evaluated the link between the data of the first stage of the argument and its conclusion, and then centred on an unmentioned rebuttal (the possibility of a "subtle numerical invariant"), suggesting that Gowers's given qualifier was inappropriate. A similar evaluation was reported in the following comment by an undergraduate student:

> The reasoning is flawed in moving from talking of experience strongly suggesting 'short sequences' occur in naturally occurring irrational numbers to saying that 'a million sevens' is likely to occur. Of course, their definition of a 'short sequence' isn't given, but I dare guess it is much fewer numbers than a million (Undergraduate Student).

Again, in this case the student criticised the strength with which the data is claimed to support the argument's conclusion, focusing on what he or she considered to be an inappropriate qualifier. In this comment, the student did not report the extent to which he or she believed that the given evidence supports the conclusion; instead they seemed to be more concerned about the relatively high qualifier given in the original argument.

In both these extracts the evaluators exhibited the hallmarks of a Type 3 evaluation; they explained that they were not persuaded by the argument as a whole because they did not accept that the given warrant (and associated backing) justifies the given qualifier (and associated rebuttal).

7.3.5 *Type 4*

A Type 4 evaluation can clearly be seen in this researcher's response:

> The argument hinges on a precise notion of randomness in the digits of pi, which may be plausible, but hasn't been proven. If a manuscript that made an analogous argument came to me for refereeing, I'd recommend it be rejected for lack of mathematical rigour. However, if someone wanted to generate 'good pseudorandom' bits from the digits of pi for a casual computer program (i.e., not one on which lives or property crucially depend), I'd say Gowers's argument would justify the strategy (Researcher).

Here the evaluator suggested that in the context of an academic mathematics journal he or she would deem the argument to be inadmissible, but that in a different context, where one merely needed to generate some random numbers, it would be admissible. This position was clarified still further by noting that in yet another context—where the random numbers were a matter of life or death—then perhaps the argument would again struggle to meet the requirements of admissibility.

Within educational contexts Type 4 evaluations are very important, and an ability to understand successfully the different rules of admissibility for different contexts may be a hard skill for students to develop. These rules undoubtedly vary between educational levels—the type of justification required at school level mathematics is typically very different from that required at university level—but may also vary between courses at a single level. For example, the types of argument which are admissible to justify the rules of integration may be very different if the notion of integral is studied during a real analysis course compared with during an applied fluid dynamics course. In the former a formal derivation from the definition of, for example, the Riemann integral, may be necessary, whereas in the latter a statement of the result might be sufficient.

7.3.6 Mixing the Types

Sometimes participants in empirical research studies may give evaluations of different types in the same response: multiple interpretations of the question may lead to answers with multiple layers. This, for example, is how one mathematics researcher responded when asked to evaluate Gowers's argument:

> Purely logically on the basis of the evidence presented, I am not persuaded at all. However, I am aware that there is a substantial body of research (rather more formal than the waffle above) specifically addressing equidistribution of digit sequences of pi. So I moved from the most sceptical to the next category by way of combining that knowledge with the information above (Researcher).

Here the evaluator explicitly noted that he or she was not persuaded by the data, warrant and backing of the argument: they were completely Type 2 unpersuaded. However, the researcher claimed that they were somewhat Type 1 persuaded on account of his or her background knowledge about the digit sequences of π. Understandably, these differing interpretations gave this participant some difficulty when asked to rate his or her level of persuasion on a Likert scale. However, the proposed typology can help us to make sense of this participant's multi-layered written comment.

7.4 One Question, Five Ways of Answering

Earlier researchers have studied two different ways of evaluating a given argument and the two corresponding levels of persuasiveness reported by their participants. This led them to establish a distinction between a private and a public, or internal and external, sense of conviction. Raman (2003, p. 320), for example, differentiated between private and public arguments and their corresponding senses of conviction (see also Raman 2002):

> By 'private argument' I mean, 'an argument which engenders understanding', and by 'public' I mean 'an argument with sufficient rigor for a particular mathematical community'.

In our terms, Raman was noting the differences between Type 4 persuasion and persuasion of Types 0–3. The usefulness of this two-way distinction in mathematics education lies in the importance of Type 4 evaluations in mathematical argumentation.[3] Mathematical proof seems to set argument admissibility in mathematical practice aside from admissibility standards in other contexts, making students' beliefs of what constitutes a valid mathematical proof, and the ways in which these beliefs influence their reported level of persuasion in a given argument, an interesting topic of study among mathematics educators.

However, we suggest that a finer typology of persuasiveness may be helpful: whereas earlier researchers have spoken only of a 'private' sense of conviction, we have demonstrated that, in the case of argument validation, there are (at least) four different ways in which such an evaluation may be conducted in a 'private' fashion. Similarly, there is not simply one variety of 'public', or Type 4, persuasion. Each particular context brings with it its own particular rules of admissibility, and these rules vary greatly between contexts. Even in the particular context of mathematics, requirements for rigour alter greatly according to educational level, mathematical subject and other particular circumstances of each evaluation.

Segal (1999) used two different questions to study this distinction between a private and public sense of conviction. Following Mason et al. (1982), she asked her participants whether or not a given argument convinced them personally, and whether or not the argument would persuade "one's enemies (as opposed to one's friends, or oneself)" (Segal, 1999, p. 199). It is unclear whether a participant's response to the first question involves an evaluation of Type 0, 1, 2 or 3. Furthermore, a participant's response to the second question (arguably related to Type 4 evaluations) is, of course, dependant upon in which context the participant situates their enemy, and what type of evaluation is expected from that enemy.

Therefore, this finer typology may be used by both teachers and researchers not only to better assess students' and participants' reported levels of persuasion in a given mathematical argument, but also to design specific questioning strategies to incline students and participants towards making the teacher's or researcher's desired type of evaluation.

7.5   Using the Typology: An Example from the Literature

In this section we give a specific example of a piece of analysis from the mathematics education literature to illustrate the utility of the typology proposed here. During their study of the proof conceptions of school children, Coe and Ruthven (1994) looked at students' investigative and problem solving strategies.

---

[3]Indeed, it has recently been demonstrated that even minor linguistic changes to task instructions can influence whether a Type 4 evaluation is conducted. Mejía-Ramos and Inglis (2011) found that responses to a convincing (but informal) visual argument differed between two groups who were respectively asked "does the argument prove the claim?" and "is the argument a proof of the claim?" The latter question seemed to privilege Type 4 evaluations and the former evaluations of Type 2.

Here we concentrate on one particular extract from an interview transcript reported by Coe and Ruthven. Whilst working on a problem regarding the sums of diagonals in a number square, Bill, a 17 year old student, checked that a statement he was investigating was true for six cases, and then said that it was "safe to make a conjecture". The interviewer pressed him by asking "what sort of percentage certainty would you put behind that, say, if I forced you on that?" Bill replied by estimating he had a "percentage certainty" in the "high nineties".

What has happened here? Bill and the interviewer were discussing the persuasiveness of an argument with an inductive warrant (which consisted of the numerical evaluation of six examples). The interviewer pressed Bill with an ambiguous question, by asking the "percentage certainty" he would be willing to "put behind that". One interpretation is that, when pressed, Bill conducted a Type 2 evaluation. He evaluated what sort of qualifier he was willing to pair with the data, warrant and conclusion of the given (self-constructed) argument, and made the decision that he was willing to deploy a high (but non-absolute) qualifier (Seen in this light, Bill's behaviour closely matches that of the highly successful research students interviewed by Inglis et al. 2007). Coe and Ruthven (1994), however, appear to have interpreted Bill's response in a different way, as a Type 4 evaluation. They wrote that Bill's "*certainty* appears to be gained just by checking a relatively small number of cases" (p. 50), and used the episode as evidence for the claim that "students' *proof* strategies were primarily and predominantly empirical" (p. 52, our emphasis). Of course, it may well be that Bill was conducting a Type 4 evaluation of his self-constructed argument: there are certainly many studies which corroborate Coe and Ruthven's claim that students often think empirical evidence can form admissible proofs (e.g. Balacheff 1987; Harel and Sowder 1998; but see Weber 2010). However, when seen within the typology set out in this paper, Coe and Ruthven's interpretation of this interview evidence is, at best, arguable.

We suggest that an awareness of the typology presented in this paper could help researchers conducting studies on mathematical conviction to deploy careful questioning strategies to increase the likelihood of accurately interpreting their interviewee's behaviour.

7.6 Concluding Remarks

Researchers interested in assessment have, for some time, been aware that there may be a gap between test designers' interpretation of a given question and the interpretation of those who respond to the question. In their study of 11 and 12 year old children's responses to national test items, Cooper and Harries (2002) found that students interpretations of how much realism to use in their answers when answering 'realistic' mathematical questions differ from those of the questions' designers. We suggest that teachers and researchers who are interested in what types of mathematical argument students find persuasive need to have awareness of the differing ways in which their questions may be interpreted. Similarly, the many researchers who have studied the manner in which students and mathematicians validate proofs (e.g. Selden and Selden 2003; Weber 2008) also need to be aware that requests to 'evaluate' a purported proof may lead to differing interpretations from different participants.

In this chapter we have proposed that there are (at least) five different ways in which participants in research studies can reasonably interpret a request to evaluate their level of conviction or persuasion in an argument. The first two types that we have described revolve around a participant evaluating a particular part of the argument (data or conclusion) and paying little or no attention to the other parts; two other types involve the participant evaluating the core part of the argument and either completing it with what they believe is an appropriate qualifier, or assessing whether or not the given qualifier is appropriate; a fifth type is related to one particular context in which the argument may take place and the participant's evaluation of whether or not the given argument would be admissible in such context. By using Toulmin's full scheme, it is possible to distinguish clearly between these different types of argument evaluation. Given these different ways, we suggest that the empirical researcher must design their methodological instruments carefully to determine which question participants are responding to, and take into account these different types of evaluations in their theorisation of students' reported levels of persuasion.

Acknowledgements We are grateful to Janet Ainley, Paola Iannone, Andreas Moutsios-Rentzos, Elena Nardi, Dave Pratt, Keith Weber and two anonymous reviewers for helpful comments on earlier drafts of this manuscript.

This article was originally published in *Research in Mathematics Education* (vol. 10, pp. 119–133). It is copyright © British Society for Research into Learning Mathematics, and is reprinted by permission of Taylor & Francis Ltd (http://www.tandf.co.uk/journals) on behalf of the British Society for Research into Learning Mathematics.

References

Balacheff, N. (1987). Processus de preuves et situations de validation. *Educational Studies in Mathematics, 18,* 147–176.
Cabassut, R. (2005). Argumentation and proof in examples taken from French and German textbooks. In *Proceedings of the fourth congress of the European Society for Research in Mathematics Education.* Sant Feliu de Guíxols: ERME.
Coe, R., & Ruthven, K. (1994). Proof practices and constructs of advanced mathematical students. *British Educational Research Journal, 20,* 41–53.
Cooper, B., & Harries, T. (2002). Children's responses to contrasting 'realistic' mathematics problems: Just how realistic are children ready to be? *Educational Studies in Mathematics, 49,* 1–23.
Evens, H., & Houssart, J. (2004). Categorizing pupils' written answers to a mathematics test question: 'I know but I can't explain'. *Educational Research, 46,* 269–282.
Gowers, W. T. (2006). Does mathematics need a philosophy? In R. Hersh (Ed.), *18 unconventional essays on the nature of mathematics* (pp. 182–200). New York: Springer.

Harel, G., & Sowder, L. (1998). Students' proof schemes: Results from exploratory studies. In A. H. Schoenfeld, J. Kaput, & E. Dubinsky (Eds.), *Research in collegiate mathematics III* (pp. 234–282). Providence, RI: American Mathematical Society.

Healy, L., & Hoyles, C. (2000). A study of proof conceptions in algebra. *Journal for Research in Mathematics Education, 31*, 396–428.

Hoyles, C., & Küchemann, D. (2002). Students' understanding of logical implication. *Educational Studies in Mathematics, 51*, 193–223.

Inglis, M., & Mejia-Ramos, J. P. (2009). The effect of authority on the persuasiveness of mathematical arguments. *Cognition and Instruction, 27*, 25–50.

Inglis, M., Mejia-Ramos, J. P., & Simpson, A. (2007). Modelling mathematical argumentation: The importance of qualification. *Educational Studies in Mathematics, 66*, 3–21.

Krantz, J. H., & Dalal, R. (2000). Validity of web-based psychological research. In M. H. Birnbaum (Ed.), *Psychological experiments on the internet* (pp. 35–60). San Diego, CA: Academic.

Krummheuer, G. (1995). The ethnology of argumentation. In P. Cobb, & H. Bauersfeld (Eds.), *The emergence of mathematical meaning: Interaction in classroom cultures* (pp. 229–269). Hillsdale, NJ: Erlbaum.

Mason, J., Burton, L., & Stacey, K. (1982). *Thinking mathematically*. London: Addison-Wesley.

Mejía-Ramos, J. P., & Inglis, M. (2011). Semantic contamination and mathematical proof: Can a non-proof prove? *Journal of Mathematical Behavior, 30*, 19–29.

Mejía-Ramos, J. P., & Tall, D. O. (2005). Personal and public aspects of formal proof: A theory and a single-case study. In D. Hewitt & A. Noyes (Eds.), *Proceedings of the sixth British Congress of Mathematics Education* (pp. 97–104). Coventry: BSRLM.

Pedemonte, B. (2005). Quelques outils pour l'analyse cognitive du rapport entre argumentation et démonstration. *Recherches en Didactique des Mathématiques, 25*, 313–348.

Raman, M. (2002). *Proof and justification in collegiate calculus*. PhD thesis, University of California, Berkeley.

Raman, M. (2003). Key ideas: What are they and how can they help us understand how people view proof? *Educational Studies in Mathematics, 52*, 319–325.

Rodd, M. M. (2000). On mathematical warrants: Proof does not always warrant, and a warrant may be other than a proof. *Mathematical Thinking and Learning, 2*, 221–244.

Segal, J. (1999). Learning about mathematical proof: Conviction and validity. *Journal of Mathematical Behavior, 18*, 191–210.

Selden, A., & Selden, J. (2003). Validations of proofs considered as texts: Can undergraduates tell whether an argument proves a theorem? *Journal for Research in Mathematics Education, 34*, 4–36.

Stanovich, K. E., & West, R. F. (2006). Natural myside bias is independent of cognitive ability. *Thinking and Reasoning, 13*, 225–247.

Stephan, M., & Rasmussen, C. (2002). Classroom mathematical practices in differential equations. *Journal of Mathematical Behavior, 21*, 459–490.

Toulmin, S. (1958). *The uses of argument*. Cambridge: Cambridge University Press.

Weber, K. (2008). How mathematicians determine if an argument is a valid proof. *Journal for Research in Mathematics Education, 39*, 431–459.

Weber, K. (2010). Mathematics majors' perceptions of conviction, validity and proof. *Mathematical Thinking and Learning, 12*, 306–336.

Weber, K., & Alcock, L. (2005). Using warranted implications to understand and validate proofs. *For the Learning of Mathematics, 25*(1), 34–38.

Yackel, E. (2001). Explanation, justification and argumentation in mathematics classrooms. In van den Heuvel-Panhuizen, M. (Ed.), *Proceedings of the 25th international conference on the psychology of mathematics education* (Vol. 1, pp. 9–23). Utrecht: IGPME.

Chapter 8
Revealing Structures of Argumentations in Classroom Proving Processes

Christine Knipping and David Reid

8.1 Introduction

This paper seeks to describe a method for revealing the rationality of arguments that are produced during proving processes in the mathematics classroom. As proving processes in the classroom follow their own peculiar rationale, the problem is to reconstruct and analyze the complex argumentative structure in classroom conversations. The rationale of these conversations is of interest here.

Formal mathematical logic alone cannot capture this rationale, as will be obvious from the examples discussed in this paper. We have to find other ways to reveal these structures and the specific types of arguments that occur in classrooms. Understanding their rationale and the contextual constraints that shape these argumentations can help us to improve our efforts in teaching proof.

The importance of alternatives to formal logic in argumentation analysis in everyday contexts has been recognized, beginning with Toulmin's (1958) work and continuing through, for example, Walton's (1989, 1998) dialectic approach, van Eemeren and Grootendorst's (2004) pragma-dialectical approach and Snoeck-Henkemans's (1992) work on complex argumentation. Argumentation analysis in mathematical contexts has a shorter history, arguably beginning with the work of Alcolea (1998).

C. Knipping (✉)
FB3: Didaktik der Mathematik, Universität Bremen, Bibliotheksstraße 1, 28359 Bremen, Germany
e-mail: knipping@math.uni-bremen.de

D. Reid
FB3: Didaktik der Mathematik, Universität Bremen, Bibliotheksstraße 1, 28359 Bremen, Germany

School of Education, Acadia University, Wolfville, NS B4P 2R6, Canada
e-mail: david.reid@acadiau.ca

A. Aberdein and I.J. Dove (eds.), *The Argument of Mathematics*, Logic, Epistemology, and the Unity of Science 30, DOI 10.1007/978-94-007-6534-4_8,
© Springer Science+Business Media Dordrecht 2013

This paper suggests a method by which complex argumentations in proving processes can be reconstructed and analyzed. A three stage process is proposed: reconstructing the sequencing and meaning of classroom talk; analyzing arguments and argumentation structures; and finally comparing these argumentation structures and revealing their rationale. In this paper the emphasis is on the second stage, arguments and argumentation structures. This involves two moves, first analyzing local arguments on the basis of Toulmin's functional model of argumentation, and second analyzing the global argumentation structure of the proving process. To illustrate patterns in the global analysis of the classroom talk a schematic representation of the overall argumentative structure is used.

The paper is organized as follows. We first discuss not only why an alternative conception of 'rational argument' (see Toulmin, 1958), distinct from the one in formal logic, is important for understanding proving processes in the mathematical classroom, but also how such a conception leads to a different reconstruction of arguments found in classroom proving processes. This provides an alternate perspective on arguments in the context of classroom proof and proving. We then describe the theoretical and methodological grounds of Knipping's (2008) method of reconstructing and analyzing argumentation structures, before presenting the method itself. Examples of how we apply the argumentation analyses to real data will illustrate each individual stage of the method. We then compare global argumentation structures reconstructed from a German and a Canadian classroom to show how this method can reveal differences in the rationale of proving processes.

8.2 The Importance of Understanding Proving Practices in the Classroom

Teaching proof is considered to be challenging. Numerous empirical studies have documented that students up to university level have difficulties in recognizing different types of reasoning and producing mathematical proofs (e.g., Harel and Sowder, 1998; Healy and Hoyles, 1998; Reiss et al., 2001).[1] Students find proof difficult and often do not understand why so much emphasis is put on mathematical proof (Moore, 1994). Some research has investigated these difficulties (Chazan, 1993; Reid, 1995; Pedemonte, 2002a,b, 2007) and has explained aspects of the problem from an individual student's point of view. Researchers have offered alternative ways of teaching proof (e.g., Garuti et al., 1998; Mariotti et al., 1997; Jahnke, 1978), but only a few have documented students' proving processes in alternative teaching environments (Balacheff, 1988, 1991).

Very little research so far has looked at these difficulties with a focus on proving practices in the classroom (Sekiguchi, 1991; Herbst, 1998, 2002a; Knipping, 2003). Such a focus is important for two reasons:

---

[1] For an overview of these studies see as well Harel and Sowder (1998) and Hanna (2000).

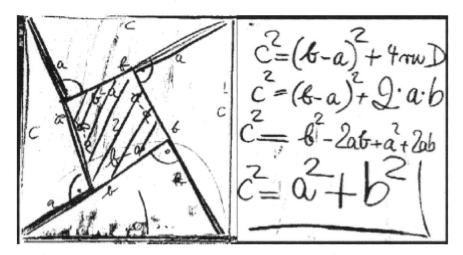

Fig. 8.1 Written proof of the Pythagorean Theorem in a grade 9 class in Germany, 1998. "4rwD" means 4 *rechtwinklige Dreiecke*, i.e., 4 right angled triangles)

- Teachers' approaches to proof are not guided solely by logical considerations; pedagogical and practical considerations are also important.
- Written proofs, whether produced in classrooms or presented in texts, do not reflect the process of their creation, which is itself worthy of study.

Sekiguchi (1991) explores the social nature of proof in the mathematics classroom using ethnographic methods. He confirms that proof practice in the classroom does not follow "the patterns of formal mathematics", but that pedagogical and practical motivations shape practices in the classroom. He finds various forms of practices called "proof" by the participants, but his research focuses on the standard form for writing proofs in US classrooms, the two-column proof format (Sekiguchi, 1991). Herbst (2002b) describes how this proving custom developed at the end of the nineteenth-century, in a historical context where the demand arose that every student should be able to *do* proofs. Herbst helps us to understand that the practice of producing two-column proofs was a reaction to this demand and the difficulty "to organize classrooms where students can be expected to produce arguments and proofs" (Herbst, 2002b, 284). Thus a way of writing proofs that is peculiar to the school context evolved, not out of logical requirements, but out of a specific historical context, students' difficulties, and teachers' efforts to address the difficulties and challenges of engaging students in proofs.

The two-column proof format is not the only format that can be found in mathematics classrooms. In France and Germany (Knipping, 2002, 2003, 2004), other formats for written proofs have evolved. Whatever format is used in presenting a proof, in the classroom context writing proofs is only part of, and more a means to engage students in, proving (or the illusion of doing proofs), as Sekiguchi and Herbst demonstrate. In classrooms the process itself, embedded in the constraints of teaching, seems to be of major importance, but arguments produced in these processes are far different from the written forms that are produced. For example, Fig. 8.1 shows a photograph of a blackboard at the end of a lesson on the Pythagorean Theorem, in a German grade 9 upper stream secondary school ("*Gymnasium*") class. The teacher and probably the students considered this a proof of the Pythagorean Theorem. The proof was the product of processes during an entire lesson. Diagrams like that presented on the left side of the board are classic illustrations of the Pythagorean Theorem and its proof. They are recognized by many people, and can be found throughout the history of mathematics in many cultures. However, our focus in this chapter is on the proof written on the right side of the blackboard and the argument that led to it. The algebraic equations written there are a very condensed form of argument, which leaves out many steps that have been developed in the oral part of the classroom proving process we will discuss later.

Because classroom proving processes are guided by more than logical considerations and because the written proofs that result from classroom proving processes are incomplete as records of those processes, reconstructing classroom proving processes and their overall structure requires a model that acknowledges their context dependence. Research within and outside of mathematics education has addressed this concern that there is no universal way to describe and formulate context dependent arguments. We will discuss this next.

8.3   Approaches to Describing Arguments

The arguments that occur in classroom proving processes cannot be analysed using formal logic. Instead approaches are needed that allow for the context and the nature of students' rationality. Toulmin's functional model of argument can be extended to describe both local arguments and global argumentation structures. The need for such an approach and the basics of it are the topic of this section.

8.3.1   The Inadequacy of Logical Analysis for Reconstructing Proving Processes in Classrooms

In mathematics considerable attention has been paid to the nature of proofs. As proofs can be seen as the end product of the work of mathematicians, i.e., as explanations which are accepted by the mathematical community as proofs (Balacheff, 1987) one might expect that the process by which proofs come to be can be analyzed on the basis of what proofs are. Further, the final goal of teaching proof is to bring students to an understanding of the logic behind mathematical proofs and to accept the same kinds of explanations as proofs as are accepted by mathematicians. Therefore, the misunderstanding can arise that the analysis of proof teaching in the classroom can be based on logical analysis of classroom arguments. This is not true for several reasons.

First, as discussed earlier, Sekiguchi (1991) and Herbst (1998, 2002a) illustrate that when teachers teach proof they do not follow "the patterns of formal mathematics", but pedagogical and practical motivations shape their practices in the classroom. Their practices in the classroom are a complex response to the challenge of teaching proofs and of engaging students in proving. In these multi-faceted processes classroom conversations with complex argumentation structures occur that might appear "illogical" to a mathematician. Revealing these complex structures is necessary to better understand the complexity of teaching proof and proving, but this cannot be undertaken by means of formal logic alone.

Second, students are in the process of developing logical thinking patterns, and so the thinking they express in classrooms will include many elements which a logical analysis would simply describe as "illogical" but which are nevertheless important to the future development of their thinking. As learning necessarily depends on the students' thinking at the time, a method of analysis that cannot go beyond dismissing it as "illogical" is not helpful. What is needed is a conception of 'rational argument' that does not cut off students' rationality.

Third, logical rationality, which has historically decontextualized intellectual and practical rationality, is widely questioned by philosophers and historians of science. Toulmin (1990), for example, deconstructs the logical ideal of reason as a historical project of Modernity that we have to appreciate, but that has been overcome by the facts of twentieth-century science. He argues that "the decontextualizing of problems so typical of High Modernity is no longer a serious option" (Toulmin, 1990, 201). Instead science came to a "renewed acceptance of practice" (1990, 192) and to a reconceptualization of rationality that does not cut "the subject off from practical considerations" (1990, 201) in the way that formal logic does. In his earlier work Toulmin (1958) had enquired into different fields of argument and addressed the question of "What things about the forms and merits of our arguments are *field-invariant* and what things about them are *field-dependent*?" (1958, 15, emphasis in original). Contrasting arguments in jurisprudence to formal logic he suggests that there is a layout of arguments that is field invariant, but that allows the characterization of arguments related to their context of use. This layout or functional model is the topic of the next section.

8.3.2 Toulmin's Functional Model of Argument

The model that Toulmin developed to reconstruct arguments in different fields has become popular across disciplines. As noted above, logical analysis is inadequate for argumentations in mathematics classrooms, which perhaps bear more resemblance to argumentations in other domains such as law, where public discussion of facts and the relationships between them are important. Toulmin's model is intended to be applicable to arguments in any field. For this reason the work of

Fig. 8.2 Toulmin model

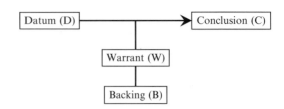

Toulmin has provided researchers in mathematics education with a useful tool for research, including formal and informal arguments in classrooms (Krummheuer, 1995; Knipping, 2002) and individual students' proving processes (Pedemonte, 2002a,b).

Rejecting a mathematical logical model Toulmin (1958) investigates the functional structure of rational arguments in general. Therefore he asks "What, then, is involved in establishing conclusions by the production of arguments?" (1958, 97). Toulmin's first answer is that facts (data) might be cited to support the conclusion. He illustrates this by the following example. If we assert that 'Harry's hair is not black', we might ground this on "our personal knowledge that it is in fact red" (1958, 97). We produce a datum that we consider as an evident fact to justify our assertion (conclusion). If this is accepted, this very simple step, datum–conclusion, can represent a rational argument.

But this step, its nature and justification, can be challenged, actually or potentially, and therefore it is often explicitly justified. Instead of additional information, an explanation of a more general style, by rules, principles or inference-licenses has to be formulated (1958, 98). Toulmin's second answer addresses this type of challenge. A 'warrant' might be given to establish the "bearing on the conclusion of the data already produced" (1958, 98). These warrants "act as bridges, and authorize the sort of step to which our particular argument commits us" (1958, 98). In the example above the implicit warrant of the argument is 'If anything is red, it will not also be black." (1958, 98). While Toulmin acknowledges that the distinction between data and warrants may not always be clear, their functions are distinct, "in one situation to convey a piece of information, in another to authorise a step in an argument" (1958, 99). In fact, the same statement might serve as either datum or warrant or both at once, depending on context (1958, 99), but according to Toulmin the distinction between datum, warrant, and the conclusion or claim provides the elements for the "skeleton of a pattern for analyzing arguments" (1958, 99); see Fig. 8.2. In the following we will use "claim" in cases where data and warrants have not yet been provided, and "conclusion" when they have been.

Toulmin adds several other elements to this skeleton, only one of which will be discussed here. Both the datum and the warrant of an argument can be questioned. If a datum requires support, a new argument in which it is the conclusion can be developed. If a warrant is in doubt, a statement Toulmin calls a "backing" can be offered to support it.

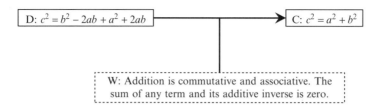

Fig. 8.3 Datum, warrant and conclusion for the final step in the written proof

Looking back at the proof in Fig. 8.1, we can analyze the last step in the argument in terms of Toulmin's model (see Fig. 8.3). In it we can see an important characteristic of many arguments: warrants are often left implicit. In this case our awareness that the warrant is not included in the written proof helps us to ask the question "Was the warrant ever stated as the proof was developed, or was it always implicit?", showing once more the importance of analyzing the classroom proving process.

Toulmin states, "The data we cite if a claim is challenged depend on the warrants we are prepared to operate with in that field, and the warrants to which we commit ourselves are implicit in the particular steps from data to claims we are prepared to take and to admit." (1958, 100). Therefore careful analyses of the types of warrants (and backings) that are employed explicitly or implicitly in concrete classroom situations allow us to reconstruct the kinds of mathematical justifications students and teacher together operate on. In particular, the comparison of warrants and backings in different arguments can reveal what sort of argument types are used in proving processes in mathematics classrooms.

For example in Fig. 8.3, we have supplied an implicit warrant based on mathematical properties of addition. In a different context the warrant for this argument might have been geometrical, interpreting $2ab$ as the area of a rectangle (or two triangles), or syntactical, not interpreting the symbols at all, but operating on them purely formally. Any of these types of warrants (and backings) could occur in a classroom and indicate the *field* of justifications in which the students and teacher operate.

Other researchers (e.g., Inglis et al., 2007) have made use of other elements in Toulmin's model, including "modal qualifiers" and "rebuttal". Many arguments do not establish their conclusions with complete certainty, and in such arguments we find qualifiers like "probably" and "possibly" as well as rebuttals that identify cases where the conclusion does not hold. Inglis, Mejía-Ramos and Simpson consider the arguments of postgraduate university students in mathematics and find that modal qualifiers play an important role in their mathematical argumentations. In our work in schools, however, we have found that the mathematical argumentations produced are quite different from what advanced mathematics students produce, and we have not found it necessary to make use of any elements in the Toulmin model beyond data, conclusions, warrants and backings. We have added one element, however, which we call "refutation". A refutation differs from a rebuttal in that a rebuttal is

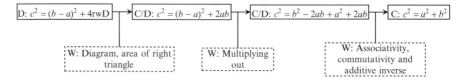

Fig. 8.4 Functional reconstruction of the written proof presented in Fig. 8.1

local to a step in an argument and specifies exceptions to the conclusion. A refutation completely negates some part of the argument. In a finished argumentation refuted conclusions would have no place, but as we are concerned with representing the entire argumentation that occurred, it is important for us to include refutations and the arguments they refute, as part of the context of the remainder of the argumentation, even if there is no direct link to be made between the refuted argument and other parts of the argumentation. Aberdein (2006) proposes extending Toulmin's rebuttal element to encompass refutations, but for our purposes we prefer to limit rebuttals to Toulmin's original role, of specifying circumstances where the conclusion does not hold.

An important way in which we have used the Toulmin model, that extends it significantly, is our application of it not only to single steps in argumentations, but also as a tool to explore the global structure of an argumentation. In the next section we will describe this distinction in more detail.

8.3.3 Local and Global Arguments

Toulmin (1958) notes "an argument is like an organism. It has both a gross, anatomical structure and a finer, as-it-were physiological one" (94). Toulmin's aim is to explore the fine structure, but in considering classroom argumentations both argumentative forms must be reconstructed. Toulmin's model is useful for reconstructing a step of an argument, which allows us to single out distinct arguments in the proving process (for example as in Fig. 8.3). We will call these "argumentation steps" or *local arguments*. But it is also necessary to lay out the structure of the argument as a whole (the anatomical structure), which we will call *global argument* or the argumentation "structure" of the proving process.

Global arguments in classrooms can be quite complex (as will be shown later). The written proof in the right hand side of Fig. 8.1 provides a simple example. As we noted earlier, a single step of that proof is shown in Fig. 8.3. The global argument presented on the blackboard, reconstructed as a chain of argumentation steps is shown in Fig. 8.4. The final conclusion (C: $c^2 = a^2 + b^2$), a formulation of the Pythagorean Theorem, is the target conclusion of the global argument. The argument can be reconstructed as a simple chain of conclusions beginning with a datum "$c^2 = (b-a)^2 + 4$rwD" that has been taken from the drawing on the blackboard. This datum leads to a conclusion: $c^2 = (b-a)^2 + 2ab$, but no warrant is explicitly given to support this inference. The information in the diagram (adjacent sides of the right triangle are a and b) and implicit calculations of the area of the four right triangles implicitly support this claim. The next two steps are also based on implicit warrants. In Fig. 8.4 we have reconstructed possible implicit warrants for each step; they are marked by a box with a dashed line. Note that the statement "$c^2 = (b-a)^2 + 2ab$" is not only the conclusion of one step but also the datum of another. Finally the target conclusion: $c^2 = a^2 + b^2$ is established.

This type of argument can be characterized as a chain of statements, each one deduced from the preceding one on logical and mathematical grounds. This has been described by Duval as "Recyclage" (Duval, 1995, 246–248). Once a statement has been established as a conclusion it functions as a datum, an established true fact, in the next step. Aberdein (2006) calls this way of combining single steps "Sequential" and he describes four other ways steps could be combined. As we will see in the following, our empirical research on classroom argumentation provides examples of Aberdein's ways of combining steps, as well as other ways.

8.4 A Method for Reconstructing Arguments in Classrooms

As mentioned above, for reconstructing arguments in classrooms a three stage process is followed:

- reconstructing the sequencing and meaning of classroom talk (including identifying episodes and interpreting the transcripts);
- analyzing arguments and argumentation structures (reconstructing steps of local arguments and short sequences of steps that form "streams"; reconstructing the global structure); and
- comparing local argumentations and comparing global argumentation structures, and revealing their rationale.

Each of these stages is illustrated in the following, but the emphasis is on the second stage. We will do so by discussing episodes of the proving process that lead to the written proof of the Pythagorean Theorem that we have presented above (see Fig. 8.1). The teacher (T) in this class will be referred to as Mrs Nissen, references (e.g. ⟨6–21⟩) indicate lines of the transcript of this lesson. "N5" indicates this is taken from the fifth lesson observed in Mrs Nissen's class. Additional data and analysis can be found in (Knipping, 2003).

The choice of what part of the lesson to analyze was based on the participants' own identification of what classroom conversations were seen as being proving, through explicit labeling of them as such. The protocols and transcripts show that it is generally the teacher who labels a proving phase.

8.4.1 Reconstructing the Sequencing and Meaning of Classroom Talk

In the following we will first describe how we reconstructed sequencing and meaning of classroom talk in proof and proving.

8.4.1.1 Layout of Episodes

The first step is dividing the proving process into episodes. This means that the general topics emerging in the classroom talk are identified and their sequencing is reconstructed. This allows one to get an overview of the different steps in the argumentation. Proving process in classrooms can occur over long periods of time, from 20 to 40 min or longer. Laying out different episodes of the process helps to make the argumentations in these episodes more accessible to analysis. Once the flow and sequencing of the emerging topics is made visible the reconstruction of the arguments can start. For example, the following topic episodes could be identified in lesson N5:

1. Sketch of the proof diagram $\langle 6$–$21 \rangle$
2. Goal of the proof $\langle 21$–$28 \rangle$
3. Meanings of a^2, b^2, c^2 $\langle 28$–$69 \rangle$
4. Calculating sub-areas of c^2 $\langle 69$–$100 \rangle$
5. Sascha's Conjecture $\langle 101$–$129 \rangle$
6. The area of the right triangles $\langle 129$–$155 \rangle$
7. A mistake on the board $\langle 156$–$175 \rangle$
8. Transforming the equations found $\langle 175$–$200 \rangle$

Mrs Nissen starts the lesson by sketching a drawing on the blackboard (episode 1, see also Fig. 8.1). The class then determines the goal of the proof of the Pythagorean Theorem (episode 2) and discusses the meanings of a^2, b^2, c^2, expressions that are used to state the Pythagorean Theorem and that are related to the drawing (episode 3). In episode 4 the teacher asks the students to calculate the sub-areas of the big square c^2. Sascha supposes (episode 5) that any two triangles form a square, but his conjecture is refuted by his peers and the teacher. Instead the class calculates the area of two right triangles in a general way (episode 6). By accident the teacher writes in the second line $(b - a^2)$ on the board, but a student points out the mistake which the teacher gratefully corrects into $(b - a)^2$ (episode 7). Together, the teacher and the students transform the equations found and deduce $c^2 = a^2 + b^2$ (episode 8).

8.4.1.2 Turn by Turn Analyses

Argumentations in classroom processes are mostly expressed orally and by a group of participants. Generally arguments are produced by several students together, guided by the teacher. As Herbst showed (2002a), it is the teacher who mostly takes responsibility for the structure and correctness of the argument, but students contribute to the argument, so there is a division of labour in the class. Argumentations are co-produced; the teacher and the students together produce the overall argument. Their turns are mutually dependant on each other; their public meanings evolve in response to each other. The argument forms in relation to these emerging meanings. So, in order to reconstruct the structure of an argument first the meanings of each individual turn put forward in class have to be reconstructed. As Krummheuer and Brandt state:

> Expressions do not a priori have a meaning that is shared by all participants, rather they only get this meaning through interaction. In concrete situations of negotiation the participants search for a shared semantic platform (Krummheuer and Brandt, 2001, 14, our translation).[2]

Because meanings emerge through interaction, reconstructing meanings necessarily involves some reconstruction of the process by which they emerge. Generally statements of classroom talk are incomplete, ambiguous and marked by deictic[3] terms. Deictic terms are replaced as much as possible in the reconstruction of the argumentation. For example, in Excerpt 1 the term "das" ("that") can be replaced with '$(b-a)$' because its meaning is apparent from earlier utterances.

| 89 | Jens: *Das* ist also die eine Seitenlänge von diesem mittleren Quadrat da. Aber trotz # | Jens: So *that* is the side length of that middle square there. But, # |

Excerpt 1. Example of an ambiguous utterance marked by a deictic term

Because the focus of the analysis is the argumentative structure of the classroom talk the reconstruction of the meanings of statements in the turn by turn analysis must consider the argumentative function of the statements: datum, conclusion, warrant, etc. These functions will be identified in the next step of analysis. Utterances are primarily reconstructed according to their function within the collectively emerging argumentation, not with respect to subjective intentions and meanings as in interaction analyses.

---

[2]Äußerungen besitzen "a priori keine von allen Beteiligten geteilte gemeinsame Bedeutung, sondern erhalten diese erst in der Interaktion. In konkreten Situationen des Verhandelns bzw. Aushandelns wird nach einer solchen gemeinsamen semantischen Bedeutungsplattform gesucht".

[3]In linguistics, a deictic term is an expression, for example a pronoun, that gets its meaning from its context. The meaning of "this" depends on what is being pointed to. The meaning of "I" depends on who is speaking. In philosophy the word "indexical" is used to express the same idea.

8.4.2 Analyzing Arguments and Argumentation Structures

In the following we will describe in detail the moves in the reconstruction of local arguments, then of intermediate argumentation streams, and then of global argumentation structures. This method for reconstructing arguments, argumentation streams and argumentation structures was developed by Knipping (2003, 2008).

8.4.2.1 Functional Reconstruction of the Argumentation

Recall that arguments have in Toulmin's model a general structure of data leading to conclusions, supported by warrants, which in turn can be supported by backings. Statements are characterized as having different functions within an argumentation, and functional analysis can help to reveal the structure of the argumentation.

Analyzing students' and teachers' utterances in the class according to this functional model allows us to reconstruct argumentations evolving in the classroom talk. In our analyses only utterances that are publicly (in the class) accepted or constituted as a statement are taken into account. The teacher's attention to some utterances and deferment of others can play a major role in this. This is not surprising given Herbst's findings that in general only the teacher takes responsibility for the truth of statements (Herbst, 2002a). Where alternative argumentations or attempts at an argument are publicly acknowledged, they are also considered in our analyses, although the focus is on the main stream of the argumentation. The issue of alternative argumentations will be discussed in more detail in Sect. 8.4.3, where types of argumentation structures are compared.

In Knipping's (2003, 2004) analyses of classroom processes, focusing first on conclusions turned out to be an effective step in reconstructing argumentations. It is helpful to begin by identifying what statement the participants are trying to justify, the claim that will gain the status of a conclusion by their argument. So, before actually analyzing the complete argument we look for conclusions and claims. The following example illustrates such a functional reconstruction of a conclusion, or in fact two conclusions.

The following excerpt marks the beginning of the proving process in class. The teacher has sketched the drawing presented in Fig. 8.1 on the blackboard and seeks to develop a proof of the Pythagorean Theorem together with the class. As yet, no written proof has been developed. The teacher asks the students to interpret the given drawing. Jens starts.

89 Jens: Das ist also die eine Seitenlänge von diesem mittleren Quadrat da. Aber trotz #	Jens: So this is the side length of that middle square there. But, #
90 L: # Also, Adam sagt, hier soll ich immer *b* minus *a* dran schreiben, oder soetwas ähnliches.	T: # So, Adam says, I should write *b* minus *a* on here, or something like that.
91 L: Er sagt diese vier Seiten sind alle *b* minus *a* lang. Wie kommt er denn darauf? *b* minus *a* lang.	T: He says that these four sides are all *b* minus *a* long. So, how did he get that? *b* minus *a* long.
92 L.: Srike.	T: Srike.
93 Srike: Wir haben ja die eine Seitenlänge vom Dreieck *b*, eh die Kathete. Und dann ist die, also	Srike: Well, we know the length *b* of one side of the triangle, in fact the adjacent side. And then this is therefore
94 vom anderen Dreieck die andere Kathete *a*, die wird ja davon abgezogen und dann ist das,	the other adjacent side *a* of the other triangle, which of course gets taken away from it and then that is
95 was übrig bleibt dieses *b* minus *a*.	what is left, this *b* minus *a*.
96 L: Einverstanden? Immer wird von der grünen Strecke 'ne gelbe abgeschnitten und es bleibt	T: Agreed? A yellow segment is always cut off the green, and only the remainder
97 von *b* nur noch der Rest über. *b* minus *a*. Minus *a* heißt das gelbe weg. So *b* minus *a*, ha.	of *b* is still left. *b* minus *a*. Minus *a* means remove the yellow. Thus *b* minus *a*, ha.

Excerpt 2. Transcript of the argumentation in episode 4 of lesson N5

Previously Adam has offered $(b - a)^2$ as a label for the inner square in the figure that is on the board (see Fig. 8.1). Jens, after first being confused about this expression, makes a connection with the side length of the middle square (89). Given this context we interpret the "das/that" as "$(b - a)$". The teacher reinforces Jens's interpretation and endorses Adam's earlier claim (90–91). Jens does not provide any data or warrant for his claim, on the basis of the drawing on the blackboard he seems to consider this as a "matter of fact", a datum. The teacher asks for an explanation, "How did he get that? b minus a long." (91), and reinterprets this "matter of fact" as a claim that needs justification. Srike provides a justification by relating $(b - a)$ to the difference of the lengths of the two legs of the right triangle in the drawing and explains why $b - a$ is the difference in length.

Although Jens proposed his statement as a datum, its status depends on its role in the public argumentation, not on the intention of its proposer. The teacher expects an explanation and Srike provides one, but for a different conclusion. However Srike's explanation is accepted as justifying both her conclusion and Jens's original claim. In this situation a single argumentation has two conclusions (see Fig. 8.5).

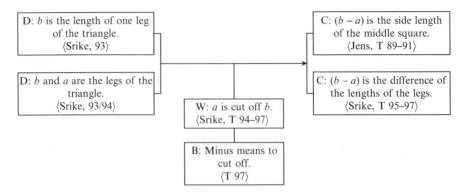

Fig. 8.5 Functional reconstruction of the argumentation in episode 4 of lesson N5, example of a step of an argumentation (*D* datum, *C* conclusion, *W* warrant, *B* backing)

Srike's argument is based on the data that "*b* and *a* are the legs of the triangle" (93/94) and that "*b* is the length of one leg" (93). She argues that because "*a* is cut off *b*" (94–97), "$(b - a)$ is the difference of the lengths of the legs" (95–97). The teacher reinforces Srike's attempt at an explanation, confirms her conclusion (94–97) and supports her warrant by the backing "Minus means to cut off" (97). In the given episode the teacher and the students together construct an argument that, with the help of the Toulmin pattern, may be reconstructed as in Fig. 8.5. Our representation is somewhat different, but this argument can been seen as combining what Aberdein (2006, 214) calls a "linked" layout, where two or more data are included, with a variant of his "divergent" layout in which one datum supports two conclusions. We see our empirical approach as offering a useful complement to theoretical approaches like Aberdein's, in that our work reveals which of Aberdein's layouts occur in classroom argumentations, and in what combinations.

It is interesting that in this case the warrant and the backing are given explicitly. Typically, reconstructed arguments in classroom proving processes are often incomplete, as was the case with the written proof analyzed above. The warrant and backing are often not mentioned explicitly. In most cases the warrant can be assumed or taken as implicit, as an inference can only be constituted with a warrant. The transition from datum to conclusion must be justified somehow. In our argumentation analyses we usually do not add implicit warrants (W(i)), but leave them implicit in the reconstruction. This is meant to illustrate the implicitness of both the argumentation and warrant. This allows the comparison of the degree of explicitness in different argumentation structures. In cases where we do want to talk about an implicit warrant we place it in a dotted box (as in Fig. 8.6).

We have occasionally come across arguments where the datum has been left implicit. In such cases the warrant is present, however, so in the reconstruction the datum is left implicit, and the argument consists of the warrant and the conclusion. As for the conclusions of arguments we have found that these are often formulated as questions. In the reconstruction of the argument these are represented by statements,

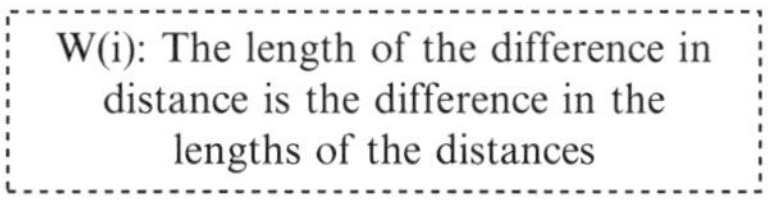

W(i): The length of the difference in
distance is the difference in the
lengths of the distances

Fig. 8.6 Example of the representation of an implicit warrant, episode 4 of lesson N5

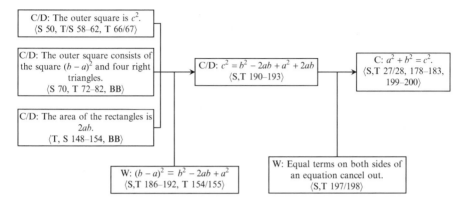

Fig. 8.7 Example of the second half of the argumentation in lesson N5. "BB" means black board. Transcript line numbers refer to (Knipping, 2003)

so that their grammatical form is no longer visible, but their function in the argument is clearer. In the descriptions and comments on the argument this is noted and discussed.

In cases where statements are questioned or doubted they are often justified in more than one argumentation step. To deal with such situations Knipping introduced another term. She called a chain of argumentation steps by which a target conclusion is justified an "argumentation stream" (AS). The target conclusion (the final conclusion of the argumentation) is often marked by the teacher as a goal of one stage within the global argumentation. The example in Fig. 8.7 illustrates the last argumentation stream of the proving process in lesson N5. It involves both "linked" and "sequential" ways to combine steps.

8.4.2.2   Reconstructing the Argumentation Structure of Proving Processes in Class

As discussed earlier in this paper, the functional model of Toulmin, which is helpful for reconstructing argumentation steps and streams, is not adequate for more complex argumentation structures. Analyzing proving processes in classrooms requires a different model for capturing the global structure of the argumentations developed there. Knipping developed a schematic representation in order to illustrate the complex argumentation structures of this type of classroom talk, which we will present in the following.

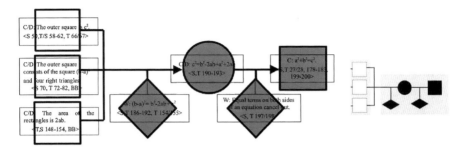

Fig. 8.8 The method of reconstructing a global argumentation

Fig. 8.9 Symbols used in argumentation structure diagrams

The argumentation structure of a classroom proving process is generally complex. Argumentation streams can be parallel, as well as nested into each other. For example, if the argumentation builds on certain statements more than on others, these will be justified and explained in more detail, leading to multiple arguments in support of them embedded within the larger argument to which they are important.

To address this complexity Knipping developed a schematic representation that allows the description of argumentations at different levels of detail. This approach differs from Aberdein's (2006) as he reduces the complexity of the argumentation by a process of folding that results in a single step that includes all the assumptions (initial data and warrants) of the full argumentation, but which hides the relationships between these assumptions. Knipping's approach also differs from that taken by Van Eemeren et al. (1987) who developed two different ways of representing the structures of everyday written argumentations, in that she makes the role of warrants more visible. We will illustrate below how Knipping's method makes the global argumentation visible while preserving the relationships in the local steps.

In the schematic representation shown in Fig. 8.8 all statements in the overall argumentation are represented by rectangles, circles and diamonds (see Fig. 8.9). The different symbols not only represent the different functions of the statements (datum, conclusion, warrant), which are also indicated by the letters D, C, W, B in front of the statements, but also the status that the statements have within the global structure of the argumentation. For example, the target conclusion is represented by a black rectangle. White rectangles represent target conclusions of intermediate stages within the global argumentation; they indicate end points of stages. These can become starting points, therefore data (D), in the next stage of the argumentation.

Three statements (the three white rectangles in Fig. 8.8) have the status of data in the argumentation stream shown, but they are at the same time conclusions (C) of earlier argumentation streams. Once their truth was established they became data for a subsequent argumentation stream, the one being presented in Fig. 8.8. Conclusions or data which do not have the status of an intermediate target are represented by circles. Warrants (W) and backings (B) are symbolized by diamonds.

In Knipping (2003) the overall structures of argumentations were analyzed by means of such schemes. She compared argumentation structures in different proving processes on the basis of the schematic representations and attempted to reconstruct their peculiar rationale. In the next section we will use some results of our research to illustrate the utility of this method for describing complex argumentations and their rationale. The processes that we have studied occurred in junior high school contexts where proof was an explicit goal of the lessons observed (see Knipping, 2003; Reid and Knipping, 2010).

8.4.3 Comparing Global Argumentation Structures

There is no established theoretical framework for investigating classroom proving processes. Therefore no model for the explanation of these processes can be formulated before researching the empirical field, but still a sound methodology of discovery is necessary. Analyzing argumentation in the primary mathematics classroom, Krummheuer considers comparison as a methodological principle that provides a reliable method of this sort and that can give "direction to a novel theoretical construction" (Krummheuer, 2007, 71). This methodological principle underlies the comparisons that we undertake in our research on proving processes in classrooms (Knipping, 2003, 2008; Reid and Knipping, 2010).

As with Glaser and Strauss (1967), for Krummheuer comparative analysis represents a central activity that allows empirical control of the heuristic generation of theory. In this approach comparisons occur continuously, "the comparison of interpretations of different observed parts of reality represents a main activity on nearly every level of analysis: from the first interpreting approach to the later more theoretical reflection" (Krummheuer, 2007, 71, describing Strauss and Corbin, 1990). The aim of these comparisons is "conceptual representativeness" (see Strauss and Corbin, 1990), that is, to ground theoretical concepts within the data. This concept differs from the one in quantitative research, where representativeness on the level of the sampling is the goal.

Knipping (2003) compares argumentations at two levels. Local argumentations are compared by analyzing and classifying the warrants (and backings) used according to the *field* of justification they belong to. Global argumentations are compared according to their overall structures. Here we will focus on comparing global argumentation structures, to show how they reveal elements of the rationale underlying the proving process. Two types of structures from our research will be described here as examples. We call them *source*-structure and *spiral*-structure.

(See Knipping, 2003 and Reid and Knipping, 2010, for descriptions of two other structures: the gathering-structure and the reservoir-structure. The four structures identified so far in our work only begin to describe the variety possible in classroom proving processes.) While the structures themselves are grounded in the data, the metaphors we use to describe them reflect our later interpretations.

8.4.3.1 Source-Structure

In proving discourses with a source-like argumentation structure, arguments and ideas arise from a variety of origins, like water welling up from many springs. The teacher encourages the students to formulate conjectures that are examined together in class. In some cases this means that students propose conjectures that are unconnected to the overall structure. More than one justification of a statement is appreciated and encouraged by the teacher. This diversity of justifications results in an argumentation structure with parallel streams in which intermediate statements are justified in various ways. False conjectures are eventually refuted, but they are valued as fruitful in the meantime. In argumentations with a source-structure a funneling effect becomes apparent. Towards the end of the argumentation only one chain of statements is developed in contrast to the beginning where many parallel arguments are considered.

This is illustrated by Fig. 8.10a, which is the global argumentation structure for lesson N5 from Mrs Nissen's class (Knipping, 2003). The structure has the following characteristic features:

- Parallel arguments for the same conclusion (AS-1 and AS-2).
- Argumentation steps that have more than one datum, each of which is the conclusion of an argumentation stream (AS-8).
- The presence of refutations in the argumentation structure (AS-3, AS-6).

In Knipping's analyses of proving processes she found that a single target conclusion might be legitimized by several parallel argumentation streams, as occurs here in AS-1 and AS-2. By a *parallel argumentation* we mean argumentation streams in a proving process in which different arguments can be found supporting the same conclusion. This happens, for example, if substantially different arguments are produced for the same conclusion. Aberdein (2006) describes parallel argumentation steps as "convergent". Both AS-1 and AS-2 support the conclusion that the side of the outer square is c (see Fig. 8.1). In AS-1 the argument is based on the conclusion that the inner quadrilateral is a square, with the drawing of the proof figure as a warrant. In AS-2 it is argued that the triangles make up the outer shape, again based on the drawing.

Several argumentation streams can support a single conclusion without being parallel. The conclusion of each one can act as a datum for a subsequent step that requires the data from all of them. Aberdein (2006) describes argumentation steps with more than one datum as "linked" and we are especially interested in noting cases where the data in a linked argumentation step are conclusions from a number

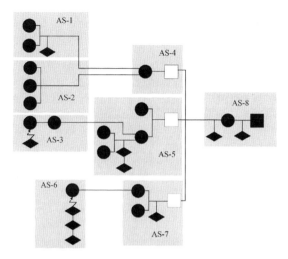

Fig. 8.10a *Source*-structure

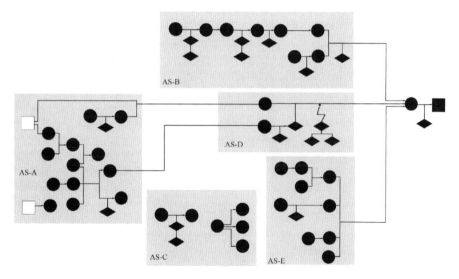

Fig. 8.10b *Spiral*-structure

of argumentation streams. For example, the conclusions of AS-4, AS-5 and AS-7 act as data for AS-8. Details of these three conclusions and AS-8 are shown in Figs. 8.7 and 8.8. In a sense, AS-4, AS-5 and AS-7 prove lemmas that make the proof of the final conclusion straightforward. Once it is known that the area of the outer square is c^2, that it is made up of the inner square (with area $(b-a)^2$) and four right triangles, and that the area of the two rectangles formed by the four right triangles is $2ab$, then the conclusion $a^2 + b^2 = c^2$ follows by a simple algebraic manipulation.

There are two interesting refutations within this argumentation structure. In AS-3 Maren, in the process of describing the area of the outer square c^2, assumes that the area of the inner square is b^2. The teacher contradicts her, refuting Maren's suggestion visually. The teacher then develops together with the class an argument that the side length of the inner square is $b - a$ (AS-5) and therefore the inner square's area must be $(b-a)^2$. AS-5 and also Maren's suggestion (and its refutation) are the source of a justification that c^2 consists of a square with side $b - a$ and four congruent right triangles, which is noted on the blackboard as follows: $c^2 = (b-a)^2 + 4$rwD (see Fig. 8.1).

Sascha's conjecture (AS-6) is the other example of a refutation. Sascha claims that two of the triangles form a square which the teacher refutes by having the class put together the cut-out triangles. However, she says "I really like your idea, I think ideas that lead to the right result in detours are wonderful" giving value to Sascha's conjecture even though she refuted it.

In Knipping's (2003) other examples of the source structure, there are argumentation streams that are completely disconnected from the main structure. AS-6 in Fig. 8.10a is an imperfect example of this. The refutation of Sascha's conjecture is connected to the main argument only in the loose sense that its negation, the area of each triangle is neither $a^2/2$ nor $b^2/2$, is part of the *context* for AS-7 in which it is concluded that the area of the two rectangles formed by the four right triangles is $2ab$.

The source-structure is also characterized by argumentation steps that lack explicit warrants. Argumentation steps without explicit warrants are evident in AS-2, AS-3, AS-4 and AS-5. While this also occurs in other types of argumentation structures it is more frequent in the source-structure. We speculate that this is because of the encouraging in the source structure of conjectures, which may be offered with some supporting data, but which are not further developed. A similar phenomenon occurs in another structure we have identified, the *gathering*-structure (Reid and Knipping, 2010). That structure involves the gathering of a large amount of data to support several related conclusions. Again the emphasis is on collecting information (data and conclusions) rather than on the connections between them. In the other structures we have studied (see below) there is much more emphasis on the transitions, and so there are more explicit warrants.

We noted above that the source-structure has four characteristic features, which we have described in the argumentation structure of Mrs Nissen's lesson 5. In another classroom we have observed lessons that have another structure, the spiral-structure, which shares the same four characteristic features, but differs in the way they occur in the global argument.

8.4.3.2 Spiral-Structure

In a proving process with a spiral argumentation structure the final conclusion is proven in many ways. First one approach is taken, then another and another. Each approach can stand on its own, independent of the others. Students suggest some

Fig. 8.11 Diagram from Mrs James's classroom (reconstruction)

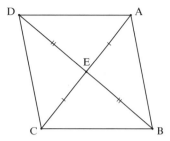

approaches and the teacher proposes others. Students' contributions that do not lead to the conclusion result in disconnected argumentation streams and refutations.

The global argumentation structure depicted in Fig. 8.10b shows a spiral argumentation structure from Mrs James's grade 9 (age 14–15 years) classroom in Canada (see Reid and Knipping, 2010). The class was trying to explain why two diagonals that are perpendicular and bisect each other define a rhombus (see Fig. 8.11). The students had discovered and verified empirically that the quadrilateral produced is a rhombus using dynamic geometry software and the proving process led by the teacher was framed as an attempt to explain this finding using triangle congruence properties.

Several features characteristic of the spiral argumentation structure are evident in Fig. 8.10b:

- Parallel arguments for the same conclusion (AS-B, AS-D, AS-E).
- Argumentation steps that have more than one datum, each of which is the conclusion of an argumentation stream (the final conclusions of AS-B and AS-E).
- The presence of refutations in the argumentation structure (AS-D).
- Argumentation streams that do not connect to the main structure (AS-C).

In the argumentation structure there are three parallel arguments AS-B, AS-D, and AS-E. They all lead to one conclusion, that the four sides are congruent, which acts as the datum for the final conclusion that the quadrilateral is a rhombus. In AS-B the congruency of the sides is shown by showing that the four triangles formed by the diagonals are congruent. In AS-D a student offers an alternative argument, based on the idea that the quadrilateral cannot be shown to be a square (see below). This argument is listened to attentively by the teacher, who eventually refutes it. Finally, in AS-E the teacher offers an alternative argument implicitly based on using the Pythagorean Theorem instead of triangle congruency to establish that the four sides are equal.

In the argumentation structure from Mrs James's classroom there are two argumentation streams that involve steps that have more than one datum, which are in turn conclusions of argumentation streams. They are AS-E and AS-B.

In AS-E four data combine to establish that the sides are congruent. First, two arbitrary numbers are assigned to the legs of triangle AEB (see Fig. 8.11). EB is assigned a length of 10 and AE a length of 3. The teacher then asks how one could find the length of the hypotenuse AB, and the students suggest squaring and adding. Mrs James supplies the final step of taking the square root. It is clear that the Pythagorean theorem is being used, but no one ever names it. It is the implicit warrant for most of the arguments in AS-E. The first argument concludes with the first datum that will be used later: that the length of AB is $\sqrt{10^2 + 3^2}$. The same procedure is then used to find the length of BC, and to conclude that AB=BC on that basis. The third argument consists of the datum that the same procedure could be used for CED and the conclusion that CD is $\sqrt{10^2 + 3^2}$. The final datum is the fact that all the triangles are right triangles. These four data are taken together to conclude that the four sides AB, BC, CD, and DA are congruent, without any explicit warrant.

In AS-B two data are used to establish the conclusion that the sides are congruent. The first is the culmination of a chain of data/conclusions leading to the conclusion that the four triangles AEB, BEC, CED, and DEA are congruent. The second is the conclusion of a shorter argument, establishing that the sides of the rhombus are corresponding sides of the congruent triangles. Once these two data are concluded, they are combined to conclude that the four sides are congruent based on the explicit warrant that corresponding sides of congruent triangles are congruent.

Simple arguments involving more than one datum also occur in AS-A, in which the known information, that the diagonals meet at right angles and bisect each other, is represented on the diagram. For example, that segment EB should be marked with two dashes rather than one depends on two data: that it is the same length as another segment (and so needs to be marked) but not the same length as segments AE and EC that have already been marked with single dashes (see Fig. 8.11). However, these data are not themselves conclusions of argumentation streams, but rather simple facts read off the diagram or from the given conditions.

The single refutation in this argumentation structure occurs in AS-D. Unlike the refutation we have discussed in the source-structure, above, what is refuted is not a datum or a warrant, but rather the applicability of the warrant to the argument (see Reid et al., 2008, for details; Verheij, 2006, discusses analogous types of rebuttals). A student, Kaylee, asserts that the diagonals must define a rhombus because in order to define a square the diagonals would have to be the same length. Mrs James does not refute the fact that if the diagonals were the same length, as well as being perpendicular and bisecting each other, they would define a square. Instead she points out that other quadrilaterals might also be possible that have not been considered and excluded. She mentions rectangles and kites as examples of quadrilaterals that share some of the same diagonal properties that the class is considering, though she also notes that they do not share all the properties. She uses these examples to back up her refutation, which is based on the fact that other quadrilaterals, other than squares and rhombuses, exist. Hence, excluding squares is not sufficient to guarantee the shape must be a rhombus. Here the refutation is directed at the warrant, but does not refute it (as it is correct). Instead it suggests that the warrant is insufficient in the logic the teacher expects mathematical arguments to follow. Although she refutes Kaylee's argument, Mrs James values it, commenting, "You're on the start, but I'm not sure that you've clinched it. I'm not sure you've got that final part, but … you're three quarters of the way there my dear."

Finally, this argumentation structure includes two disconnected arguments, in AS-C. Both are focused on labeling. One concerns whether the segment DE should be labeled DE or ED. The other concerns the difference between the Angle-Side-Angle congruency property (abbreviated ASA) and the Angle-Angle-Side congruency property (abbreviated AAS). While these points are important to clarify in the context of Mrs James's mathematics lesson, they do not result in conclusions that are used later.

While representing the argumentation structures of proving processes is useful as a way of identifying and structuring the important elements in it, we gain more from comparisons of argumentation structures. For example, Knipping (2002, 2003, 2004) found that in the six classrooms she observed, the proving processes either had argumentations with the source-structure, or another structure that she calls the *reservoir*-structure. The reservoir-structure differs from the source-structure in many ways. Most notably, in the reservoir-structure the reasoning sometimes moves backwards in the logical structure and then forward again. Initial deductions lead to desired conclusions that then demand further support by data. This need is made explicit by identifying possible data that, if they could be established, would lead to the desired conclusion. Also, because transitions are a focus in the proving process, the reservoir-structure has more explicit warrants.

The different structures in the lessons Knipping observed revealed interesting differences in the nature of proof teaching in the two contexts in which they were found. In the next section we will compare the source-structure with the spiral structure.

8.4.3.3 Comparing Source and Spiral Argumentation Structures

Both the source-structure and the spiral-structure were observed in proving processes in which the teacher took a prominent role in guiding the process. Arguments were co-produced by teachers and students, but the teacher was in control of the emerging overall structure. Therefore it is not surprising that these argumentation structures have several similar characteristic features including parallel arguments, argumentation steps that have more than one datum, the presence of refutations, and argumentation streams that do not connect to the main structure. However, they differ in how these features play out in the global structure.

One of the main distinctions between the spiral-structure and the source-structure is the *location* of the parallel arguments. In the source-structure the parallel arguments occur at the start of the proving process (AS-1 and AS-2 in Fig. 8.10a). The teacher invites input at this stage, but once the basis for the proof is established, the teacher guides the class to the conclusion through an argumentation that no longer has parallel arguments. In the spiral-structure, however, the conclusions of the parallel arguments are almost the final conclusion in the entire structure. In fact, two of the three parallel arguments in Fig. 8.10b (AS-B and AS-E) could stand alone as proofs of the conclusion. Having proven the result in one way, the teacher goes back and proves it again in a different way. And she values students' attempts to prove the conclusion using other approaches.

The source-structure and the spiral-structure differ also in the kinds of refutations they involve and in the inclusion or omission of warrants. Recall that in the source type argumentation structure shown in Fig. 8.10a, the refutations are refutations of data. When interpreting the figure on the blackboard students propose statements that the teacher refutes. In contrast, the refutation in the spiral type argumentation structure shown in Fig. 8.10b is a refutation of an argument. The data and warrant are accepted but their adequacy to justify the conclusion is refuted. In the source structure we see also many steps that omit warrants, while most steps in the spiral structure include warrants. The notable exceptions are in AS-A and AS-E. We will discuss some possible reasons for this later.

Although the source and spiral argumentation structures have the same four characteristic features, they differ in the placement of the parallel arguments and in the nature of the refutations they include. These differences result in fundamentally different argumentation structures. It is through comparisons of structures that such differences become apparent. While reconstructing argumentation structures and comparing them allows us to identify important features and differences, it does not explain why they occur. To try to explain these differences, we need to return to the data and consider the nature of the local arguments that make up the global structure.

In the proving process in Mrs Nissen's classroom, the figure on the blackboard is the starting point. Almost all the steps in the argument consist of establishing data that is depicted in the figure. For example, the part of AS-5 shown in Fig. 8.5 establishes that the side of the inner square is $b - a$. The three sub-conclusions of AS-4, AS-5 and AS-7 are all present implicitly in the figure, and unpacking that data is the main focus of the proving process. This explains the focus on data, which we see in the lack of warrants and in the refutation of inaccurate data. Once all the necessary data has been unpacked, the final steps of the argument are straightforward and algebraic. They are what is written on the blackboard at the end, as the written proof, and notably, the warrants that are expressed verbally in the class are not included when the proof is written down. Again the focus is on the data, rather than on the arguments and their warrants. When we consider the nature of the omitted warrants this makes sense, as they are either visual (based on features of the figure on the blackboard) or algebraic procedures well known to everyone in the classroom. In the first case it is hard to imagine how the warrants could be formulated in words or symbols, and in the second case the teacher is simply modeling the standard mathematical practice of omitting from proofs any warrants that the reader can be expected to provide.

In Mrs James's lesson, the argumentation begins with the given data: the diagonals are perpendicular and meet at their midpoints. From this data the figure is constructed. There are not many explicit warrants in this argumentation stream (AS-A). As in Mrs Nissen's lesson, if the warrants were made explicit they would either be visual or refer to conventions well known to the students. In the other streams, however, there is a focus on the arguments themselves and warrants become more explicit. Duval's "recyclage" is evident in AS-B where almost all the data are recycled conclusions of previous steps, and in which most steps have explicit warrants. In AS-D we see the focus on the argument in the refutation of it, as opposed to the refutation of data in the source structure. Even in AS-E, where there are many omitted warrants and a configuration similar to the source structure overall, with the final conclusion depending on more than one datum, the fact that the teacher offers this alternative approach to establishing the final conclusion indicates a focus on arguments rather than data and conclusions.

Examining the argumentation structures in these two classrooms allows us to describe their characteristic features, and by comparing them we can understand the different ways these features occur. We see the parallel arguments, refutations and omitted warrants in both, but we see these features occurring differently. Looking more closely at the features of the local arguments helps to explain these differences, and reveals an importance distinction between the rationales of the proving processes taking place. In Mrs Nissen's class we find in the local arguments a focus on interpreting the given figure. The activity is essentially one of unpacking the data in the figure and expressing it verbally. It is not clear how this could be transferred to proving another theorem, unless a similar complex figure were provided. We suspect this is inevitable in a class focussing on the Pythagorean Theorem. Historically, mathematics educators have struggled with the problem of using mathematically significant theorems as a context for learning proving. The proofs of such theorems are usually sufficiently complex that it is unreasonable to expect students in schools to discover them, unless the teacher provides so much guidance that the students' contributions are limited to activities such as unpacking a diagram, as in Mrs Nissen's class. Herbst (2002b) describes how this struggle affected the evolution of proof teaching in the US.

In contrast, in Mrs James's class the focus is more on proving. The result itself is relatively uninteresting, but the recycling of conclusions as data, the provision of warrants, the fact that the same result can be proven in different ways, and bringing different prior knowledge to bear, are all important. Student contributions are valued, even when flawed, and the argumentation, especially in AS-B, served as a model for the students when proving similar claims in subsequent lessons.

The source structure and the spiral structure are interesting to compare because they have many characteristic features in common, including parallel arguments, argumentation steps that have more than one datum, refutations, and unconnected argumentation streams. There are differences in how these features play out in the global structures, however, and to explain these we focus again on local arguments.

8.5 Conclusion

Toulmin's functional model of argument allows us to reconstruct arguments in mathematics classrooms at the local level, and also to assemble local arguments together into a global structure. By examining argumentation structures (the metaphoric anatomies of proving processes) we can describe their characteristic features, and by comparing structures we can understand the different ways these features occur in an argumentation structure. To better understand the differences we observe, and to shed light on the rationales of the proving processes we return to the local arguments, to physiology in our metaphor. Attention to both the local and the global levels are essential to understanding proving processes in the classroom. While we are still, as Hoyles (1997) noted, "a long way from understanding how students acquire proof", examining argumentation structures provides a tool to better understand the different ways in which teachers teach proof in actual classrooms and how students in those classrooms come to an understanding of proof and proving.

References

Aberdein, A. (2006). Managing informal mathematical knowledge: Techniques from informal logic. In J. M. Borwein & W. M. Farmer (Eds.), *MKM 2006* (pp. 208–221), number 4108 in LNAI. Berlin: Springer.

Alcolea Banegas, J. (1998). L'argumentació en matemàtiques. In E. Casaban i Moya (Ed.), *XIIè Congrés Valencià de Filosofia* (pp. 135–147). Valencià. [Trans.: A. Aberdein & I.J. Dove (Eds.), The argument of mathematics (pp. 47–60). Dordrecht: Springer].

Balacheff, N. (1987). Processus de preuve et situations de validation. *Educational Studies in Mathematics, 18*(2), 147–176.

Balacheff, N. (1988). *Une étude des processus de preuve en mathématique chez les élèves de collège*. PhD thesis, Université Joseph Fourier, Grenoble.

Balacheff, N. (1991). The benefits and limits of social interaction: The case of mathematical proof. In A. Bishop, E., Mellin-Olsen, & J. van Dormolen (Eds.), *Mathematical knowledge: Its growth through teaching* (pp. 175–192). Dordrecht: Kluwer.

Chazan, D. (1993). High school geometry: Student's justification for their views of empirical evidence and mathematical proof. *Educational Studies in Mathematics, 24*(4), 359–387.

Duval, R. (1995). *Sémiosis et pensée humaine: Registres sémiotiques et apprentissages intellectuels*. Bern: Peter Lang.

Garuti, R., Boero, P., & Lemut, E. (1998). Cognitive unity of theorems and difficulty of proof. In A. Olivier & K. Newstead (Eds.), *Proceedings of the 22nd conference of the international group for the psychology of mathematics education* (Vol. 2, pp. 345–352). Stellenbosch: University of Stellenbosch.

Glaser, B., & Strauss, A. (1967). *The discovery of grounded theory: Strategies for qualitative research*. Chicago, IL: Aldine.

Hanna, G. (2000). Proof, explanation and exploration: An overview. *Educational Studies in Mathematics, 44*(1–2), 5–23.

Harel, G., & Sowder, L. (1998). Students' proof schemes: Results from exploratory studies. In A. H. Schoenfeld, J. Kaput, & E. Dubinsky (Eds.), *Research on collegiate mathematics education* (Vol. 3, pp .234–283). Providence, RI: American Mathematical Society.

Healy, L., & Hoyles, C. (1998). *Justifying and proving in school mathematics: Technical report on the nationwide survey*. London: Institute of Education.

Herbst, P. G. (1998). *What works as proof in the mathematics class*. PhD thesis, University of Georgia, Athens.

Herbst, P. G. (2002a). Engaging students in proving: A double bind on the teacher. *Journal for Research in Mathematics Education, 33*(3), 176–203.

Herbst, P. G. (2002b). Establishing a custom of proving in American school geometry: Evolution of the two-column proof in the early twentieth century. *Educational Studies in Mathematics, 49*(3), 283–312.

Hoyles, C. (1997). The curricular shaping of students' approaches to proof. *For the Learning of Mathematics, 17*(1), 7–16.

Inglis, M., Mejía-Ramos, J. P., & Simpson, A. (2007). Modelling mathematical argumentation: The importance of qualification. *Educational Studies in Mathematics, 66*(1), 3–21.

Jahnke, H. N. (1978). *Zum Verhältnis von Wissensentwicklung und Begründung in der Mathematik-Beweisen als didaktisches Problem.* PhD thesis, Institut für Didaktik der Mathematik, Bielefeld.

Knipping, C. (2002). Proof and proving processes: Teaching geometry in france and germany. In H.-G. Weigand, et al. (Eds.), *Developments in mathematics education in German-speaking countries: Selected papers from the annual conference on didactics of mathematics, Bern 1999* (pp. 44–54). Hildesheim: Franzbecker Verlag.

Knipping, C. (2003). *Beweisprozesse in der Unterrichtspraxis: Vergleichende Analysen von Mathematikunterricht in Deutschland und Frankreich.* Hildesheim: Franzbecker Verlag.

Knipping, C. (2004). Argumentations in proving discourses in mathematics classrooms. In G. Törner, R. Bruder, A. Peter-Koop, N. Neill, H.-G. Weigand, & B. Wollring (Eds.), *Developments in mathematics education in German-speaking countries: Selected papers from the annual conference on didactics of mathematics*, Ludwigsburg, March 5–9, 2001 (pp. 73–84). Münster: Gesellschaft für Didaktik der Mathematik.

Knipping, C. (2008). A method for revealing structures of argumentations in classroom proving processes. *ZDM Mathematics Education, 40*, 427–441.

Krummheuer, G. (1995). The ethnography of argumentation. In P. Cobb & H. Bauersfeld (Eds.), *The emergence of mathematical meaning: Interaction in classroom cultures* (pp. 229–269). Hillsdale, NJ: Lawrence Erlbaum Associates.

Krummheuer, G. (2007). Argumentation and participation in the primary mathematics classroom: Two episodes and related theoretical abductions. *Journal of Mathematical Behavior, 26*(1), 60–82.

Krummheuer, G., & Brandt, B. (2001). *Paraphrase und Traduktion. Partizipationstheoretische Elemente einer Interaktionstheorie des Mathematiklernens in der Grundschule.* Weinheim: Beltz.

Mariotti, M. A., Bartolini, B. M., Boero, P., Franca, F. F., & Rossella, G. M. (1997). Approaching geometry theorems in contexts: From history and epistemology to cognition. In E. Pehkonnen (Ed.), *Proceedings of the 21st conference of the international group for the psychology of mathematics education* (pp. 180–195). Lathi: University of Helsinki.

Moore, R. (1994). Making the transition to formal proof. *Educational Studies in Mathematics, 27*(3), 249–266.

Pedemonte, B. (2002a). *Étude didactique et cognitive des rapports de l'argumentation et de la démonstration.* Grenoble: Leibniz, IMAG.

Pedemonte, B. (2002b). Relation between argumentation and proof in mathematics: Cognitive unity or break? In J. Novotná (Ed.), *Proceedings of the 2nd conference of the European society for research in mathematics education* (pp. 70–80). Marienbad: ERME.

Pedemonte, B. (2007). How can the relationship between argumentation and proof be analysed? *Educational Studies in Mathematics, 66*(1), 23–41.

Reid, D., & Knipping, C. (2010). *Proof in mathematics education: Research, learning and teaching.* Rotterdam: Sense.

Reid, D., Knipping, C., & Crosby, M. (2008). Refutations and the logic of practice. In O. Figueras, J. Cortina, S. Alatorre, T. Rojano, & A. Sepúlveda (Eds.), *Proceedings of the 32nd annual conference of the international group for the psychology of mathematics education* (Vol. 4, pp. 169–176). México: Cinvestav-UMSNH.

Reid, D. A. (1995). Proving to explain. In L. Meira & D. Carraher (Eds.), *Proceedings of the 19th annual conference of the international group for the psychology of mathematics education* (pp. 137–144). Brasilia: Centro de Filosofia e Ciencias Humanos.

Reiss, K., Klieme, E., & Heinze, A. (2001). Prerequisites for the understanding of proofs in the geometry classroom. In M. van der Heuvel-Panhuizen (Ed.), *Proceedings of the 25th conference of the international group for the psychology of mathematics education* (pp. 97–104). Nottingham: PME Proceedings.

Sekiguchi, Y. (1991). *An investigation on proofs and refutations in the mathematics classroom.* PhD thesis, University of Georgia, Athens.

Snoeck Henkemans, A. F. (1992). *Analyzing complex argumentation: The reconstruction of multiple and coordinatively compound argumentation in a critical discussion.* Amsterdam: Sic Sat.

Strauss, A., & Corbin, J. (1990). *Basics of qualitative research: Grounded theory procedures and techniques.* Newbury Park, CA: Sage.

Toulmin, S. (1958). *The uses of argument.* Cambridge: Cambridge University Press.

Toulmin, S. (1990). *Cosmopolis: The hidden agenda of modernity.* Chicago, IL: University of Chicago Press.

Van Eemeren, F. H., & Grootendorst, R. (2004). *A systematic theory of argumentation: The pragma-dialectical approach.* Cambridge: Cambridge University Press.

Van Eemeren, F. H., Grootendorst, R., & Kruiger, T. (1987). *Handbook of argumentation theory: A critical survey of classical backgrounds and modern studies.* Dordrecht: Foris.

Verheij, B. (2006). Evaluating arguments based on Toulmin's scheme. In D. Hitchcock & B. Verheij (Eds.), *Arguing on the Toulmin model: New essays in argument analysis and evaluation* (pp. 181–202). Dordrecht: Springer.

Walton, D. N. (1989). *Informal logic: A handbook for critical argumentation.* Cambridge: Cambridge University Press.

Walton, D. N. (1998). *The new dialectic: Conversational contexts of argument.* Toronto, ON: University of Toronto Press.

Chapter 9
Checking Proofs

Jesse Alama and Reinhard Kahle

9.1 Introduction

Argumentative practice in mathematics evidently takes a number of shapes. An important part of understanding mathematical argumentation, putting aside its special subject matters (numbers, shapes, spaces, sets, functions, etc.), is that mathematical argument often tends toward formality, and it often has superlative epistemic goals: often the aim of a piece of mathematical argumentation is to *prove* that such-and-such a statement is *logically true* or a *logically valid consequence* of some assumptions; a proved statement thus seems to be *indubitable*, *certain*, or *irrefutable*. These aims generally do not depend on the subject matter of what is being argued about; whether one discusses functions, numbers, spaces, shapes, sets, arrangements, flows, figures, or fields, mathematical argumentation, in its final, published, form (and even in ordinary mathematical conversation) tends to be formal and self-consciously explicit about its own argumentative structure. The problem, then, is to better understand the notion of *mathematical proof*. We are interested in this chapter in the phenomenon of mathematical proof considered as a species of argumentative practice in mathematics.

For us in this chapter the central feature of mathematical argumentation—specifically, mathematical arguments that are put forward with the intention of showing that a certain proposition is true or validly derived—is its in-principle formalizability. By the in-principle formalizability of an argument we understand that there exists a formal derivation in some conventionally accepted formalism

J. Alama (✉)
Faculty of Science and Technology, Center for Artificial Intelligence, Universidade Nova de Lisboa, P-2829-516 Caparica, Portugal
e-mail: j.alama@fct.unl.pt

R. Kahle
CENTRIA and DM, FCT, Universidade Nova de Lisboa, P-2829-516 Caparica, Portugal
e-mail: kahle@mat.uc.pt

A. Aberdein and I.J. Dove (eds.), *The Argument of Mathematics*, Logic, Epistemology, and the Unity of Science 30, DOI 10.1007/978-94-007-6534-4_9,
© Springer Science+Business Media Dordrecht 2013

suited for mathematical reasoning of the proposition that commences from some conventional set of foundational axioms in a gap-free way all the way to the (formalized version of the) proposition.

Even a modest exposure to the practice of producing formal proofs common in introductory courses in logic, mathematics, computer science, law, linguistics, etc., quickly makes clear that the conventional formalisms for reconstructing an argument formally tend to be practically unsuited to the task for which they were designed. Without claiming to offer a complete list, we have

- the method of truth tables,
- Aristotle's syllogisms and variants thereof,
- analytic and semantic tableaux (e.g., Smullyan-style or Jeffrey-style),
- statement-justification tables (as one often sees in elementary courses of geometry),
- Euler/Venn diagrams (for expressing relationships of inclusion and exclusion),
- natural deduction (in various forms: Gentzen, Fitch, Jáskowski, Suppes, ...),
- Hilbert-style calculi (a linear format preceding from axioms and generally using a handful of rules, e.g., modus ponens),
- Toulmin diagrams (in which the various roles of statements are represented, so that not all statements are simply bald "premises"),
- sequent calculi (à la Gentzen).

The list is quite incomplete; the reader is invited to recall other formats for representing argumentation formally. The point of our list is to suggest to the reader that there are numerous formalisms available for representing or reconstructing arguments, especially mathematical ones.

It is one thing to formalize a piece of syllogistic reasoning or to use a truth table or tableau method for showing that a short propositional statement such as $p \wedge q \rightarrow p$ is a tautology. But for arguments of any complexity, one sees that reconstructing the argument formally quickly becomes tedious: the formalized argument is often much longer, in an everyday sense, than the argument that it is intended to formalize. One loses the thread of the formalized argument, since most formalisms mandate that one spell out all steps, significant or insignificant. The formal reconstruction takes too long, and the reward at the end (if one has enough patience!) pales in comparison to the cost of the formalization.

If one insists on writing mathematical arguments entirely in accordance with the demands of, say, a standard Hilbert-style calculus (where modus ponens is the only rule of inference),[1] then checking a formalized mathematical argument is indeed exceedingly routine, but also exceedingly time-consuming. One wonders

---

[1] It is not always the case that modus ponens is the only rule of inference available in a Hilbert-style system. In certain systems of modal logic, for example, one typically finds a rule of necessitation as part of a Hilbert-style formalism. But for classical propositional and predicate logic, as well as for others, it is standard to assume that in a Hilbert-style calculus modus ponens is the only rule of inference. The main feature of Hilbert-style calculi is that they have very few rules, placing the deductive burden of the formalism on its axioms, which are formulas, rather than rules of inference.

what payoff might be had if one were to formalize one's arguments. Lakatos, asking what one can discover in a formalized mathematical theory gives one answer: "One can discover the solution to problems which a suitably programmed Turing machine could solve in a finite time (such as: is a certain alleged proof a proof or not?). No mathematician is interested in following out the dreary mechanical 'method' prescribed by such decision procedures" (Lakatos, 1976, 4). It seems that these formalisms in general and the formalisms in which one can reconstruct "informal" arguments, whatever virtues they have (e.g., the soundness and completeness of various systems for expressing derivations), are simply not a practical tool. Those who stress formalisms for writing proofs seem to be overpromising what those formalisms can deliver. It seems that, even if we are interested in the formal side of mathematical reasoning, we need to rest content with its in-principle formalizability. Those who are not so interested in formalization might even suspect that the impracticality of formalizing interesting mathematical arguments constitutes a reductio of the notion of the in-principle formalizability of mathematical proof. If the "in-principle" part of "in-principle formalizability" is so essential, perhaps there is something wrong with the notion of formalizability in the first place.

We do not disagree with the view that formalizing interesting arguments (let alone mathematical proofs) is often tedious. We would like to defend, though, the in-principle formalizability of mathematical proofs as one of their important features by explaining their in-practice formalizability. We are interested, specifically, in the problem of *checking a proof*, because the ordinary discussion of in-principle formalizability does not take full account of an important capability that can be brought to bear when formalizing arguments. Instead of formalizing mathematical arguments in our heads or with pencil-and-paper, why not use computers to assist us in the task? Computers are obviously capable of doing symbolic computation at a faster rate, and with less regard for tedium, than humans have when working only in their heads or with pencil and paper. We are now in the possession of a wealth of tools that help us to practically reconstruct an informal argument in a formal setting, check whether the formalization is a valid argument and, if not, what defects it has.

In our view these technological developments are important for argumentation theory because, on the one hand, they shed new light on classical topics such as reconstructing and appraising arguments (especially mathematical ones), and, on the other hand, the developments suggest formal treatment of topics that might be thought to be essentially informal.

Our view is closely related to Azzouni's (2004) so-called derivation-indicator view about mathematical proofs (briefly, that ordinary "informal" mathematical proofs serve as indicators of corresponding formal derivations). The success of computer-assisted (formal) theorem proving projects as discussed here bears on Azzouni's view by showing how formal derivations in his sense do occur 'in the wild' and are indicated by informal proofs. Derivations for many mainstream mathematical theorems are now available; the days when one could only dream of live derivations for any substantial mathematical theorem are long over. Of course, the empirical success of computer-assisted formal theorem proving projects does not show that Azzouni is right. Our view is also compatible with that of his critics, such as Rav (2007). We do not offer formal proof construction and checking as a replacement for ordinary mathematical proof practices, nor are we implicitly suggesting that formal proofs ought to be a central object of interest in argumentation theory of mathematics. Nor, finally, can we recommend formal proofs for everyone; it requires, certainly, a friendly (or at least patient) attitude toward formalization, which not everyone has.

This chapter stands out from several of the others in this volume by its focus on mathematical proof and its connection with formal logic. A fair amount of argumentation theory can be seen as trying to escape from the musty chains of formal reasoning by asking questions about argumentation that are ignored, spurned, or untreatable by the old tools of deductive, valid, formal logic. From this perspective, our contribution might appear to be an unwelcome throwback. Although we focus on such "traditional" matters, we are by no means claiming that in-principle formalizability exhausts the interests of an argumentation theorist looking at mathematics. Further, we do not claim that the subject of *proof* exhausts the study of mathematical argumentation. Proof is clearly but one aspect of a multifarious phenomenon, as other contributors to this volume can testify. (See, for example, van Bendegem, 1988.) And although our interests are clearly "traditional" or "foundational", our focus on formalizability stems from the same desire among argumentation theorists looking at mathematics to find out what formal logic does *not* account for. Our modest suggestion is that, thanks to developments in automated reasoning systems (Portoraro, 2008, is a useful survey), new light is shed on the notion of in-principle formalizability and new problems suggested that, we believe, will be of interest to argumentation theorists.

Our focus in the present chapter is on evaluating formalized mathematical arguments, or, rather, checking proofs. We illustrate how this apparently "dreary mechanical method" that Lakatos referred to can in fact offer insight into a formalized mathematical argument. Our discussion is based on modern computer implementations of the process of checking formalized proofs, in the guise of so-called *interactive theorem provers* (also known as *proof assistants*). We will focus on the MIZAR interactive theorem prover,[2] which is based on classical first-order logic, set theory, and natural deduction. There are many actively maintained interactive theorem provers now available:

- COQ[3]
- ISABELLE[4]
- HOL[5] and some variants, such as HOL LIGHT[6] and HOL ZERO[7]

---

[2]http://mizar.org

[3]http://coq.inria.fr

[4]http://www.cl.cam.ac.uk/research/hvg/isabelle/

[5]http://hol.sourceforge.net/

[6]http://www.cl.cam.ac.uk/~jrh13/hol-light/

[7]http://proof-technologies.com/holzero.html

From among these we chose MIZAR for its relatively straightforward proof syntax, which is most likely to be immediately accessible to an unfamiliar reader. We are not interested in defending a claim about which of the great variety of interactive theorems provers now available is "best". For lack of space, we cannot provide a comprehensive introduction to MIZAR; we refer the reader to (Grabowski et al., 2010). We will explain the relevant parts of MIZAR's language for representing mathematical proofs as necessary.

9.2   Computer-Assisted Proof Construction

The problem of formalizing mathematical arguments is rather old. The roots of formalization arguably go back to Euclid, if not earlier (Netz, 2003). For lack of space, we have to ignore a rich history, doing much injustice to many intellectual forbears, and skip ahead, past the invention of the modern computer in the 1930s. Some of the earliest research in what we now call artificial intelligence was on the formalization of mathematical arguments, specifically, the task of using computers to search autonomously for formal proofs of mathematical claims (e.g. Wang, 1960). Wang's groundbreaking research led to automatically found proofs of many theorems of *Principia Mathematica*. His remarks, made in 1960, have proved to be rather prescient:

> The time is ripe for a new branch of applied logic which may be called "inferential" analysis, which treats proofs as numerical analysis does calculations. This discipline seems capable, in the not too remote future, of leading to machine proofs of difficult new theorems. An easier preparatory task is to use machines to formalize proofs of known theorems (Wang, 1960, 2).

Wang distinguishes the automated search for genuinely new mathematical results from the formalization of known theorems. We are interested in this second practice. Such work, we urge, provides a fascinating glimpse into the practice of mathematical argumentation.

Early research in the field of *theorem proving*—the search for proofs (or disproofs) of mathematical claims—has led to rather sophisticated techniques and impressive milestones. It is standard to divide automated theorem proving (in which computers search more or less autonomously for proofs, models, refutations, etc.) from interactive theorem proving (in which the emphasis is on the construction of proofs, assisted by a machine). Although there are some precursors of interactive theorem proving going back to the earliest days of modern computers, the field began to pick up steam mainly in the 1960s and 1970s: early important projects include AUTOMATH by N. G. de Bruijn in Eindhoven, the Netherlands (de Bruijn, 1980), SAD (System for Automated Deduction) in Kiev, Ukraine (Verchinine et al., 2007), and MIZAR in Białystok, Poland (Matuszewski and Rudnicki, 2005). (Our account of the early history of interactive theorem proving must, for lack of space, be cut short.) The products of these systems that are most interesting to us are their formal languages for reconstructing the mathematical vernacular, that is, the informal though highly stylized, even slightly rigid, parole used by mathematicians when communicating mathematical proofs. In the end, these reconstructions of mathematical vernacular are themselves formal languages, but they are far from simply being "raw" formalisms such as natural deduction or sequent calculus.

For a more thorough account of the state of the art, one can consult the excellent survey articles by Wiedijk (2008), Hales (2008), and Harrison (2008). Here we focus on the aspects of formal mathematical proofs that may be of interest to those working in argumentation theory and mathematics.

The use of computers in mathematical theorem proving is becoming increasingly important. One can distinguish two directions: *proof search* and *proof checking*. Proof search suffers from well-known complexity problems and has so far had only limited success solving general mathematical problems. On the other hand, while it might be very hard to find a proof, to *check* a proof for correctness is, in general, of lower complexity (although the process can be rather technical and long).

Wiedijk (2006) surveys a corner of the state of the art. He presents 17 theorem provers and evaluates them in a uniform way: to prove that $\sqrt{2}$ is irrational. Wiedijk's survey gives interesting insight into the state of the art of theorem proving, strongly emphasizing proof check (and less strongly proof search). Wiedijk presents a six line proof of the famous theorem, taken from the classical textbook of Hardy and Wright (1960, 39 f.):

The traditional proof ascribed to Pythagoras runs as follows. If $\sqrt{2}$ is rational, then the equation

$$a^2 = 2b^2$$

is soluble in integers a, b with $(a,b) = 1$. Hence a^2 is even, and therefore a is even. If $a = 2c$, then $4c^2 = 2b^2$, $2c^2 = b^2$, and b is also even, contrary to the hypothesis that $(a,b) = 1$.

This proof should be understandable by anyone with basic mathematical knowledge. Emphasizing proof checking, Wiedijk writes (2006, 3): "Ideally, a computer should be able to take this text as input and check it for its correctness." Insofar as Wiedijk means that Hardy and Wright's proof *needs* to be checked for correctness, we consider this perspective misleading. Of course, the correctness of a proof is essential, but that doesn't mean that the correctness of a proof is always in doubt, or that mathematical proofs require (repeated) verification. In fact, the proof above, as a textbook proof, has surely another objective besides simply displaying the logical correctness of its conclusion. On the one hand, it should *convince* the reader of the truth of the proven theorem; on the other hand it should provide text which is *memorable*, that can be reproduced whenever needed. For the moment, let us follow the line of proof checking, as this might be an important task when we are in doubt about the truth of a theorem or about a particular proof of a theorem. When Wiedijk presents the results which computer-aided theorem provers provide for the irrationality of $\sqrt{2}$, these results might very well be checkable, but the proofs themselves are far from acceptable for a human reader.[8] In fact, to guarantee correctness, they have to take into account too many details, details which a mathematician does not like to see exposed in the proof. This was formulated by Scott as follows:

> For verification (...) *checkable proofs* have to be generated and archived. Computers are so fast now that hundreds of pages of steps of simplifications can be recorded even for simple problems. Hence, we are faced with the questions, 'What really is a proof?' and 'How much detail is needed?' (Scott, 2006, *ix* f.)

That formalized proofs are not a good answer to the former question is argued by several authors. For instance, Avigad, working from historical case studies of proofs in elementary number theory, concludes that "Standard models of deduction currently used in mathematical logic cannot easily support the type of analysis [of proofs] we are after" (Avigad, 2006, 131).[9]

We agree with Scott that formalized or computer-checked proofs challenge our notion of proof with respect to *proof representation*. How do we represent proofs in such a way that a computer could understand them, while still being practical and useful for humans? And can we trust formal proofs in the same way that we can trust "human proofs"? (See Rehmeyer, 2008, for a further discussion.) And when is a proof really a proof, anyway?

This already holds "in the small" for the proofs of the irrationality of $\sqrt{2}$ presented in (Wiedijk, 2006). It surely also holds "in the large" when we come to the controversial case of the computer-aided proof of the four color theorem, which suffers from a huge number of case distinctions checked by computer, but which were not (and most likely could never) be checked by a human mathematician. And it also holds for the case of the alleged proof of the Kepler conjecture (Hales, 2005). But, while we said that the verification of Hardy and Wright's proof above is not an issue, the verification of these proofs *is* at issue, and is valuable. For Appel and Haken's original 1976 proof of the four color theorem (Appel and Haken, 1977) this is mentioned by Thomas, who writes in an informal explanation of the second proof:

> We have in fact tried to verify the Appel–Haken proof, but soon gave up. Checking the computer part would not only require a lot of programming, but also inputting the descriptions of 1476 graphs, and that was not even the most controversial part of the proof (Thomas, 2007).

So, Robertson, Sanders, Seymour, and Thomas came up with a new proof (Robertson et al., 1997). It is still performed by computer aid, but it reduces the cases from 1476 to 633. While this is still a number which cannot be checked "by hand", no one would deny that it is an improvement, i.e., that this proof is

---

[8]"We can also see clearly from the examples in this collection that the *notations* for input and output have to be made more human readable" (Scott, 2006, viii f., in the foreword of Wiedijk, 2006).

[9]Avigad actually proposes *methods*—which correspond, for instance, to *tactics* in the ISABELLE theorem prover—as alternatives (cf. Avigad, 2006).

clearly *better* than the original one. But with respect to the verification, now we have the possibility to verify—by computer aid—the programs involved (as has now been done: Gonthier, 2008).

The question of verification of large or otherwise controversial proofs was reignited recently by Hales when he launched a program to verify "formally", i.e., by computer aid, his proof of the Kepler conjecture. Hales is engaged in his project because the initial verification attempt *by mathematicians* of his solution to the Kepler conjecture led only to "99% certainty" about the correctness of his proof (Hales, 2005). While the situation seems to be in principle similar to the proofs of the four color theorem, the new aspect is that the author should try to convince the mathematical community of the correctness of his proof *entirely* by formal and computer verified means.

What is the difference between proving and checking a proof? When is a proof a proof, anyway? Is there a social aspect to whether a proof is a proof? (We refer the reader to Heintz, 2003, and to Löwe et al., 2010, for a discussion of further epistemological issues in formal proof.)

As a consequence of the philosophical discussion of Appel and Haken's original proof of the four color theorem (there are many sources: Tymoczko, 1979; Detlefsen and Luker, 1980; Teller, 1980; MacKenzie, 1999; Arkoudas and Bringsjord, 2007; Bassler, 2006), Prawitz stresses—with reference to Teller—the importance of distinguishing proof from its verification:

> That one has verified that a proof is a proof [...] is therefore not a part of the proof. That is not to say, of course, that it is not wise to check one's proof; as Hume rightly remarks, the confidence in a proof increases when one runs over it. But the checking does not add anything to the proof itself (Prawitz, 2008, 89).

From this perspective, it is clear that Hales's verification project should concern *only* the confidence one might have in his proof, not the proof itself. But what will be the status of the proof even after it is verified on the lines that Hales proposes?

In fact, this situation is not as new as it might look; it didn't just emerge from the use of computers. There is actually one other instance in which a modularization in numerous cases was carried out by just a large number of mathematicians: the classification of finite groups. In fact, its "proof" (which was, in part, also done by computer aid) has an interesting history with respect to its correctness and its acceptance (Aschbacher, 2004).[10] What distinguishes it most from the computer proofs mentioned above, is the fact that the different cases might be considered as interesting in their own right, i.e., for the study of a particular group (or class of groups). A study of the single cases in the proofs of the four color theorem does not provide any such mathematical surplus value.

---

[10]"I have described the Classification as a theorem, and at this time I believe that to be true. Twenty years ago I would also have described the Classification as a theorem. On the other hand, 10 years ago, while I often referred to the Classification as a theorem, I knew formally that that was not the case, since experts had by then become aware that a significant part of the proof had not been completely worked out and written down" (Aschbacher, 2004, 737 f.).

The lesson to take away from the reception of the proof of the four color theorem and the reception of the classification of finite simple groups is that mathematicians still seem to prefer proofs checked by other (human) mathematicians. Even if the proof is so complex that no single mathematician can check it, it is preferred that the mathematical community as a whole cooperates in carrying out the verification. (See also Buss, 1998, who divides proofs into formal and social.) In the application of computers to proving properties of computer programs, one can also find controversy; De Millo, Lipton, and Perlis's paper (1979) on the social nature of proof is a famous gauntlet thrown down in the discussion about these matters. MacKenzie (2004) provides a comprehensive discussion.

9.3   Checking a (Formal) Proof

The problem of checking a mathematical proof can sometimes be surprisingly complex. It seems that for many ordinary mathematical proofs the process of checking the proof occurs simultaneously with a reconstruction of the proof. Lakatos expresses this thus: "Often the checking of an *ordinary* proof is a very delicate enterprise, and to hit on a 'mistake' requires as much insight and luck as to hit on a proof" (Lakatos, 1976, 4). But checking proofs even in a formal setting can also be a delicate enterprise, as well. How?

One can view an argument for a claim as a structure that specifies how claims of the argument are justified by various moves. We begin with an initial thesis, and then make inferential moves from it, making in turn additional claims. Each step we make transforms the thesis to be proved (and possibly introduces new theses) into a different claim. We discharge some obligations and possibly introduce others along the way. The argument can be said to be successful if all our steps are the result of sound applications of rules of inference and the thesis to be proved at the end of the argument is acceptable.

Lest we slip into an infinite regress à la Carroll (1895), we need to agree to our rules of inference. Thus, if the thesis is B and we have established A and $A \to B$, we need to agree that:

- An application of modus ponens to the two claims already established (A and $A \to B$) yields B,
- Modus ponens is an acceptable rule of inference, and
- The occurrence/utterance of B that is obtained by applying modus ponens is "the same" as the B that we set out to establish.

The first item pins down the premises of an application of modus ponens. The second item is meant to rule out the possibility that the acceptability of modus ponens becomes itself a disputable issue in the argument. The third item, like the first, is meant to pin down the issue under discussion; it won't do if, establishing A and $A \to B$, we nonetheless reject the conclusion of the application of modus ponens to these two premises because the conclusion B now differs from the B that we set out to establish.[11]

To some extent, one could say that there is little at issue when it comes to the problem of checking a formal mathematical proof. We simply choose some target formalism in which to reconstruct the proofs, such as a standard Hilbert-style calculus or a Fitch-style natural deduction calculus. We might prefer calculi that are complete for the notion of logical consequence in which we are interested (often, but not always, classical first-order logic).[12] We then "formalize" the argument in the chosen formalism, producing some kind of figure d (graph, tree) representing the initial argument. The initial argument is then checked if and only if d is a legal figure according to the rules of the proof formalism that we started with. Showing that d is legal is, generally, an entirely mechanical matter.

For example, in a Hilbert-style calculus, d is a finite sequence $\langle A_1, A_2, \ldots, A_n \rangle$ of logical formulas. The question of whether d is a legal derivation of a formula ϕ from assumptions Γ consists of showing that d terminates with ϕ and that for each term A_i of the sequence, that either

- A_i is a member of Γ
- A_i is an axiom (which amounts to simple pattern matching: does A_i have the form $\phi \vee \neg \phi$? Does A_i have the form $\phi \rightarrow (\psi \rightarrow \phi)$?)
- There exist earlier terms A_{i_1} and A_{i_2} of the sequence such that A_{i_2} is $A_{i_1} \rightarrow A_i$. This is just another way of saying that A_i is obtained from A_{i_1} and A_{i_2} by modus ponens.

For other formalisms, e.g., Fitch-style natural deduction or Gentzen-style sequent calculus, checking whether some figure d is legal according to the formalism is likewise quite straightforward.

Yet it often happens that among proof formalisms, there is a balance between the complexity of verifying that a figure purporting to be a legal derivation really is legal, and the length of the proofs. Thus, in a Hilbert-style calculus, it is trivial indeed to check that a given sequence of formulas is a legal Hilbert-style derivation; but the length of the legal sequences, as one considers derivations of increasingly non-trivial mathematical results, grows quite rapidly. (For a thorough systematic discussion of this and related issues of proof complexity, see Orevkov, 1993.) It seems, moreover, that such proofs are unsatisfactory because they diverge significantly from mathematical practice. What is wanted is a formalism that tries

---

[11]This is not to say that such phenomena are not worth studying. One way of coming to grasp the meaning of a statement is by arguing with it; we may find, for example, that if we have reached an unacceptable conclusion through sound reasoning from premises that we accept, we find ourselves having reached a better understanding of the conclusion. Thus the B we reach at the end is, in some sense, different from the B (in $A \rightarrow B$) from which the argument commenced. Such a phenomenon might be understood as argument-based discovery of meaning. Such argumentation—which might be seen as fallacious—is present in mathematics, but we shall not consider it here.

[12]"Classical" means that the law of excluded middle is assumed to be valid: the disjunction $\phi \vee \neg \phi$ is assumed to hold for any formula ϕ.

to be more faithful to the practice of mathematical argumentation, while still being sufficiently delimited that one can compute with the formalizations.

There are, we submit, such formalisms, balancing ease of use with the practical need that checking whether a figure is a legal derivation is fast. Later we shall see some example proofs written in one of them.

One kind of argument appraisal that is available in the formal setting that is not easily available in the informal setting is the question of what an argument depends on. Essentially any contentful argument takes something for granted. It appeals explicitly to some background knowledge, or perhaps makes certain assumptions implicitly or carries out parts of the argument without any justification. Even certain mathematical definitions might take something for granted. For example, the definition of the real number π as the ratio of a circumference of a circle to its radius, on the surface, simply defines a function from *particular* circles to the set of real numbers. No one doubts that the notions "circumference of a circle" and "diameter of a circle" vary from circle to circle (that is, distinct circles can have distinct circumferences and radii), so without any further analysis all the definition of π gives us is yet another quantity that varies from circle to circle:

$$\forall \gamma (\text{circle}(\gamma) \to (\pi(\gamma) = (\text{circumference}(\gamma)/\text{diameter}(\gamma)))).$$

But in fact in standard Euclidean geometry the quantity $\pi(\gamma)$ does *not* vary with γ. Thus, in a fully formal treatment of Euclidean geometry, one would have to establish the theorem

$$\forall \gamma, \gamma' [\pi(\gamma) = \pi(\gamma')].$$

Thus, in a fully formal development of Euclidean geometry, one would expose this implicit dependency of a definition on some of the axioms of Euclidean geometry.

Nonetheless, modern interactive theorem provers generally provide some kind of support for omitted inferences. Which inferences are omitted? We see an interesting formal analogue of the subject taken up by Fallis, concerning intentional gaps in mathematical proofs (Fallis, 2003). In what sense are there gaps in a formal, computer-checked proof?

At the end of formalization, one typically has, at least in principle, an utterly formal proof of a theorem, logically correct down to all details, down to the axioms of whatever background theory one is working with. However, a completely formal proof, for a theorem of any mathematical substance, would be unmanageable. Let us be clear that when working with interactive theorem provers, one does not typically work with "totally" formal proofs that, say, proceed *only* by introduction and elimination rules for quantifiers and connectives. (This is in accordance with the usual notion of *analytic proof*, which proceeds by an analysis of the structure of the claim to be proved and the structures of the assumptions used to prove it. By contrast, a *synthetic proof* brings in some new ingredients that are not formally contained in the statement to be proved.) It is well know that these are simply too big. One uses the computer not to assist in the drudgery of simply storing a large derivation figure in its memory and manipulating it with somewhat greater facility than would be the case were one to just use pencil and paper. Instead, the standard practice is that one works with a formal language that sits above a "totally" formal language. One writes proofs in the intermediate formal language that, in some sense, can be compiled into a totally formal proof. Thus, modern interactive theorem provers typically provide mechanisms for suppressing some inferences. One sees here a computable version of Azzouni's derivation-indicator view. An "informal" or ordinary mathematical proof is said to be an indicator of a formal derivation. Likewise, proofs conducted with modern interactive theorem provers, even though they are rather more formal than informal proofs, can likewise be seen as indicators of totally formal derivations.

This does not mean that a proof constructed with an interactive theorem prover is, in some interesting sense, *informal*. Rather, it is formal, but with some gaps that can be, as it were, computably traversed. That is, gaps generally represent proof search problems; the traversability of a gap means that there is a solution to the proof search problem. Consider, for example, a proof in the MIZAR system for the following elementary set-theoretic fact[13]:

```
for x, X, Y being set holds
  x in X \+\ Y iff not (x in X iff x in Y)
proof
  let x, X, Y be set;
  x in X \+\ Y iff x in X \ Y or x in Y \ X
    by DefDisjointUnion;
  hence thesis
    by DefRelativeComplement;
end;
```

Here the claim is that a set x is in the disjoint union of two sets X and Y (x in X \+\ Y) iff it is not the case that x is in X iff x is in Y (not (x in X iff x in Y)). The proof uses the definition of disjoint union (DefDisjointUnion),

```
definition
  let X, Y be set ;
  func
    X \/ Y -> set
  means :DefDisjointUnion:
  for x being set holds x in it iff (x in X or x in Y);
```

which is defined in terms of relative complement (X\Y), defined as

```
definition
  let X, Y be set ;
  func X \+\ Y -> set
    equals
    (X \ Y) \/ (Y \ X);
```

---

[13] http://mizar.org/version/current/html/xboole_0.html#T1

The disjoint union of X and Y is the union $(X \setminus Y) \cup (Y \setminus X)$ of the relative complement of X from Y and Y from X), and the definition of relative complement itself, which is

```
definition
  let X, Y be set ;
  func X \+\ Y -> set
    equals
  (X \ Y) \/ (Y \ X);
```

The proof is three steps long: the initial "let" statement, the inference from the definition of disjoint union, and the final inference (thesis) from the definition of relative complement. One might wonder what the trouble is all about; see the discussion of what counts as an "obvious" inference in Sect. 9.4 for an explanation of why it's necessary, at least for the case of MIZAR, to spell out an argument. Taking for granted the need to articulate these steps in the proof, one might wonder whether this is *really* a formal proof. Indeed, it may fail to adhere to, say, the requirements of a Fitch-style natural deduction system. The "let" of the first three steps introduces three variables (x, X, and Y) all at once, in one step, rather than one at a time. Another way in which the MIZAR proof could fail to strictly adhere to the requirements of a totally formal proof is that it fails to start with the definition of disjoint union, considered as the universal claim

$$\forall X \forall Y [X \oplus Y = (X \setminus Y) \cup (Y \setminus X)],$$

and instantiate it for the terms of interest, and then apply a rule of equality to transform the given claim in terms of \+\ into one involving union and relative complement. A complete Fitch-style natural deduction proof of the claim would proceed as in Fig. 9.1. This Fitch-style deduction takes dozens of steps. We cannot claim that there is no shorter proof. Still, it is implausible to us that there is a Fitch-style deduction of the same theorem from the same premises that is much shorter than this one. The Fitch proof even economizes in some ways: we took as an axiom an instance of the propositional tautology

$$\neg(p \leftrightarrow q) \rightarrow [(p \wedge \neg q) \vee (\neg p \wedge q)]$$

in step 1 and an instance of

$$(p \wedge q) \rightarrow (q \wedge p)$$

in step 2 as premises. The principle assumptions are of course 3, 4, and 5, which are definitions of three mathematical concepts. In a Fitch-style natural deduction formalism in which (instances of) this formula are axioms, then no further work is needed. If this formula is not an axiom, then of course a proof of it must be given, so the proof would need to be even longer. We have also economized by

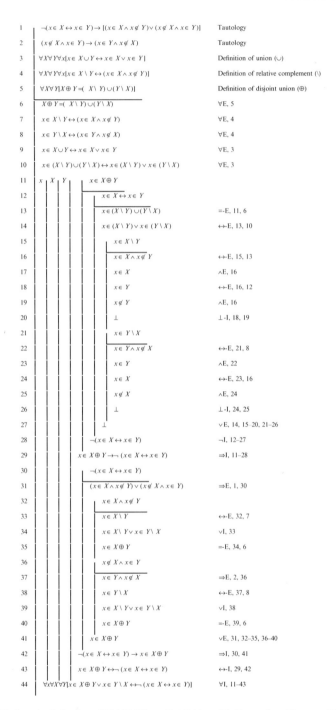

Fig. 9.1 Fitch-style deduction of $\forall x \forall X \forall Y [x \in X \oplus Y \lor x \in Y \setminus X \leftrightarrow \neg(x \in X \leftrightarrow x \in Y)]$

allowing multiple-variable instantiations for universal formulas, that is, permitting the inference of

$$\phi[x_1,x_2,\ldots,x_n := t_1,t_2,\ldots,t_n]$$

from

$$\forall x_1 \forall x_2 \cdots \forall x_n \phi,$$

where $n \geq 1$, in a single step (where $\phi[x_1,\ldots,x_n := t_1,\ldots,t_n]$ denotes the simultaneous substitution of the term t_k for the variable x_k, $1 \leq k \leq n$). Thus, depending on precisely how restricted the natural deduction system is, the deduction easily approaches 50 steps.

Contrast the 3-step formal proof earlier, from which the Fitch-style derivation came, with the totally formal Fitch-style derivation. The point is that there is no need to write down all those 40+ steps of the Fitch-style derivation, because most of them can be *computed*. The above MIZAR proof might even be just what we want, since, apart from the first step of instantiating the variables, the two steps are, essentially:

1. Apply the definition of disjoint union, then
2. Apply the definition of relative complement.

This seems to be the heart of the matter, and the proof does not diverge from that. By contrast, it is not apparent what the "heart" of the corresponding Fitch-style proof is, since we had to carry out various instantiations of universal premises and even take a detour through a propositional logic.

9.4 What Inferences Are "Obvious"?

When giving a mathematical proof, one has to decide what to say and what can go without saying. Learning the norms for communicating proofs is an important part of acquiring knowledge of mathematics. (One might even see mathematics as an instance of Toulmin's fields: Toulmin, 2003.) It is a recurring question that teachers of mathematics face: "Which steps should be included? What steps can I omit?" Rarely (if ever) do we see all steps of a mathematical proof, if by "step" we understand a single application of a rule of inference in some conventionally accepted formalism akin to the Fitch-style derivation given in the previous section. Indeed, if a student were to give a totally formal derivation as a solution to a mathematical problem, we would rightly feel that the student has, in some sense, failed, despite the inarguable validity of his solution. In the opposite extreme, teachers of mathematics can surely recount occasions where a student simply asserts a complicated statement that, by reasonable lights, cannot be simply asserted, since it needs justification, at least in the context of instruction.

At issue is the problem of distinguishing, in a context, which mathematical claims need to be justified and which can be simply granted. (A related topic is the problem of characterizing persuasiveness of certain mathematical moves. For more on the subject, see the contribution by Inglis and Mejía-Ramos in this volume.) A full answer evidently requires a classification of the various contexts in which mathematical claims are made. A mathematics teacher might reject a student's unjustified claim of an equation, while accepting, only 5 min later, the very same equation put forward by a colleague in the mathematics department. The teacher is not being duplicitous because the contexts of mathematical acceptability in the two situations are different.

What about in the formal context? The choice of formalism determines which claims count as justified and which require justification. The result is extreme: *anything that is not an axiom requires justification.* It is not so bad that non-trivial propositions require proof. What is worse is that it often happens that even "trivial" statements, so long as they are not axioms, require proof. And sometimes trivial statements turn out to be not so trivial after all, in that they apparently require unexpectedly long formal proofs. Moreover, the definition of formal derivation requires that every step in the proof that is not an axiom be the result of an application of a rule of inference.

If we are working cooperatively with a computer, though, the answer to the question of which inferences need justification can be deferred somewhat to the computer. The strength of the mechanisms for doing automated reasoning will have a powerful influence on the proofs. Rudnicki presents the situation thus:

> The core of an automatic proof-checking system is a decision procedure for accepting/ rejecting presented inferences. A rejection does not mean that an inference is logically invalid, it simply mirrors the fact that the proof-checker was unable to certify the inference's correctness. Certainly, an invalid inference has to be rejected. Using a proof-checker is similar to a discussion between humans. One admits that one does not see why a conclusion follows from premises (even if it does in fact), but one agrees quickly that the adversary is right after being given additional explanation. The criterion for acceptance/rejection of valid logical inferences in a proof-checking system is said to define a class of 'obvious' inferences in the system (Rudnicki, 1987, 383).

We shall now follow the terminology used by Rudnicki (who is reusing a term introduced by Davis) and discuss the problem of *what inferences are obvious?* The problem is to try to give a formal or computable account of an informal notion. We will, of course, not succeed in full, but we believe that we can learn about the notion of obviousness and attendant features of mathematical argumentation by studying it through a formal lens.

As Rudnicki suggests, we are interested in *sound* obvious inference. (To precisely specify what we mean by "sound", we should specify a logic, and there seems to be room for non-classical logics. For our purposes, let us stick to classical first-order logic.) Just because something is judged to be non-obvious does not mean that it is false or invalid. Moreover, we are all too aware of claims put forward as "obvious" but which turn out to be false or otherwise unacceptable. Calling something "obvious" can sometimes run the risk of being nothing more than a thinly-veiled appeal to authority, argument by intellectual boasting or belittlement (it's obvious to me—why isn't it obvious to you?), or premature abandonment ("it's obvious" might just amount to saying that we don't have time to argue or that the arguer isn't willing to argue).

The extreme solution of stipulating that all (valid) inferences are obvious seems unacceptable. With this understanding of obviousness, any mathematical inference whatsoever would be acceptable without justification. This extreme solution might not even be well-defined. After all, what notion of validity should we use? Should we use classical logic or non-classical logic? What is the characterization of validity? Should the notion of validity be syntactic or semantic? If we were worried about consistency (is it possible that incompatible claims are both obvious?), we might try to restrict obviousness so that only true conclusions are accepted—but what claims are true? Evidently not all true claims are obvious and at least some require proof; indeed, often we don't even *know* that a mathematical claim is false until we try to prove it, tentatively accepting it as true, and realize only in our failure to prove it that it is not even true.

We have already discussed the other extreme, in which nothing is obvious except what is formally an axiom. The result here is likewise unacceptable, because we then have that only the most brutally obvious claims are accepted, and we would give up quickly. One could take as axioms an extremely large set of principles; but then we collapse to the other extreme in which essentially everything is obvious.

What is wanted is a notion of obvious inference that allows us to omit *some* (valid) reasoning, but not too much. The notion should also be practical: if we are interested in doing much mathematical reasoning formally, checking whether a step in the proof is an obvious inference should be done quickly. The notion of obviousness should therefore have a low computational complexity.

Davis has taken on the problem of characterizing, proof-theoretically, the notion of an "obvious" inference (Davis, 1981). The context from which Davis devised his characterization of obvious inference was a project of natural deduction at Stanford. Natural deduction arguments à la Gentzen, Fitch, or Suppes (as with any serious proof formalism) can be rather tedious. The heart of an informal argument is often obscured or diffused if one adheres strictly to the requirements of the formalism. In formal contexts one wants a rule of inference that would allow one to dispense with certain tedious details.

Here is a precise reformulation of Davis's proof-theoretic definition of "obvious inference" in first-order classical logic:

Definition 1. A logical formula ϕ is an **obvious logical consequence** of assumptions Γ if there is a Herbrand proof of ϕ from Γ in which each quantified formula of Γ is instantiated at most once.

The precise definition of Herbrand proof will not be given here; see Davis (1981) or Harrison (2009). The idea is that in drawing an obvious logical inference, it is ruled out to use multiple instances of quantified formulas in Γ. Once one has chosen, for each quantified formula α in Γ, an instance α^* (one may elect not to instantiate

α at all), then one has to formally derive ϕ from Γ and the instances $\alpha_n^*, \ldots$ using only a "light" form of first-order reasoning and propositional calculus.

The notion of obvious logical inference, as defined by Davis, clearly does not characterize how we use the term "obvious" in the context of ordinary reasoning. A consequence of the proposal is that quite a lot of propositional inferences get classified as "obvious". This seems to accord with our intuitions in cases such as

$$q \text{ because } p \text{ and } p \rightarrow q$$

but fail for cases such as

$$((p \rightarrow q) \rightarrow p) \rightarrow p$$

(a famous classical validity known as Peirce's formula), or

$$p \text{ because } X \text{ is unsatisfiable,}$$

where X is a large, complex unsatisfiable set of propositional formulas. Davis's proposal also fails to account for such first-order inferences as:

$$\phi(f(f(a))) \text{ because } \phi(a) \text{ and } \forall x[\phi(x) \rightarrow \phi(f(x))],$$

because we need to instantiate the universal premise $\forall x[\phi(x) \rightarrow \phi(f(x))]$ twice (once with a, and again with $f(a)$).[14]

Still, it is a valuable proposal because of its conceptual simplicity and practical efficiency. In general, if we have some quantified formulas Γ and a conclusion ϕ, deciding which instances of formulas in Γ to choose can quickly lead to a vast number of possibilities. Davis's proposal limits the search for instances: *choose at most one instance each time.* When a conclusion ϕ is not an obvious logical consequence of background assumptions Γ, then the inference is "too complicated" to be checked by a machine, and the human formalizer needs to supply more information.

Following on from Davis's work, Rudnicki (1987) has outlined some problems with Davis's proposal and has offered a second mathematical characterization of the notion of obvious inference.

As mentioned earlier, the problem faced by the designer of an interactive theorem proving system is to sail between an extremely dense but extremely fast notion of acceptable inference, as opposed to trying to make as much as possible obvious by devoting arbitrary computational resources to determining whether an inference is obvious. The effect of this decision is that the class of proofs will tend to be extremely long on the one hand, and maximally concise on the other.

One might say that the preference here is clear: one should go for the most concise proofs possible! The problem is that computer proof search for interesting

---

[14]This example appears in Rudnicki (1987).

logics is a rather difficult computational task. The search spaces for theorem proving problems tend to be extremely large, even intractable, so that devoting increasing amounts of computational power often has surprisingly little effect. It often happens, because of the sheer size of the search spaces involved, that if a theorem proving problem cannot be solved in 5 min, then it cannot be solved in 1 h either.[15] We are thus forced by complexity to keep our aims modest.

If we vary the strength of the checker for obvious inferences, can we detect a dividing line between what is obvious, in the everyday sense of the term, from what is not obvious? Surely there must come some point (though we concede that it seems likely that there could be Sorites-type paradoxes here). For at the extreme end we could have complicated theorems of mathematics that are known to be valid consequences of some recursively enumerable set of axioms, such as those of Zermelo-Fraenkel set theory (ZF); given arbitrary computational resources to check whether an inference is "obvious", we would find that, say, the fundamental theorem of calculus gets deemed as an obvious inference from some finite set of axioms of ZF, though we would surely balk at judgment.[16]

The result is that varying the strength of the mechanism that certifies obviousness has an effect on the intuitive *explanatory value* of the proofs. (For a further discussion of explanation in mathematics, see Mancosu, 2000.) This is intuitively clear. Earlier we presented "$\phi(a)$ and $\forall x[\phi(x) \to \phi(f(x))]$ therefore $\phi(f(f(a)))$" as a case where Davis's notion of obvious inference fails. This case would be counted as obvious if we strengthened Davis's notion and permitted two instances of universal premises to be selected.

Here is a slightly more mathematical example of the same phenomenon, showing the effect that such a strengthening has on live mathematical proofs coming from the MIZAR Mathematical Library, the curated body of formalized mathematical knowledge that has been reconstructed in the MIZAR interactive theorem prover.[17] The proofs in the MIZAR Mathematical Library are governed by a notion of obviousness similar to Davis's notion. (A precise definition of the class of MIZAR-obvious inferences is not needed.)

Consider the following formal theorem:

```
reserve X, Y, Z for set;

Lemma1: X c= Y & Y c= Z implies X c= Z;

Lemma2: X c= X \/ Y;

Lemma3: X c= Y implies X \/ Z c= Y \/ Z;
```

---

[15]There are many counterexamples to this general outlook. The solution, by an automated theorem prover, of the long-outstanding Robbins problem required 8 days of continuous computation (McCune, 1997).

[16]If one is not satisfied with this example, we could replace the fundamental theorem of calculus by some other significant mathematical fact, perhaps even one that has not yet been discovered.

[17]We thank Artur Korniłowicz for this example.

```
X c= Y implies X c= Z \/ Y
proof
  assume X c= Y;
  then A1: Z \/ X c= Z \/ Y by Lemma3;
  X c= Z \/ X by Lemma2;
  hence X c= Z \/ Y by A1,Lemma1;
end;
```

The first line says that in what follows, the variables X, Y, and Z are sets (more precisely, they are assigned the type set). The next three statements are background lemmas (assigned the labels Lemma1, Lemma2, and Lemma3).

Lemma 1 (Lemma1) expresses the transitivity of the subset relation: if $X \subseteq Y$ (X c= Y) and $Y \subseteq Z$ (Y c= Z), then $X \subseteq Z$ (X c= Z).

Lemma 2 (Lemma2) expresses the fact that X is always a subset c= of the union of X with any other set Y (X \/ Y).

Lemma 3 (Lemma3) expresses the fact that if X is a subset of Y (X c= Y), then the union $X \cup Z$ is a subset of the union $Y \cup Z$ (X \/ Z c= Y \/ Z).

Lemmas 1, 2 and 3 will be taken for granted, for the sake of discussion. That is, the text above is not acceptable to MIZAR as written, because all three lemmas are not MIZAR-obvious and therefore require MIZAR-proof. Our interest is the final result (the theorem) of the MIZAR text fragment, which expresses the simple result that if X is a subset of Y (X c= Y), then X is a subset of the union $Z \cup Y$ for any set Z (X c= Z \/ Y). Unlike the case for the three lemmas, a proof of this fact is provided. Let us proceed through it.

We are carrying out a natural deduction-style proof of an implication (X c= Y implies X c= Z \/ Y); the first step, naturally, is to assume the antecedent (assume X c= Y). From this assumption we get, using Lemma 3, the inclusion Z \/ X c= Z \/ Y. (MIZAR is also implicitly using the commutativity of the binary union operation—note that Lemma 3 puts Z on the right-hand side of the union, but in the conclusion just drawn, it appears on the left-hand side of the union.) The notation A1: is simply assigning a label to the statement just concluded; we will use the formula later in the argument by appealing to its label. The third step in the argument simply applies Lemma 2; we have from it that $X \subseteq Y \cup X$. We do not use the hypothesis of the theorem nor the previously concluded statement (whose label is A1) to infer the result; this follows immediately from Lemma 2 alone. The final step of the proof (followed by hence) is the desired conclusion: we have that $X \subseteq Z \cup Y$ because of

- $X \subseteq Z \cup X$ (from the previous line)
- the formula labeled A1, i.e., $Z \cup X \subseteq Z \cup Y$
- the formula labeled Lemma1, i.e., $X \subseteq Y \wedge Y \subseteq Z \rightarrow X \subseteq Z$.

This argument is optimal in the sense that no step can be removed. The MIZAR proof checker rejects all possible compressions of the argument. In the maximal compression, one justifies the final theorem by simply declaring, without proof, that it follows from the lemmas, i.e.,

```
X c= Y implies X c= Z \/ Y by Lemma1, Lemma2, Lemma3;
```

one finds that the MIZAR proof checker rejects the inference. Other kinds of attempted compression, such as removing the intermediate statement A1, viz.

```
X c= Y implies X c= Z \/ Y
proof
 assume X c= Y;
 hence X c= Z \/ Y by Lemma1, Lemma2, Lemma3;
end;
```

or dropping the application of Lemma 2, viz.

```
X c= Y implies X c= Z \/ Y
proof
 assume X c= Y;
 then A1: Z \/ X c= Z \/ Y by Lemma3;
 hence X c= Z \/ Y by A1, Lemma1, Lemma2;
end;
```

are all rejected by the MIZAR proof checker. They are rejected because they require that multiple instances of background universal premises be taken, and this is precisely what is ruled out by the definition of obvious inference.

The preceding MIZAR proof is constructed according to the MIZAR notion of "obvious inference": to verify any purported inference, at most an instance of any universal premise is used. Moreover, no inference in the proof can be compressed or omitted while preserving this property.

To illustrate how the strength of the notion of obvious inference has an effect on the proofs one writes, let us consider a strengthening of the MIZAR-obviousness by permitting not one but two instances of universal premises to be used. We might call this notion "2-obviousness". (The old notion of obvious would be understood as "1-obviousness".) The result is that the above proof can indeed be compressed:

```
X c= Y implies X c= Z \/ Y by Lemma1, Lemma2;
```

This is a fascinating compression. We don't even need to articulate a proof any longer; for the MIZAR proof checker, our claim is a 2-obvious consequence of Lemmas 1 and 2 alone. We don't even need the help of Lemma 3, which was essential before when we were operating under the constraint that all inferences must be 1-obvious.

To sum up, we argue that mathematical proofs should be evaluated with respect to their ability to provide answers—answers in a "digestible" form. Computer proofs are often just "indigestible", not only due to excessive information, but also because they can't be addressed by our questions.

9.5   Appraising and Improving Formal Proofs

After completing a formal proof, one can return to it and improve it in various ways. We again see some advantages of the formal approach to mathematical proof and its implementation in an interactive theorem prover: we can evaluate an argument in ways that might be more difficult were we to insist on working with informal proofs.

For example, when constructing a formal argument, it can happen that parts of the argument play no role in the final conclusion. This can be automatically detected, thus providing a kind of mechanical detection of irrelevant reasoning.

It can also happen that in the inferences appearing in a proof, some can be safely omitted. One might say that this is an appraisal of *redundant information*. That is, a proof is valid, but it also remains valid if one takes away parts of some of the justifications. Moreover, a formal mathematical proof can also be more verbose than necessary for a proof checker. Consider:

```
Step1: A by Lemma1;
MiddleStep: B by Theorem1, Step1;
Step2: C by Lemma2, Theorem2, MiddleStep;
```

It can happen that we can eliminate the middle step and combine the justifications:

```
Step1: A by Lemma1;
Step2: C by Theorem1, Lemma2, Theorem2, Step1;
```

The first step (Step1) is unchanged. We moved justification of the middle step (MiddleStep) to the second step (Step2). The proof is now compressed by one step.

Interestingly, even when such compressions are possible, we may not always wish to carry them out, because intermediate steps might be important for our understanding of the proof; deleting such intermediate steps may decrease the comprehensibility of the proof.

9.6   Conclusion

We intended to explain how the formal view of mathematical proof, far from being a stale, old subject, gets new life breathed into it thanks to technological progress coming from automated reasoning. The problem of what inferences count as acceptable without any further justification—which we have called "obvious inferences"—might be thought to be inherently informal, but appears quite naturally in the setting of automated reasoning. Those interested in mathematical practice and argumentation can, we hope, see that a formal approach to mathematical proof is not at odds with their analyses but complements them.

Acknowledgements Both authors were partially supported by the ESF research project Dialogical Foundations of Semantics within the ESF Eurocores programme 'LogICCC', LogICCC/0001/2007, and the project 'The Notion of Mathematical Proof', PTDC/MHC-FIL/5363/2012, both funded by the Portuguese Science Foundation FCT. Alama's research was conducted in part as a visiting fellow at the Isaac Newton Institute for the Mathematical Sciences, Cambridge, in the programme 'Semantics & Syntax'. Kahle was partially supported by the FCT project 'Hilbert's Legacy in the Philosophy of Mathematics', PTDC/FIL-FCI/109991/2009.

References

Appel, K., & Haken, W. (1977). Every planar map is four-colorable. *Illinois Journal of Mathematics, 21*, 439–567.

Arkoudas, K., & Bringsjord, S. (2007). Computers, justification, and mathematical knowledge. *Minds and Machines, 17*(2), 185–202.

Aschbacher, M. (2004). The status of the classification of the finite simple groups. *Notices of the American Mathematical Society, 51*(7), 736–740.

Avigad, J. (2006). Mathematical method and proof. *Synthese, 153*, 105–149.

Azzouni, J. (2004). The derivation-indicator view of mathematical practice. *Philosophia Mathematica, 12*, 81–106.

Bassler, O. B. (2006). The surveyability of mathematical proof: A historical perspective. *Synthese, 148*, 99–133.

Buss, S. (1998). An introduction to proof theory. In S. Buss (Ed.), *Handbook of proof theory, volume 137 of studies in logic and the foundations of mathematics* (pp. 1–78). Amsterdam: Elsevier.

Carroll, L. (1895). What the tortoise said to Achilles. *Mind, 4*(14), 278–280.

Davis, M. (1981). Obvious logical inferences. In *Proceedings of the 7th international joint conference on artificial intelligence (IJCAI)* (pp. 530–531). Los Angeles, CA: William Kaufmann.

de Bruijn, N. (1980). A survey of the project AUTOMATH. In J. R. Hindley & J. P. Seldin (Eds.), *To H. B. Curry: Essays on combinatory logic, lambda calculus and formalism*. London: Academic Press.

De Millo, R., Lipton, R. J., & Perlis, A. J. (1979). Social processes and proofs of theorems and programs. *Communications of the ACM, 22*(5), 271–280.

Detlefsen, M., & Luker, M. (1980). The four-color theorem and mathematical proof. *Journal of Philosophy, 77*(12), 803–820.

Fallis, D. (2003). Intentional gaps in mathematical proofs. *Synthese, 134*, 45–69.

Gonthier, G. (2008). Formal proof—The four color theorem. *Notices of the American Mathematical Society, 55*(11), 1382–1393.

Grabowski, A., Korniłowicz, A., & Naumowicz, A. (2010). Mizar in a nutshell. *Journal of Formalized Reasoning, 3*(2), 153–245.

Hales, T. C. (2005). A proof of the Kepler conjecture. *Annals of Mathematics, 162*(3), 1063–1185.

Hales, T. C. (2008). Formal proof. *Notices of the American Mathematical Society, 55*(11), 1370–1380.

Hardy, G. H., & Wright, E. M. (1960). *An introduction to the theory of numbers* (4th ed.). Oxford: Oxford University Press.

Harrison, J. (2008). Formal proof—Theory and practice. *Notices of the American Mathematical Society, 55*(11), 1395–1406.

Harrison, J. (2009). *Handbook of practical logic and automated reasoning*. Cambridge: Cambridge University Press.

Heintz, B. (2003). When is a proof a proof? *Social Studies of Science, 33*(6), 929–943.

Lakatos, I. (1976). *Proofs and refutations: The logic of mathematical discovery* (edited by J. Worrall & E. Zahar). Cambridge: Cambridge University Press.

Löwe, B., Müller, T., & Müller-Hill, E. (2010). Mathematical knowledge as a case study in empirical philosophy of mathematics. In B. van Kerkhove, J. de Vuyst, & J. P. van Bendegem, (Eds.), *Philosophical perspectives on mathematical practice*. London: College Publications.

MacKenzie, D. (1999). Slaying the Kraken: The sociohistory of a mathematical proof. *Social Studies of Science, 29*(1), 7–60.

MacKenzie, D. (2004). *Mechanizing proof: Computing, risk, and trust*. Cambridge, MA: MIT Press.

Mancosu, P. (2000). On mathematical explanation. In E. Grosholz & H. Berger (Eds.), *Growth of mathematical knowledge* (pp. 103–119). Dordrecht: Kluwer.

Matuszewski, R., & Rudnicki, P. (2005). MIZAR: The first 30 years. *Mechanized Mathematics and Its Applications, 4,* 3–24.

McCune, W. (1997). Solution of the Robbins problem. *Journal of Automated Reasoning, 19*(3), 263–276.

Netz, R. (2003). *The shaping of deduction in Greek mathematics: A study in cognitive history.* Cambridge: Cambridge University Press.

Orevkov, V. P. (1993). *Complexity of proofs and their transformations in axiomatic theories, volume 128 of translations of mathematical monographs* (A. Bochman, Trans. from the original Russian manuscript, translation edited by D. Louvish). Providence, RI: American Mathematical Society.

Portoraro, F. (2008). Automated reasoning. In E. N. Zalta (Ed.), *Stanford encyclopedia of philosophy.* Fall 2008 Edition. http://plato.stanford.edu/archives/fall2008/entries/reasoning-automated/.

Prawitz, D. (2008). Proofs verifying programs and programs producing proofs: A conceptual analysis. In R. Lupacchini & G. Corsi (Ed.), *Deduction, computation, experiment: Exploring the effectiveness of proof* (pp. 81–94). Milan: Springer.

Rav, Y. (2007). A critique of a formalist-mechanist version of the justification of arguments in mathematicians' proof practices. *Philosophia Mathematica, 15*(3), 291–320.

Rehmeyer, J. (2008). How to (really) trust a mathematical proof. *Science News.* Accessed May 2013. http://www.sciencenews.org/view/generic/id/38623/title/How_to_(really)_trust_a_mathematical_proof.

Robertson, N., Sanders, D. P., Seymour, P. D., & Thomas, R. (1997). The four colour theorem. *Journal of Combinatorial Theory. Series B, 70,* 2–44.

Rudnicki, P. (1987). Obvious inferences. *Journal of Automated Reasoning, 3*(4), 383–393.

Scott, D. (2006). Foreword. In F. Wiedijk (Ed.), *The seventeen provers of the world, volume 3600 of lecture notes in computer science* (pp. vii–xii). Berlin: Springer.

Teller, P. (1980). Computer proof. *Journal of Philosophy, 77*(12), 797–803.

Thomas, R. (2007). The four color theorem. Accessed May 2013. http://www.math.gatech.edu/~thomas/FC/fourcolor.html.

Toulmin, S. E. (2003). *The uses of argument* (updated ed.). Cambridge: Cambridge University Press.

Tymoczko, T. (1979). The four-color problem and its philosophical significance. *Journal of Philosophy, 76*(2), 57–83.

van Bendegem, J. P. (1988). Non-formal properties of real mathematical proofs. In J. Leplin, A. Fine, & M. Forbes (Eds.), *PSA: Proceedings of the biennial meeting of the philosophy of science association* (Vol. 1, pp. 249–254). East Lansing, MI: Philosophy of Science Association (Contributed papers).

Verchinine, K., Lyaletski, A. V., & Paskevich, A. (2007). System for automated deduction (SAD): A tool for proof verification. In F. Pfenning (Ed.), *CADE, volume 4603 of lecture notes in computer science* (pp. 398–403). Berlin: Springer.

Wang, H. (1960). Toward mechanical mathematics. *IBM Journal of Research and Development, 4*(1), 2–22.

Wiedijk, F. (Ed.). (2006). *The seventeen provers of the world, volume 3600 of lecture notes in computer science.* Berlin: Springer.

Wiedijk, F. (2008). Formal proof—Getting started. *Notices of the American Mathematical Society, 55*(11), 1408–1414.

Part III
Mathematics as a Testbed
for Argumentation Theory

Chapter 10
Dividing by Zero—and Other Mathematical Fallacies

Lawrence H. Powers

In this paper I shall discuss a fallacy involving dividing by zero. And then I shall more briefly discuss fallacies involving misdrawn diagrams and a fallacy involving mathematical induction.

I discuss these particular fallacies because each of them seems at first—and seemed to me myself at one time—to be a counterexample to a theory of mine. The One Fallacy Theory says that every real fallacy is a fallacy of equivocation, of playing on some sort of ambiguity. But these particular fallacies do not seem to involve ambiguity, and yet they do seem to be real fallacies.[1]

Let me begin with the dividing-by-zero fallacy.

It goes as follows:

1. Let $a = b$
2. So $a^2 = ab$ (multiply each side by a)
3. So $a^2 - b^2 = ab - b^2$ (subtract b^2 from each side)
4. So $(a+b)(a-b) = b(a-b)$ (factoring)
5. So $a + b = b$ (cancelling $(a-b)$ on each side)
6. So $2b = b$ (since $a = b$)
7. So $2 = 1$ (cancelling b on each side).

Now this argument appears to be a counterexample to my theory. Each step is stated in unambiguous algebraic terminology. The invalid move takes us from an unambiguously true equation at line 4 to an unambiguously false equation at line 5 by a move of cancelling $a - b$ which is unambiguously though not obviously a division by zero. There seems to be no ambiguity.

---

[1] My thanks to R. De Souza, who chided me about holding a theory to which I seemed to know counterexamples. His comments led me to explore these examples more thoroughly.

L.H. Powers (✉)
Department of Philosophy, Wayne State University, 5057 Woodward Avenue, Detroit, MI 48202, USA
e-mail: ab3406@wayne.edu

A. Aberdein and I.J. Dove (eds.), *The Argument of Mathematics*, Logic, Epistemology, and the Unity of Science 30, DOI 10.1007/978-94-007-6534-4_10,
© Springer Science+Business Media Dordrecht 2013

My theory then seems to imply that there is no real *fallacy*; we do not have an invalid step which appears, by virtue of a covering ambiguity, to be valid, but rather a naked mistake with no appearance of goodness. A naked mistake is not a true *fallacy*.

But surely, the argument is a real fallacy. For it passes the phenomenological test. The first time I myself saw this argument in a book, I went through it carefully looking for the wrong step. And *I could not find it*, at least not just by going through the argument step by step. It looked like a proof *to me*, and at a time when I *knew* there had to be *something* wrong and was, in an intellectually serious way, looking for the mistake!

So clearly the argument is a real fallacy. It therefore seems a counterexample to my theory.

Now in trying to defend my theory, I think as follows. If a serious person is taken in by an invalid argument $A/ \therefore B$ and 'A' and 'B' are not ambiguous, perhaps there is some other reasoning in the person's mind. Perhaps he thinks that A implies C and C implies B, and it is the interpolated term C which is ambiguous. Another person who accepts $A/ \therefore B$ may accept it for a different reason, using a different confusion, say $A/ \therefore D/ \therefore B$.

I therefore ask: Why did I *myself* think the argument dividing by zero was valid step by step?

It is often said that people divide by zero, as in our example, because you can usually divide and people just forget about the special case of zero. I have never liked this kind of explanation. How can one *just forget* about special cases? If the rule is that you can always divide unless the would-be divisor is zero, how can one apply this rule without determining whether the would-be divisor is zero?

At any event, the explanation about forgetting the special case did not apply to me. I didn't *forget* the special case. I had never heard of any such special case. I learned from studying this very fallacy that one can't divide by zero. I was astounded to find that one couldn't always divide! I thought that you could *always* divide and that I knew you could always divide.

Now here too there is a popular explanation about why people think they can always divide. The explanation is that people overgeneralize: since you can almost always divide, we overgeneralize and think we can divide in the case of zero also. I do not like this explanation. Such inductive reasoning could easily lead a rational person to think that one can always divide, probably. However, mathematical knowledge is not about what is probably true but about what is *proven*. I thought I *knew* that one can always divide, that I had seen a *proof* of this.

Now after examining the argument and not finding the mistaken step, I substituted the concrete number 5 for a and b. The equations then became: Let $5 = 5$. So $25 = 25$. So $25 - 25 = 25 - 25$. Upon factoring, $10 \cdot 0 = 5 \cdot 0$. Cancelling, $10 = 5$. So $10 = 5$. So $2 = 1$. And here it is obvious where the mistake is. The equation $10 \cdot 0 = 5 \cdot 0$ balances, but $10 = 5$ doesn't.

And reflecting on this wrong move, we see that its general form is $x \cdot 0 = y \cdot 0/ \therefore x = y$. So if we can divide by zero, then all numbers are equal. This proves that we cannot divide by zero. Of course, when I saw this, I distrusted my reasoning and went and looked in a math book to assure myself that it was really true that you can't divide by zero.

Having thus decided that you can't divide by zero, I started to consider my reasons for thinking you can divide by zero.

How can it be that we can't divide by zero? After all, I first thought, multiplication is always well-defined. But division is defined as the inverse of multiplication. Doesn't it follow that division is always well-defined as well?

I knew immediately that there was something wrong with this reasoning. In the natural numbers, it is always possible to add but one cannot always subtract, say, 7 from 3. Yet subtraction is defined as the inverse of addition. How then can it be that one can't always subtract?

To understand this fallacy more clearly, let me state my argument in more sophisticated terminology. In modern logic, definite descriptions are well-formed expressions whether they refer or not. Thus, in a Russellian sense, 'the king of France' is a well-defined expression. And so, for any x, is '$(\imath z)(x = z \cdot 0)$'. But the latter is the definition of '$x/0$', which is thus well-defined, in a Russellian sense. For *Frege*, however, a referring expression is not well-defined unless it is proven that it actually succeeds in *referring* to something. Mathematicians speak of functions as being 'well-defined' in *Frege's*, not Russell's sense. If $x/0$ were well-defined in Frege's sense, then division by zero would be possible. So my argument involved an equivocation, on two different meanings of 'well-defined'.

When, years ago, I fell into the dividing-by-zero fallacy, I found that one can't divide by zero, and asked myself 'how can that be?' I then went through the 'well-defined' problem as just rehearsed. However, when I saw that there were two different concepts of 'well-defined' involved, I did not feel that this point really addressed my perplexity, for I thought I had somewhere seen a proof that division always *was* well-defined, even in Frege's sense. Hadn't I seen a proof that you *can* always divide?

Before looking at the proof I had in mind at that time, it is convenient here to consider another possible supposed proof.

In a book, *Lapses in Mathematical Reasoning*, the authors, Russian mathematicians, mention fallacies in which a true mathematical law is applied but in the wrong field of numbers (Bradis et al., 1959, 14). It is interesting that fallacies involving dividing by zero can be thought of as a subclass of those applying a true law in the wrong field of numbers, and these in turn are a subclass of fallacies of ambiguity.

When we learn about numbers in our school years, we learn to use the word 'number' ambiguously. At first the teacher says that *numbers* are those things you count with: 1, 2, 3, 4, etc. So we learn to use 'number' to mean a *natural* or *whole* number, a positive integer. In this sense of 'number,' we learn that we can always add and always multiply, but we cannot always subtract or always divide. For instance, we cannot subtract 7 from 3 or divide 3 into 7. But then later the teacher told us that, after all, we could *always* divide as well as always add or multiply, though we still could not always subtract. We could now always divide because, the teacher said, 'there are more numbers than you yet know about.'

Even as a youngster, I was rather hyper about ambiguity, and I said—though to myself, not out loud—'Come on, teacher, there aren't more *numbers* than we know about. The truth is: you're going to change the meaning of the word 'number'. And so it happened. *Now* 'number' meant *positive rational*, the fractions were numbers, and we could always add, multiply, and *divide*. With 'number' in this meaning, any number whatsoever could be divided by any number whatsoever, without any exception whatsoever.

Later the term 'number' will be extended again, from the positive rationals to the rationals generally. Now subtraction will always be possible, as well as addition and multiplication but division by zero will not be possible.

And so one fallacious way of dividing by zero would be to apply the true law that division is always possible—true in the positive rationals, but to apply this law wrongly to rationals generally. This way of dividing by zero would involve equivocation on the term 'number' and so would be in accord with the One Fallacy Theory.

Still, when I myself divided by zero, I did not do it in this way, I believe. I knew that 'number' was ambiguous. I knew that when you extend the number system, as from positive rationals to rationals generally, in order to make a new operation, as subtraction, always possible, you have to *recheck* the previously always possible operations—addition, multiplication, and division—to make sure they are *still* always possible. But 1 thought I had seen in my readings just such a rechecking, a proof that these operations were always possible in the rationals generally.

So I recalled the argument in question. Take addition. Addition was always possible in the positive rationals and subtraction is now always possible. So let a and b be positive. Then $a + b$ always exists. But $a + (-b)$ is $a - b$ and also always exists. And $(-a) + (+b)$ is $b - a$ and always exists. And, finally, $(-a) + (-b) = -(a+b)$, a negative, and always exists. So addition is always possible, it seems.

But the exact same argument can be given for multiplication and division. So they are all always possible.

Of course, the mistake in this argument becomes clear when we look at the version concerning division. But it is already there in the argument for addition. By considering a and b and $-a$ and $-b$, I consider the positive numbers and the negative numbers but I forget to consider *zero* What about zero!?

But this seems rather embarrassing, I said at the outset that I didn't like the explanation that people divide by zero because they simply forget the special case of zero. Yet here I seem to have done precisely that! I just forgot about zero. How could I just forget about zero??

If there are *three* kinds of numbers, the positive, the negative, and zero, then in order to prove something about *all* numbers, you have to prove it about all three kinds, and not just about two. If there are three people, Arthur, Barbara, and Carl, in a room and I argue that *all* the people in the room are tall because Arthur is tall and Barbara is tall and I just forget about Carl, who is short, then that argument is not a *fallacy*; it is just a stupidity, Surely I couldn't have just *forgotten* about zero!

Actually, I don't think I just forgot about zero in the above reasoning, rather I vaguely thought I had covered zero twice over, though in fact my reasoning was not valid for zero. For I tend to use the terms 'positive' and 'negative' both strictly, excluding zero, and loosely, including it. So by proving something for *all* positives and negatives, I vaguely felt I had proven it for zero.

First zero seems positive in some ways. It is a square equal to 0^2. It is its own absolute value. It is the end point of the positive half of the real line. By the familiar end point ambiguity an end point seems both to be and not to be a point of the line segment whose end point it is. Also the positive and negative segments are two halves of the real line, and two halves seem to complete the whole. And if zero seems to be positive, then -0, which is also 0, seems also to be negative.

Given that 0 seems in some ways to be positive and negative, the basic reason I tend to use these two terms ambiguously is because it is convenient. We wish to prove results about an infinite class of things, the numbers. We cannot prove results about numbers one by one, so we divide them into large classes, the positives and the negatives. If it happens that there are cases, such as zero, which do not exactly fit into these large classes, we tend to include or exclude the special cases into the large classes. For the purpose of one proof, we think of zero as positive, for another, as negative, for another as both or neither.

We have a tendency to stretch and contract the more general class terms to include and exclude the special cases, as convenience dictates. This, I think, is why the argument that we could always divide in the rationals generally sounded correct to me. As I said, when I proved the result for *all* positives and *all* negatives, I vaguely thought I had covered zero twice over. This general sort of fallacy, shuffling the special case in and out of the general classes, I shall call the 'special case fallacy.'

It turns out that a variant of this fallacy is used in the remaining two fallacies I wish to discuss.

Misdrawn diagram fallacies in geometry seem at first to be counterexamples to my theory. The problem is in the misdrawn diagram, not in any ambiguity in the language used in discussing the diagram. Yet I clearly remember being shown an argument involving a misdrawn diagram and being unable to see the error in it.

However I shall argue that the diagram itself is a representation and therefore can be ambiguous. In other words, the diagrams are not really *misdrawn* so much as *misinterpreted*.

In looking over various examples of this sort of fallacy in the *Lapses* book, I did not find one simple enough to present here in detail. However the ones I looked at generally had a common form. In the givens we are told that there is a point with property P. Call this point A. We are told also that there is a point with property Q. Call this point B. We represent this by drawing two representing points, labelled 'A' and 'B'. In the reasoning which follows, we are asked to consider the line from point A to point B. We show that this line has property X. Then we show it has property *not-X*. We seem to have proven a contradiction.

The solution is that point A and point B are the *same* point. So there is no line from A to B (Bradis et al., 1959, 22).

Fig. 10.1

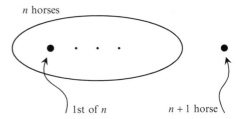

The fallacy can be thought of as an example of the special case fallacy. When we originally draw the representing points '*A*' and '*B*', these are floating points which may or may not coalesce. They represent that there is an *A* and a *B*, which may or may not be identical. Given *A*, then *B* may be the same as *A*, the special case, or anywhere else, the more general subcase. So the representation assimilates the special case to the general case; the two points, so to speak, may be one. But then, when we agree to draw a line from *A* to *B*, we misinterpret the representation as representing that *A* and *B* are different, two strictly, the more general subcase excluding the special case.

Therefore it is a fallacy of ambiguity, after all: the ambiguity of the representing diagram.

A very similar analysis can be given for the last fallacy I want to look at. Here we set out to prove that all horses are the same color.[2] We 'prove' this by 'proving' by mathematical induction that, for any *n*, any *n*-membered set of horses has the *same-color property*, namely the property that all its members are the same color. The 'theorem' is obvious for *n* = 1, for any set of only one horse has all its members the same color. So we need to prove the inductive step: if every *n* membered set has the same-color property, so does any *n* + 1 membered set. We illustrate the argument for *n* = 5, *n* + 1 = 6, but this case is to stand in for general *n* and *n* + 1. We have a set of 5 horses and a sixth horse. All the 5 horses are the same color. Remove the first of the 5 and consider all the remaining horses. These again are 5 horses and all have the same color. Therefore all 6 horses are the same color. QED. So all horses are the same color.

Now the mistake in this 'proof' is that the argument for the inductive step works for any *n* and *n* + 1 with *n* more than 1, but does not work when *n* is 1 and *n* + 1 is 2. We do not notice this because, I think, we abstract from the 5 and 6 case a mental picture which plays the role of a misleading diagram.

This picture looks like Fig. 10.1, then Fig. 10.2.

Here the first big dot is the first horse. The second is the *n* + 1 horse. The three dots represent whatever is left of the *n* horses, the first excluded.

The ambiguity in this representation is in the meaning of the three dots. It originally represents all but the first of the initial *n* members, *if* there *are* any but the

---

[2]I believe my colleague Bob Titiev first suggested this fallacy as a possible counterexample to my theory.

Fig. 10.2

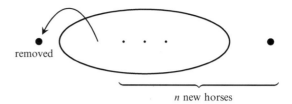

n new horses

first. It is then misinterpreted as meaning that there *are* such remaining members. Initially the special case of there being *no* remaining members is included, but then it is excluded. So here again we have a special case fallacy, and we also have a misdrawn—or really misinterpreted—diagram fallacy, although now the diagram is not actually drawn, but is a mental picture.

In this paper, I have considered three mathematical fallacies which at one time I thought were counterexamples to my One Fallacy Theory. In each case, I have argued that these fallacies can be analyzed as fallacies of ambiguity after all.

Acknowledgements   Originally published in F.H. van Eemeren, Rob Grootendorst, J.A. Blair, and Charles A. Willard, editors, *Proceedings of the Fourth Conference of the International Society for the Study of Argumentation*, pp. 655–657. Sic Sat, Amsterdam, 1999.

Reference

Bradis, V. M., Minkovskii, V. L., & Kharcheva, A. K. (1999 [1959]). *Lapses in mathematical reasoning*. Mineola, NY: Dover.

Chapter 11
Strategic Maneuvering in Mathematical Proofs

Erik C.W. Krabbe

11.1 Introduction

The purport of this chapter is to show that concepts from argumentation theory can be fruitfully applied to contexts of mathematical proof. As a source for concepts to be tested I turn to pragma-dialectics: both to the Standard Theory and to the Extended Theory.

The Standard Theory has been around for a long time and achieved its final formulation in 2004 (Van Eemeren and Grootendorst, 1984, 1992, 2004). According to the Standard Theory argumentation is a communication process aiming at the resolution of a difference of opinion (which could consist of an opinion held by one party, but not by the other, though the other does not adhere to the opposite point of view either). Argumentative communication takes place in explicit dialogical interaction or in monological texts or speeches, which are however regarded as containing an implicit dialogue between author and addressee. Thus the primary model for argumentation, used both for analysis and evaluation, is a dialogical model: the model of critical discussion. Ideally, a discussion should contain four stages (which may be repeated in subdiscussions). In the *confrontation stage* the difference of opinion is defined and clarified. In the *opening stage* it is determined what common ground is available to start the *argumentation stage*, which consists of an exchange of arguments and criticisms. In the *concluding stage* the result of the discussion is determined. The procedure is further regulated by a number of rather technical rules (Van Eemeren and Grootendorst, 1984, 2004) the effect of which can be summarized in a practical code of conduct consisting of ten commandments (Van Eemeren and Grootendorst, 1992, 2004). Violations of these rules are fallacies; they threaten the resolution process. Among the fallacies thus defined, one finds

E.C.W. Krabbe (✉)
Faculteit Wijsbegeerte, University of Groningen, Oude Boteringestraat 52, 9712 GL Groningen,
The Netherlands
e-mail: e.c.w.krabbe@rug.nl

A. Aberdein and I.J. Dove (eds.), *The Argument of Mathematics*, Logic, Epistemology, and the Unity of Science 30, DOI 10.1007/978-94-007-6534-4_11,
© Springer Science+Business Media Dordrecht 2013

not only logical fallacies (infringements of the *Validity Rule*: "Reasoning that in an argumentation is presented as formally conclusive may not be invalid in a logical sense," 2004, 193), but also many other disruptions of the resolution process, such as failing to take on one's burden of proof, or arguing for something different than the point of view one is supposed to argue for (*ignoratio elenchi*), or substituting one's authority for argument (*argumentum ad verecundiam*). In order to apply the Standard Theory to mathematical proofs, it must first be established that, at least in some contexts, proofs can be regarded as arguments; a matter that will be briefly discussed in Sect. 11.2.

The Extended Theory (Van Eemeren and Houtlosser, 1999a,b, 2002, 2004, 2005) starts from the observation that, usually, people argue not only with the objective of resolving a difference of opinion, but also with that of resolving it in their own favor: they pursue the rhetorical[1] objective of having their point of view accepted by the other (1999a, 164) . When arguers attempt to pursue both objectives at the same time, this leads to strategic maneuvering. To get a better grasp of the analysis of argumentative discourse, the Extended Theory extends the Standard Theory by rhetorical considerations, all the same maintaining the norms of the Standard Theory where evaluation is concerned. When strategic maneuvering oversteps these bounds, it is said to have been derailed, and a fallacy has been committed (2002, 141–143). Strategic maneuvering that remains within the bounds of the Standard Theory can still be suboptimal with respect to the second objective. In that case, no fallacies are committed, but one is confronted by flaws or blunders (2002, 142; Walton and Krabbe, 1995, 25).

The point of view of the Extended Theory can be further extended by introducing yet other objectives that an arguer may simultaneously pursue. For mathematical proofs (presuming them to be arguments) one may think of epistemic and didactic objectives that figure in various contexts of proof and might interfere with the two objectives mentioned above. This will be explored in Sect. 11.3.

After having thus obtained a rough notion of what various arguers try to achieve when they set out to prove a theorem, one will find in Sect. 11.4 a number of examples of strategies, derailments, and blunders, grouped according to the four stages mentioned above. Derailments will be shown to occur both within and beyond the bounds of what may be called a "mathematical argument." Sect. 11.5 contains some conclusions and adds some remarks on the concept of proof.

11.2   Proofs and Arguments

Is there any strategic maneuvering in mathematical proofs? If, on the one hand, we use the term 'strategic maneuvering' in the way it is used in the Extended Theory and, on the other hand, stick to the notion of mathematical proof as a

---

[1]Elsewhere I argued that this objective can be seen as a (secondary) *dialectical* objective (Krabbe, 2004).

conclusive establishment of a theorem, flawless and unassailable for all ages—
what I shall call 'the austere concept of proof'—then the two seem miles apart.
The habitat of strategic maneuvering is a context of controversy and critical testing
where one party tries to steer the resolution process so as to serve his personal
aims. According to the austere notion of proof, there are no personal aims, and
establishing the truth is the only goal in sight. Johnson (2000, 232) claims that
mathematical proofs are no arguments. If that is so, there would be in them no
place for strategic maneuvering in the sense of the Extended Theory either. But of
course, all depends on how one defines these notions of argument and proof. Dove
(2007) points out that the characteristics of proof assumed by Johnson do not concur
with mathematical practice as shown in Imre Lakatos's famous example about the
polyhedra (Lakatos, 1976). Lakatos's conception of proof is argumentative and
admits, even requires, a dialectical tier. So there are conceptions of proof, connected
with mathematical practice, that place proofs within the scope of argumentation
theory and therefore, potentially, within the scope of strategic maneuvering. In an
earlier paper (1991; 1997, reprinted in this volume) I also maintained that ordinary
informal mathematical proofs are arguments. It should be noted that mathematical
proofs, even the axiomatic ones, are normally informal. Formal proofs (written in
a formal language and constructed according to formalized derivation rules, as in
formal natural deduction) are a logician's gadget.

Let it be granted that mathematical proofs are (normally[2]) arguments. But
mathematical proofs, being mathematical arguments, also display some special
features in that they are situated in a specific semi-institutionalized scientific context
in which arguments are supposedly subject to requirements that are additional
to those applying to arguments in general. Not every argument, not even every
argument about mathematical propositions, is a mathematical argument, let alone a
mathematical proof. One may cite Fermat's Last Theorem and invoke the authority
of mathematicians to argue that it is true. This may be a good argument, but it is not
a mathematical argument, and not a mathematical proof. It is not a mathematical
argument because it is not using the mathematical way of reasoning. Aristotle would
say that it is not according to the art (*kata tên technên*). In each discipline there
are certain conventions to which an argument must conform to be an argument
within the discipline. For instance, in medicine you cannot reason as in philosophy.
Aristotle's example is well-known: "if someone were to deny that it is better to
take a walk after dinner because of Zeno's argument,[3] it would not be a medical
argument" (*SE* 11, 172a8–9; Aristotle, 1965, 65). But even those arguments that are
really mathematical can fail to count as proofs. They can fail for two reasons. First,
they may contain a fallacy, and therefore be no more than a false (or invalid) proof,
and not a real proof.[4] According to Aristotle "the man who draws a false figure"

---

[2]Proofs that are not arguments include: immediate proofs (where there is no reasoning), formal
proofs, and proofs in a context where there is no difference of opinion (see Sect. 11.3).

[3]A philosophical argument that motion is impossible.

[4]On this point the grammar of the term "proof" differs from that of the term "argument," for an
invalid argument is still an argument.

(*pseudographôn*) does not prove anything, and "his process of reasoning is based on assumptions which are peculiar to the science but not true" (*Top.* I.1, 101a10, 13–15; Aristotle, 1979, 275). Thus his argument may be counted as a mathematical argument, but not as a proof. The second way in which a mathematical argument can fail to be a proof is that it fails to comply with some further additional requirements. But people differ on what these additional requirements are. Aristotle tells us that by proof or demonstration (*apodeixis*) he means deductive reasoning ('syllogism') which gives us scientific knowledge. Such demonstrative knowledge, however, can only be obtained if one proceeds by deductive reasoning "from premises which are true, primary, immediate, better known than, prior to, and causative of the conclusion" (*Anal. post.* I.2, 71b17–22; Aristotle, 1979, 31). Thus Aristotle places proof within a foundationalist philosophy of science. There are other options, some of which make the notion of proof a relative notion: a proof is a proof for a given person and perhaps not for someone else. According to John Corcoran there are "two basic issues: are the premises known to be true by the given person? And does the chain of reasoning deduce the conclusion from the premise-set for the given person?" (Corcoran, 1989, 25). In my proposal (1991, 1997, reprinted in this volume), in order that an argument be a proof for a given person, the starting points of the (underlying) critical discussion must be assertions and not merely concessions of that person, and all the possible dialectic options for that person must be followed through, not merely those that happen to occur to that person; of course there must be no fallacies, and the proponent of the theorem proved (the protagonist) must be the winner of the discussion.

For the purposes of this chapter, it will not be necessary to make a choice between the different sets of additional requirement that are meant to distinguish mathematical proofs from mere mathematical arguments. But any set of additional requirements making the notion of proof relative to the addressee seems to bring proofs even more clearly within the scope of argumentation theory, and in particular within the scope of the theory of strategic maneuvering (the Extended Theory).

11.3   Contexts of Proof

Supposing that mathematical proofs are normally arguments, we may (1) regard them as (implicit) critical discussions, and (2) investigate whether certain modes of strategic maneuvering are characteristic for such proofs and (3) how these modes may lead to blunders and derailments (fallacies). In this, one should keep in mind that mathematical proofs come in kinds as to their objectives and the audiences they are aimed at. They also occur in different types of dialogue (Walton and Krabbe, 1995), here seen as activity types (Van Eemeren and Houtlosser, 2005), and display different functions of reasoning (Krabbe and Van Laar, 2007). As a first list of contexts of proof, I propose the following:

1. thinking up a proof to convince oneself of the truth of some theorem;

2. thinking up a proof in dialogue with other people (inquiry dialogue; probative functions of reasoning);
3. presenting a proof to one's fellow discussants in an inquiry dialogue (persuasion dialogue embedded in inquiry dialogue; persuasive and probative functions of reasoning);
4. presenting a proof to other mathematicians, e.g. by publishing it in a journal (persuasion dialogue; persuasive and probative functions of reasoning)
5. presenting a proof when teaching (information-seeking and persuasion dialogue; explanatory, persuasive, and probative functions of reasoning).

These contexts of proof are all very different and a proof that succeeds in one context may well fail in some others. Persuasive functions aim at convincing the other. These functions are argumentative in the sense that, given some difference or conflict, they serve to overcome doubt of an interlocutor (Krabbe and Van Laar, 2007, 31).[5] So, whenever in a proof the reasoning displays persuasive functions, the proof can be regarded as an argument. Probative functions aim at extending the knowledge of some company or community; they are not argumentative. Explanatory functions, which are not argumentative either, aim at enhancing understanding. In all contexts listed, except the first, the reasoning in proofs has a probative function: extending available knowledge. In the first context it has not this function, but only that of extending one's personal knowledge. In all contexts, except the first two, it also has persuasive functions, mostly that of convincing others of the truth of some theorem. In the first context, however, there is no other, whereas in the second context the others participate in making the proof. Thus in these two contexts the reasoning is not argumentative, and one should say that the proofs (as speech or thought events) are not arguments, though the proofs as products could of course be used argumentatively in another context, and then be arguments. That proofs in teaching are presented in a context of information-seeking dialogue and have explanatory functions will not be surprising. But it should be stressed that in teaching there is also persuasion dialogue with persuasive and even probative functions in play.

The last point, that mathematical instruction involves persuasion dialogue, and is therefore argumentative, can be illustrated by the famous passage in Plato's *Meno* (82b–85b; Plato, 1997) where Socrates teaches a slave how, for a given square, to find the length of the side of a square with twice the area of the given square. Socrates proposes to do so without passing on any information to the slave (without information-seeking dialogue), instead he will just ask questions. Socrates's ultimate, philosophical intention is to demonstrate that so-called learning is a kind of recollection of what one knew in a life before this one. The given square (*ABCD*) has a side (*AB*) with a length of 2 ft (Fig. 11.1).

The slave is at first quite sure that in order to double the area one needs to double the length of the side (extend *AB* to get *AE*). After some leading questions by Socrates, he realizes that 4 ft is too much. Since 2 ft is too little, he opts for 3 ft

---

[5]Other argumentative functions of reasoning mentioned by Krabbe and Van Laar are the polemic functions (in eristic dialogue) and the directive functions (in negotiation).

Fig. 11.1 Doubling the
square

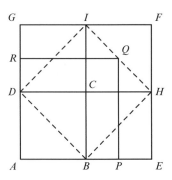

(*AP*). More questioning makes him realize that a square with a 3 ft side would have
an area of 9 ft^2 (*APQR*), one too many. Then Socrates induces the slave to draw
a figure of a square of 16 ft^2 (*AEFG*) divided in four parts of 4 ft^2 each (*ABCD*,
BEHC, *CHFI*, *DCIG*) and asks whether the diagonals of the smaller squares (*DB*,
etc.) do not divide these squares in two equal parts. Some more questioning elicits
the answer: the side of the square with an area of 8 ft^2 must have the length of
this diagonal (square *DBHI*). Though I do not think this demonstration is very
successful in demonstrating a life before this life (the questions actually pass on
lots of information), it does show how in teaching mathematics presenting a proof is
something more than passing on the information. The student is to think through the
proof for him- or herself and to become convinced of its cogency as well as of the
truth of the conclusion. Each step in the proof needs to be checked by the student,
something Socrates achieves by his questioning.

So proof in a didactic context has not just explanatory functions, but also
persuasive ones. The persuasive strategy used is that of letting the student give
wrong answers first and letting him (with some help) discover that these answers are
wrong, thus strengthening his commitment to the right answer in the future. Notice
that going through the rejection of wrong answers first is in itself not necessary for
a valid proof, but it is permissible, and here contributes to the specific aims of the
teacher. So we may say it is a case of nonfallacious strategic maneuvering which
combines the objective of valid argument with that of making a student master a
particular piece of mathematics.

11.4 Proofs in Stages

Assuming again that mathematical proofs are arguments, and that pragma-
dialectical theory enriched by the theory of strategic maneuvering applies to all
kinds of arguments, we should be able to identify in proofs the four stages of critical
discussion, and to investigate for each stage what kind of strategic maneuvering
would be typical for that stage of the proof and to what blunders and derailments
these kinds of maneuvering are typically prone.

11.4.1 The Confrontation Stage

In the confrontation stage, authors of proofs must formulate what they set out to prove. According to the rules of critical discussion they must do so in a clear and unambiguous way. In order to attract attention and to convince as many of their colleagues (or students) of the truth of the theorem it is also important for them to formulate the theorems in simple and attractive terms. To achieve this a well-known strategic maneuver is to state the theorem in terms that appear to be simple and familiar but actually refer to concepts that are quite complex. The complexities behind these terms are then covered by definitions that accompany the theorem. For instance, the completeness theorem is simply formulated as: Every valid formula is provable. A formulation that calls for extensive explanations of what is meant by 'formula,' 'valid,' and 'provable.'

This strategy is certainly commendable. It would be a strategic blunder (a forgoing of a legitimate advantage) if one did not use it. Yet, it could be a derailment if an author framed his terms and definitions so as to let his theorems appear to be relevant beyond mathematics, suggesting applications in the real world that are not really feasible. For instance mathematical concepts of space, probability, preference, and validity may suggest corresponding concepts from daily life (which may, however, diverge among themselves), and the mathematics using these terms may yield nice theorems which disclose interesting parallels between mathematical and common conceptions of space, etc., but to presume that the mathematics applies to these concepts as commonly understood would be premature. In the hands of mathematicians such concepts may also change into something quite different. As Goethe said: *Die Mathematiker sind eine Art Franzosen: redet man zu ihnen, so übersetzen sie es in ihre Sprache, und dann ist es alsobald ganz etwas anders*[6] (Goethe, 1949, 660, nr. 1279). Whether such tactics are fallacious depends, however, on further circumstances, such as the kind of audience. It would be fallacious to hide the changes in the concepts and to try to make the audience believe there are not any.

11.4.2 The Opening Stage

In the opening stage, the author should, in a context of axiomatic proof, state or indicate her axioms, or, if the context is nonaxiomatic, the general frame that is presupposed in the proof. Is the Axiom of Choice accepted? Does the theorem pertain to the real numbers or to complex numbers? Of course it would be wise to choose a frame or axiomatic setting in which the proof can be completed, but which is also attractive to the audience. The author may blunder by indicating a frame in which the theorem later appears not to hold, or commit a fallacy by sticking

---

[6]Mathematicians are a kind of Frenchmen: If one speaks to them, they will translate it into their language and then it is in no time something totally different.

to the theorem in such a frame. In a context of teaching that uses nonaxiomatic proofs, strategic maneuvering should yield starting points that are close enough to the theorem to prove it by a reasonably short proof, but distant enough to avoid triviality. Also, the starting points should be comprehensible to the students. It would, of course, be a derailment of such maneuvering to start from false principles in order to facilitate the proof.

11.4.3   The Argumentation Stage

In the argumentation stage, the author should try to make her proof neither too sketchy nor too fine-meshed. The degree of detail in the proof must be adapted to the audience. Is the proof meant to convince an expert or a student? And if a student, is it a mathematically mature student, or not? This makes a great difference: what constitutes a proof for the one does not necessarily constitute a proof for the other. Failure to adapt the proof to the audience in this respect would constitute a blunder, and in more serious cases it could be a derailment. The fallacy would then be one of evading the burden of proof, because when the steps in the reasoning are too big or too complicated for the audience to grasp no argument has been given that constitutes a proof for that audience, and the audience's implicit critical questions remain unanswered.

The author should also attempt to construct an attractive division of work between the proofs of lemmas and the proof of the theorem itself (based on the lemmas). Elements of surprise may be introduced to make the proofs more attractive, and the greatest surprise should be left for the proof of the main theorem, if possible. This part of the rhetoric of proofs reminds one of a general tactic of argumentation, recommended by Aristotle (*Top.* VIII.1), known as *krupsis* or the hiding of one's intentions about the argument's set-up. Generally, an elegant proof will be preferred over a messy one. Missing opportunities to make the proof more attractive amounts to blundering, whereas substituting elegance for validity would constitute a derailment. A serious fallacy would be committed, were one to use a theorem to establish a lemma needed for the theorem's own proof (begging the question), whereas loops in a proof that are merely superfluous, would constitute mere blunders.

Leaving routine parts of the proof to the audience constitutes a recommendable tactic since it will include the audience in the enterprise of establishing the theorem, and thus strengthen their conviction in the end. But it would be a blunder to leave tasks to the audience that are so demanding as to upset them, rather than to make them participants in proof construction. (It is a blunder relative to the objective of convincing the audience by rational means, not relative to the objective of giving the audience some training.) If a task cannot be performed at all, to ask the audience to perform this task would be a derailment of the strategy of leaving tasks to the audience. But also, if the task can in principle be performed, but arguably not by the intended audience (at least not by means of a reasonable effort and within a reasonable period of time), the strategy gets derailed. This would once more constitute a fallacy of evading the burden of proof.[7]

The well-known strategy of proof called *reductio ad absurdum* can lead to a derailment when one is too eager to disprove an hypothesis. A vulgar type of this derailment would consist in annoying one's audience by an endless chain of derivations from the hypothesis (to be refuted) leading nowhere, making them grant the absurdity of the hypothesis rather than listen any longer. This tactic, known as *reductio ad nauseam*, would not be "according to the art." Another vulgar tactic would consist in claiming to have reached an absurdity when there is none.

However, a claim of having reached an absurdity where there is none could also occur in a proof according to the art, though it uses a false assumption about what constitutes an absurdity. A serious case of this can be found in the purported proof of Euclid's fifth postulate by (Giovanni) Girolamo[8] Saccheri (1667–1733). As many before him, Saccheri wanted to show that this postulate is superfluous by proving it from the other Euclidean axioms (Saccheri, 1733). To achieve this he chose a strategy of *reductio ad absurdum*. More specifically he purported to use that of the *consequentia mirabilis* (to prove *P*, derive *P* from the supposition that *not-P*) of which he had been a champion for years (Kneale and Kneale, 1962, 345–347, 380). Thus he added the denial of the fifth postulate to the other axioms and sought to derive the fifth postulate from that.

Now the fifth postulate is, granted the other postulates and some traditional features of geometrical reasoning, equivalent to the Axiom of Parallels: given a line *l* and a point *P* not on *l*, there is exactly one line through *P* that is parallel to *l* (i.e. in the same plane and having no point in common with *l*). The denial of the Axiom of Parallels, for a given *l* and *P* not on *l*, leaves two possible suppositions: (1) No line through *P* runs parallel to *l*. (2) More than one line does.[9] Saccheri managed to derive the fifth postulate from (an equivalent of) the first supposition (his Proposition XIII; Beth, 1929, 21), thus reducing this supposition to absurdity; however, in the part of the proof that starts from (an equivalent of) the second supposition he derived a number of consequences that are absurd merely from a Euclidean point of view, but became later known as theorems of hyperbolic geometry (discovered by Bolyai, Lobachevsky, and Gauss). When reaching the consequence that there were two different straight lines approaching one another asymptotically and thus having a common perpendicular at a common point in the infinite, Saccheri decided that this contradicted the nature of straight lines and falsely inferred the absurdity of (2), thus completing the "proof" of the fifth postulate (Proposition XXXIII; Beth, 1929, 25–26). This was a mistake, but it doesn't make Saccheri's work

---

[7]It is a common experience of math students that authors of textbooks sometimes exasperate their readers by leaving not the routine parts but the more difficult parts of a proof to them.

[8]Often: Gerolamo.

[9]Clearly, for a given *P* and *l*, there are precisely three possible suppositions: no parallel, exactly one, or more than one. It should be remarked that when one of these suppositions holds, the same can be shown to hold for all other points *Q* and lines *m* (*Q* not on *m*) of the plane.

nonmathematical. His work is still conforming, I would say, to the art (in contrast to the vulgar reductions, discussed above). It was a sad mistake, for strange enough Saccheri, having actually worked in non-Euclidean geometry, failed to discover it. According to Evert Beth (1965, 12) this was "one of the most tragic occurrences in the history of science—one can without hesitation say: in the whole history of the world; tragedy does not necessarily presuppose that blood is spilt—that Saccheri let himself be carried along by his desire to prove the axiom of parallels."

Thus derailment of strategies may occur both in purported proofs that are not according to the art and in those that are so in principle, but display some flaw or other. Another example of the first kind is constituted by so-called "proofs by lack of space on the blackboard." Here the teacher scribbles all over the blackboard, until he arrives at the bottom right corner, and then skips the last and most essential part of the proof by lack of space. Nowadays, these tactics have become obsolete by lack of blackboards. Another fallacious tactic would be not to present a proof at all, but to claim that one has one at home (stored in one's desk). Of course this may, in some contexts, be a good argument for the truth of the theorem, but it does not constitute mathematical proof, since it is not a mathematical argument.[10]

Examples of the second kind (according to the art, but displaying some flaw) include cases of flaws discovered in mathematical work, among them cases of notorious errors by famous mathematicians such as Leonhard Euler's (1707–1783) elegant but incomplete proof that Fermat's Last Theorem[11] holds for exponent $n = 3$ (1770), which contained a serious gap, even though it may be that Euler actually had a proof (Edwards, 1977, 39–40, 45); or the general "proof" of Fermat's Last Theorem that Gabriel Lamé (1795–1870) presented to the Paris Academy in 1847, in which he unwarrantedly presumed the Fundamental Theorem of Arithmetic (unique factorization into primes) to hold for certain complex numbers (Edwards, 1977, 76–77)[12]; or Andrew Wiles's presentation of an alleged proof for Fermat's Last Theorem in 1993, which contained a subtle error (corrected the following year). These examples illustrate how the tension between, on the one hand, the objective of presenting proofs conforming to the norms of critical discussion (including logical requirements) as well as to the special requirements of the art of mathematics and, on the other hand, such objectives as being successful in proving a particular theorem, presenting an elegant proof, or simply presenting it before others do, can lead to derailments of strategic maneuvering.

A curious case of derailment, related by Haskell Curry (1900–1982; 1963, 88), is that of Alfred Tarski (1901–1983) who formulated in an unfamiliar notation what he thought were E. V. Huntington's axioms of Boolean algebra (Tarski, 1956, corrected version 1983). In fact the published postulates did form an axiom set,

---

[10]Moreover, even as a nonmathematical argument it could be a fallacy (*argumentum ad verecundiam*).

[11]Fermat's Last Theorem: There are no positive integers x, y, z, n ($n > 2$) such that $x^n + y^n = z^n$.

[12]The story that Ernst Kummer (1810–1893) succumbed to a similar illusion before discovering the ideal complex numbers (Bell, 1965, 522) is probably apocryphal (Edwards, 1977, 80).

Fig. 11.2 More tangents

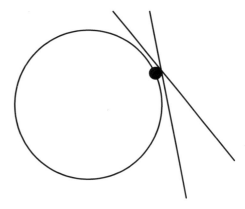

not of Boolean algebra, but of something else. This error then went unnoticed from the original Polish version into German and English translations, but was finally discovered by accident in 1956. According to Curry such a situation would have been inconceivable had the axioms been expressed in a more familiar notation. Now Tarski was very surprised that no one had made the discovery before: "It seems strange that so far I nor anybody else has noticed the mistake (though the paper has probably been read by a number of people)" (quoted by Curry, 1963, 88). The moral of the story is that one should not suppose that people read one's paper. Moreover, the paper was 127 pages long and the error consisted of just three misprinted subscripts. If this is a case of derailment, we should dub it a micro-derailment.

Among the examples presented thus far, those that reason according to the art (though making a mistake) seem to take mathematics seriously, whereas those that do not conform to the art don't. But one could be very serious in proposing a proof that is completely out of bounds. Thus I remember a student contesting the theorem that through one point P on the circumference of a circle there can be drawn only one tangent line to the circle. To show that more tangents can be drawn, he drew a very fat point on the circumference and let a line wobble on this fat point (Fig. 11.2). The student may have been serious, but his proof was not according to the art.

The same holds for some so-called false proofs (or invalid proofs) of false or nonsensical propositions, for instance the proof in *Wikipedia* that $\infty = 1/4$, can in some contexts be understood as a serious attempt to do mathematics and not just as a kind of joke. But then, by taking ∞ for a number, the alleged proof gets completely derailed and fails to be mathematical (see Appendix).

On the contrary, the proof in *Wikipedia* that $4 = 5$ and the one that any angle is zero (see Appendix) can hardly be understood as serious, but strangely enough these proofs do conform to the art (just making a mistake at some point; see Appendix). The same holds for the notorious proof by mathematical induction that all horses have the same color:

Proof. In a group consisting of only one horse all horses have the same color, there being only one horse. Suppose that in any group of *n* horses all horses have the same color. Take an arbitrary group of *n*+1 horses. Take away one horse *h*. What is left is a group of *n* horses. According to the supposition they must all have the same color *C*. It only remains to prove that horse *h* also has color *C*. Now put *h* back into the group and take away another horse. Again we have a group of *n* horses. According to the supposition they must have the same color. Since the others have color *C*, *h* must have color *C* as well. So all horses in the group of *n*+1 horses have the same color (color *C*). Since this group was arbitrarily chosen, any group of *n*+1 horses consists of horses of the same color. By mathematical induction, for each *n*, in any group of *n* horses all horses have the same color. Let the number of horses in the world be *m*, then all horses together form a group of *m* horses and must all have the same color, QED.

This proof uses mathematical induction, which in itself is conforming to the art, but in its present application contains a flaw. The puzzle only works with those acquainted with mathematical induction, but not too experienced in its application. In fact most people, being unfamiliar with this method, attack parts of the proof that are impeccable, inexperienced students may be really puzzled for some time, whereas experienced mathematicians may not be challenged. This nicely illustrates how the usability of a particular strategy depends on the audience

11.4.4   The Concluding Stage

In the concluding stage, it is advisable to clearly indicate that the proof has ended. Traditionally, this was effected by repeating the theorem, letting it be followed by 'QED', but at some point this went out of fashion. It is however crucial to indicate the end of proof, otherwise people may continue to look for a proof when the proof is actually finished, thus being likely to miss it. Nowadays, various typographical devices, such as a bold dot or a crossed square, are in use. At the same time, such a symbol stands for the proponent's claim to have resolved the difference of opinion in his favor. The proponent can derive side-benefits from this happy conclusion by adding reflections about the method of proof and the ways difficulties were dealt with, and also by stressing further results that may be derived from either the theorem (its corollaries) or the method of proof (which could be used in other cases). It would be a blunder to overdo this, and it would be a serious derailment if the proponent started bragging about his proof prematurely, without having proved what had to be proved, though perhaps having proved something else (*ignoratio elenchi*).

11.5 Conclusion

The dialectical and rhetorical view of proofs presented here contrasts with the austere concept of proof: the view of proofs as flawless and unassailable entities that are either there (success) or not (failure). According to the present view, proofs may be more or less successful, depending upon context and audience. We saw that what counts as a proof for some does not count as a proof for others. Some proofs are rather sketchy, leaving much room for questioning (or work for the reader, if the author is not there to answer questions), whereas others, by answering these questions, are developed to great dialectical depth.

But is a formalized proof not the natural limit of dialectical depth? It is true that the notion of a formalized proof is based on the idea of a proof without gaps (Gottlob Frege, 1848–1925, 1879), i.e., a proof where no questions remain, since no intermediate steps are left out. Also, formalized proofs are flawless and unassailable. But by themselves they prove nothing: for this they need to be interpreted. The proofs of mathematical logic that carry conviction are not these formal objects, but informal proofs about these objects, and normally these proofs are arguments.

Moreover, it is not true that a proof of greater dialectical depth is always the better proof. One can overdo the amount of detail, given a certain audience. This, as we saw, would constitute a blunder rather than a fallacy. But what if the details make it impossible to understand the proof?

Finally, the austere concept of proof cannot be saved by saying that the true proofs are the proofs for the experts. In contemporary mathematics the unity of the conceptual world has broken down. There is of course classical and intuitionistic mathematics, but also within classical mathematics there is a wealth of options for building an underlying set-theoretical universe (with or without the axiom of choice, with or without the continuum hypothesis, etc.). Nowadays mathematicians live in different mathematical universes (Van Dalen, 2006). So, even among experts, it is true that what constitutes a proof in someone's universe is no proof for some colleagues living in a different mathematical universe.

In constructing a proof, then, there is room for strategic maneuvering to combine one's allegiance to norms of critical discussion and to the special requirements and conventions of mathematics with other propensities, such as one's desire to complete the proof of a nice theorem and to convince the audience. Derailments correspond to dialectical fallacies, which include, but are not confined to, mathematical errors (and some types of which may be specific for attempts to present a mathematical proof), whereas blunders correspond to a suboptimal maneuvering within the bounds set by critical discussion and the art of mathematics.

Acknowledgements The chapter was first presented at the NWO-conference on "*Strategic Manoeuvring in Institutionalised Contexts*," University of Amsterdam, 26 October 2007 and has been previously published in *Argumentation* (2008) 22:453–468.

Appendix: False Proofs

Three false proofs that were mentioned in this chapter appeared in *Wikipedia* (2007). Paraphrases of these false proofs are given below.

Proof that $\infty = 1/4$

To find the distance between, say -4 and 3 (the points $(-4,0)$ and $(3,0)$ on the x-axis) one calculates $3 - (-4)$, which gives 7. Generally, for points $(-a,0)$ and $(b,0)$ the distance is equal to $b - (-a)$. The whole x-axis runs from $-\infty$ to ∞, its length is therefore equal to $\infty - (-\infty)$. The same is true for the y-axis. The area of the plane must be equal to the product of the lengths of the two axes, i.e. $(\infty - (-\infty))^2$. Since the plane is infinite, we get: $\infty = (\infty - (-\infty))^2$, which can be simplified into $\infty = (2\infty)^2$. This yields to $\infty = 4\infty^2$, from which we get $1 = 4\infty^2/\infty$, which simplifies into $1 = 4\infty$, and finally yields the value of the infinite: $\infty = 1/4$ (this can be checked by substituting $1/4$ for ∞ in the equation $\infty = (\infty - (-\infty))^2$), QED.

Proof that $4 = 5$

Nobody doubts the law of identity: $-20 = -20$. Differently expressed: $25 - 45 = 16 - 36$. Introducing factors one obtains the equivalent expression: $5^2 - 5 \times 9 = 4^2 - 4 \times 9$. Adding $81/4$ to both sides gives: $5^2 - 5 \times 9 + 81/4 = 4^2 - 4 \times 9 + 81/4$. This can also be written as: $5^2 - 2 \times 5 \times 9/2 + (9/2)^2 = 4^2 - 2 \times 4 \times 9/2 + (9/2)^2$. Now applying the algebraic formula $a^2 - 2ab + b^2 = (a - b)^2$ twice (on the left, take $a = 5$ and $b = 9/2$, on the right take $a = 4$ and $b = 9/2$), you will get: $(5 - 9/2)^2 = (4 - 9/2)^2$. Take the square root of both sides and you obtain: $5 - 9/2 = 4 - 9/2$. Adding $9/2$ on both sides gives: $5 = 4$. The law of symmetry of identities then gives $4 = 5$, QED.

Proof That Any Angle Is Zero

Let $\angle BAB'$ be any nonzero angle, where B and B' are chosen so that they satisfy $AB = AB'$ (Fig. 11.3). We shall show that, contrary to what is assumed, it must be the case that $\angle BAB' = 0$. Construct a rectangle $ABCD$ with side AB, such that the points C and D lie outside $\angle BAB'$. Draw the perpendicular bisector of CB, which intersects CB at E ($CE = EB$). Draw CB' and the perpendicular bisector of CB', which intersects CB' at F ($CF = FB'$).

Fig. 11.3 The vanishing
angle

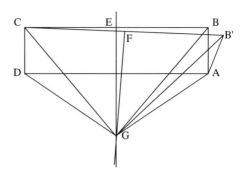

These two perpendicular bisectors cannot be parallel (check). Let them intersect at G. Now, $BE = CE$, $\angle BEG = \angle CEG$, and $EG = EG$, so triangle BEG and triangle CEG are congruent. Consequently $BG = CG$. Similarly, since EG is also the perpendicular bisector of DA, $AG = DG$. Since $AB = DC$, it follows that triangles ABG and DCG are congruent. Using triangles $B'FG$ and CFG, one proves, in a similar way, consecutively that $B'G = CG$ and that (because $AG = DG$ and $DC = AB = AB'$) triangles DCG and $AB'G$ are congruent. From these two congruencies we derive that triangles ABG and $AB'G$ are also congruent. Hence $\angle GAB = \angle GAB'$. But then $\angle BAB' = \angle GAB' - \angle GAB = 0$. Thus the initial supposition that $\angle BAB'$ be a nonzero angle leads to a contradiction and we must conclude that nonzero angles do not exist.

References

Aristotle. (1965 [1955]). *On sophistical refutations. On coming-to-be and passing-away. On the cosmos.* (E. S. Forster & D. J. Furley, Trans.). Cambridge, MA: Loeb Classical Library, Harvard University Press.

Aristotle. (1979 [1960]). *Posterior analytics. Topica.* (H. Tredennick & E. S. Forster, Trans.). Cambridge, MA: Loeb Classical Library, Harvard University Press.

Bell, E. T. (1965 [1937]). *Men of mathematics* (Vol. 2). Harmondsworth: Penguin.

Beth, E. W. (1965). *Mathematical thought.* Dordrecht: Reidel.

Beth, H. E. J. (1929). *Inleiding in de niet-Euclidische meetkunde op historischen grondslag* [Introduction to non-Euclidean geometry on a historical basis]. Groningen: Noordhoff.

Corcoran, J. (1989). Argumentations and logic. *Argumentation, 3,* 17–43.

Curry, H. B. (1963). *Foundations of mathematical logic.* New York: McGraw-Hill.

Dove, I. J. (2007). On mathematical proofs and arguments: Johnson and Lakatos. In F. H. Van Eemeren & B. Garssen (Eds.), *Proceedings of the sixth conference of the international society for the study of argumentation* (Vol. 1, pp. 346–351). Amsterdam: Sic Sat.

Edwards, H. M. (1977). *Fermat's last theorem: A genetic introduction to algebraic number theory.* New York: Springer.

Euler, L. (1770). *Vollständige Anleitung zur Algebra.* St. Petersburg: Kaiserliche Akademie der Wissenschaften.

Frege, G. (1879). *Begriffsschrift, eine der arithmetischen nachgebildete Formelsprache des reinen Denkens.* Halle: Louis Nebert.

Goethe, J. W. V. (1949). *Die Wahlverwandtschaften, Die Novellen, Die Maximen und Reflexionen*. Zürich: Artemis (Gedenkausgabe der Werke, Briefe und Gespräche 9).

Johnson, R. H. (2000). *Manifest rationality: A pragmatic theory of argument*. Mahwah, NJ: Lawrence Erlbaum Associates.

Kneale, W., & Kneale, M. (1962). *The development of logic*. Oxford: Clarendon.

Krabbe, E. C. W. (1991). Quod erat demonstrandum: Wat kan en mag een argumentatietheorie zeggen over bewijzen? [QED: What can and may a theory of argumentation say about proofs?]. In M. M. H. Bax & W. Vuijk (Eds.), *Thema's in de Taalbeheersing: Lezingen van het VIOT-taalbeheersingscongres gehouden op 19, 20 en 21 december 1990 aan de Rijksuniversiteit Groningen* (pp. 8–16). Dordrecht: ICG.

Krabbe, E. C. W. (1997). Arguments, proofs, and dialogues. In M. Astroh, D. Gerhardus, & G. Heinzmann (Eds.), *Dialogisches Handeln: Eine Festschrift für Kuno Lorenz* (pp. 63–75). Heidelberg: Spektrum Akademischer Verlag.

Krabbe, E. C. W. (2004). Strategies in dialectic and rhetoric. In H. V. Hansen, C. W. Tindale, J. A. Blair, R. H. Johnson, & R. C. Pinto (Eds.), *Argumentation and its applications* (CD-ROM). Windsor, ON: Ontario Society for the Study of Argumentation.

Krabbe, E. C. W., & van Laar, J. A. (2007). About old and new dialectic: Dialogues, fallacies, and strategies. *Informal Logic, 27*, 27–58.

Lakatos, I. (1976). *Proofs and refutations: The logic of mathematical discovery* (edited by J. Worrall & E. Zahar). Cambridge: Cambridge University Press.

Plato. (1997). Meno. In J. M. Cooper & D. S. Hutchinson (Eds.), *Complete works* (pp. 870–897). Indianapolis: Hackett.

Saccheri, G. G. (1733). *Euclides ab omni naevo vindicatus* [Euclid freed from every flaw]. Milan: Paulus Antonius Montanus.

Tarski, A. (1956). The concept of truth in formalized languages. (J. H. Woodger, Trans.). In A. Tarski, *Logic, semantics, metamathematics: Papers from 1923 to 1938* (pp. 152–278). Oxford: Clarendon.

Tarski, A. (1983). The concept of truth in formalized languages. (J. H. Woodger, Trans.). In A. Tarski, *Logic, semantics, metamathematics: Papers from 1923 to 1938* (2nd ed., edited by J. Corcoran, pp. 152–278). Indianapolis, IN: Hackett.

Van Dalen, D. (2006). *Weet u dat zeker? Over (on)zekerheden in de wiskunde* [Are you sure? About (un)certainties in mathematics]. Groningen: Johann Bernoulli Lecture, March 7th.

Van Eemeren, F. H., & Grootendorst, R. (1984). *Speech acts in argumentative discussions. A theoretical model for the analysis of discussions directed towards solving conflicts of opinion*. Dordrecht: Foris.

Van Eemeren, F. H., & Grootendorst, R. (1992). *Argumentation, communication, and fallacies: A pragma-dialectical perspective*. Mahwah: Lawrence Erlbaum Associates.

Van Eemeren, F. H., & Grootendorst, R. (2004). *A systematic theory of argumentation: The pragma-dialectical approach*. Cambridge: Cambridge University Press.

Van Eemeren, F. H., & Houtlosser, P. (1999a). Delivering the goods in critical discussion. In F. H. van Eemeren, R. Grootendorst, J. A. Blair, & C. A. Willard (Eds.), *Proceedings of the fourth conference of the international society for the study of argumentation* (pp. 163–167). Amsterdam: Sic Sat.

Van Eemeren, F. H., & Houtlosser, P. (1999b). William the Silent's argumentative discourse. In F. H. van Eemeren, R. Grootendorst, J. A. Blair, & C. A. Willard (Eds.), *Proceedings of the fourth conference of the international society for the study of argumentation* (pp. 168–171). Amsterdam: Sic Sat.

Van Eemeren, F. H., & Houtlosser, P. (2002). Strategic manoeuvring in argumentative discourse: A delicate balance. In F. H. Van Eemeren & P. Houtlosser (Eds.), *Dialectic and rhetoric: The warp and woof of argumentation analysis* (pp. 131–159). Amsterdam: Kluwer.

Van Eemeren, F. H., & Houtlosser, P. (2004). More about fallacies as derailments of strategic maneuvering: The case of *Tu Quoque*. In H. V. Hansen, C. W. Tindale, J. A. Blair, R. H. Johnson, & R. C. Pinto (Eds.), *Argumentation and its applications* (CD-ROM). Windsor, ON: Ontario Society for the Study of Argumentation.

Van Eemeren, F. H., & Houtlosser, P. (2005). Theoretical construction and argumentative reality: An analytical model of critical discussion and conventionalised types of argumentative activity. In D. Hitchcock & D. Farr (Eds.), *The uses of argument: Proceedings of a conference at McMaster University* (pp. 75–84). Hamilton, ON: Ontario Society for the Study of Argumentation.

Walton, D. N., & Krabbe, E. C. W. (1995). *Commitment in dialogue: Basic concepts of interpersonal reasoning*. Albany, NY: State University of New York Press.

Wikipedia. (2007). Invalid proof. http://en.wikipedia.org/wiki/Invalid_proof. Accessed 10 Oct 2007.

Chapter 12
Analogical Arguments in Mathematics

Paul Bartha

12.1 Introduction

Many years ago, Pólya observed that mathematicians have to be good at both
deductive and inductive reasoning. A "serious student of mathematics," he suggested, "must learn plausible reasoning; this is the kind of reasoning on which ...
creative work will depend" (Pólya, 1954). Analogy was at the top of Pólya's list of
plausible reasoning strategies.

Analogical reasoning is important in many arenas: the law, the sciences, philosophy, and certainly mathematics. Analogy has a widely recognized role as a
fallible but highly useful *heuristic* tool.[1] Striking breakthroughs have emerged from
analogies between such diverse fields as number theory and complex function
theory. Less spectacular, but far more common, is the use of humble analogies to
guide everyday problem-solving, as when a student solving a problem in multi-
variable calculus draws upon what happens in the one-variable case. The term
'heuristic' actually embraces two functions of analogies: their role in discovery,
and their role in assessing the plausibility of conjectures that have been formulated
but not yet proved or disproved.

Analogies also play a *systematizing* or *unifying* role in mathematics. Descartes'
(1637) correlation between geometry and algebra, for example, made it possible to
reduce a whole class of geometric problems to finding the roots of a polynomial
equation. Descartes provided a method for systematically handling geometrical

---

[1]Pólya (1954), Poincaré (1952) and Hadamard (1949) have all stressed the importance of analogies
in mathematical discovery.

P. Bartha (✉)
Department of Philosophy, University of British Columbia, 1866 East Mall, E370 Vancouver,
BC V6T 1Z1, Canada
e-mail: paul.bartha@ubc.ca

A. Aberdein and I.J. Dove (eds.), *The Argument of Mathematics*, Logic, Epistemology,
and the Unity of Science 30, DOI 10.1007/978-94-007-6534-4__12,
© Springer Science+Business Media Dordrecht 2013

problems that had long been recognized as analogous, since they corresponded to equations of the same degree. In the nineteenth century, the similarities between "substitution groups" (today's permutation or symmetric groups) and isometric linear transformations on crystal structures led to the concept of a group (Kline, 1972).

The heuristic and systematizing roles of analogy are related through the idea of generalization. The goal of finding a generalization drives analogical reasoning in the early stages of an investigation. By contrast, after a general result (or general framework) has been elucidated, the analogy sheds light on the fruitful connections between two previously unrelated areas. Along the way, individual analogical arguments help to guide the mathematical constructions.

In this chapter, I shall mostly be concerned with the heuristic role of analogies in mathematics. Specifically, my objective is to provide an analysis of individual analogical arguments used to show that a conjecture is plausible. I will focus on three questions:

1. What does it mean for an analogical argument to show that a conjecture is plausible?
2. What are the criteria for a good analogical argument?
3. What philosophical justification can be given for analogical reasoning?

I shall propose a model for representing and evaluating analogical arguments that gives partial answers to these questions.

The chapter proceeds as follows. Section 12.2 reviews some basic general ideas and concepts about analogical arguments. Section 12.3 answers question (1): individual analogical arguments typically aim to establish that their conclusions are worthy of further investigation (rather than to assign a probability). The bulk of the chapter is directed at question (2). Section 12.4 articulates a basic test for plausible analogical arguments in mathematics; Sect. 12.5 raises three important challenges to this simple test; Sects. 12.6–12.8 offer responses. Section 12.9 addresses question (3) by explaining how analogical arguments in mathematics that pass the test described in Sects. 12.4–12.8 genuinely do make their conclusions plausible. Finally, the conclusion (Sect. 12.10) suggests some ways in which our treatment of individual analogical arguments can be expanded to illuminate the 'systematizing' role of analogies in mathematical research.

12.2 Analogical Arguments: Basic Notions

In general, an *analogy* is a comparison of similarities between two domains, or systems of objects. This comparison is typically represented as a kind of mapping between the objects, properties and relations of the two domains, commonly referred to as the source domain and the target domain.[2] If the domains are similar in known ways, then it appears reasonable to infer that some further characteristic of the source domain, or an appropriate analog of such a characteristic, may hold in the target. An *analogical argument* is an explicit representation of this sort of reasoning, citing accepted similarities between two systems in support of the conclusion that some further similarity is plausible. Sometimes (as in legal reasoning) we rely upon multiple analogies; at other times (as in most mathematical examples) we have an individual analogy involving a single source domain. Of course, analogical reasoning is seldom represented neatly as an explicit argument. I shall assume, however, that such reasoning is capable of representation in argument form. That assumption is implicit in Pólya's popular discussion of mathematical analogies Pólya (1954); it is appropriate if we are interested in justifications that are communicable and public.

In order to fix some terminology and notation, let's begin with two simple examples.

Example 1 (Rectangles and Boxes).   Suppose that you have proved the following result: if you consider all rectangles with the same fixed perimeter, then the one with maximum area is a square. Here is an analogous conjecture in three dimensions: of all rectangular boxes with a constant fixed perimeter, the cube has maximum volume. We argue by analogy that this conjecture is, to some degree, plausible.

Example 2 (Triangles and Tetrahedra).   Suppose you have proved that the three medians of any triangle have a common intersection point. By analogy, you conjecture that the four medians of any tetrahedron—the lines joining each vertex with the center of the opposite face—have a common intersection point.

In Example 1, the source domain is the set of rectangles, the target is the set of three-dimensional boxes, and the feature being transferred is that area is maximal for a square. In Example 2, the source domain is triangles, the target is tetrahedra, and the feature being transferred to the target is that the medians are concurrent—that is, they intersect in a common point.

It is sometimes helpful to represent an analogical argument in *tabular form*, listing first the known similarities that constitute the basis of the argument and then the conjectured further similarity. We can illustrate with Example 1. Here, x and y stand for edge lengths in two dimensions, while x^*, y^* and z^* represent edge lengths for a three-dimensional box. *PER* and *PER** stand for two- and three-dimensional perimeter, *AREA* for area and *AREA** for volume. (Starred symbols are used for items of the target domain that correspond to the source domain.) Q is the known result that is analogous to the conjecture, Q^*. The information is summarized in Table 12.1.

---

[2]Chapters 2 and 3 of my book (Bartha, 2010) provide a selective survey of the extensive literature on analogies and analogical reasoning. A recent and fairly representative collection of papers on analogy, mainly from the perspective of cognitive science, is Kokinov et al. (2009).

Table 12.1 Rectangles and boxes

Two-dimensional rectangles	Three-dimensional boxes
$PER = 2(x + y)$	$PER^* = 4(x^* + y^* + z^*)$
$AREA = x \cdot y$	$AREA^* = x^* \cdot y^* \cdot z^*$
Q: for maximum $AREA$, $x = y$	Q^*: for maximum $AREA^*$, $x^* = y^* = z^*$

Table 12.2 Schema for analogical arguments (AA)

Source domain (S)	Target domain (T)	
P	P^*	Positive analogy
A	$\sim A^*$ ⎫	
$\sim B$	B^* ⎬	Negative analogy
C	C^*?	Neutral analogy
Q		
	Q^* (plausibly)	

In general: let S be the source domain and T the target domain. We can represent an analogical argument in tabular form by listing known similarities, which we refer to as the *positive analogy*, followed by the known differences or *negative analogy*, and then the conjectured further similarity. The *neutral analogy* consists of features of the source domain about whose target analogs we lack knowledge. This gives us the representation in Table 12.2.

In short: it is plausible that Q^* holds in the target domain because of certain known or accepted similarities between the domains, despite certain known or accepted differences. The analogy, if successful, persuades us to look for a rigorous proof. This representation is a little unusual in that it identifies known differences (the negative analogy) that may weaken the argument. In practice, such differences are often conveniently omitted or even suppressed.

12.3 Two Conceptions of Plausibility

Let's focus first on the conclusion in the above argument scheme. What is the meaning of "Q^* (plausibly)"? The basic idea, of course, is that we have some reason to believe the conclusion, Q^*. When we try to make this idea more precise, however, we encounter two distinct conceptions of plausibility.

First, there is a *probabilistic* conception, on which plausibility is identified with degree of belief, or more broadly with some *degree of strength* that is naturally represented as a probability. To say that a hypothesis is *highly plausible* is to say that it has high probability. To say that a hypothesis is *somewhat plausible* is to say that it has a low but appreciable probability. In general, the probability is thought to depend upon the degree of similarity between the two domains.

The probabilistic conception of plausibility has long been prominent in analyses of analogy. Hume, in his *Dialogues Concerning Natural Religion*, sees analogical arguments as possessing a degree of strength that varies with the similarity of the domains:

> Wherever you depart, in the least, from the similarity of the cases, you diminish proportionably the evidence; and may at last bring it to a very weak analogy, which is confessedly liable to error and uncertainty (Hume, 1779, 144).

Mill, in *A System of Logic*, also appears to endorse a probabilistic interpretation. Mill paints a vivid picture of analogical reasoning as a "competition between the known points of agreement and the known points of difference," which boost or lower the probability of the conclusion. He writes:

> There can be no doubt that every resemblance [not known to be irrelevant] affords some degree of probability, beyond what would otherwise exist, in favour of the conclusion [and] every dissimilarity which can be proved between them, furnishes a counter-probability of the same nature on the other side (Mill, 1843, 333).

The probabilistic conception persists in recent work on analogy and inductive logic in the Carnapian tradition. Even those writing on analogical reasoning in mathematics tend to adopt a probabilistic model. Pólya, for example, states that the probability calculus is the most perspicuous way "to render more precise our views on plausible reasoning" (Pólya, 1954, 116).

Second, and less familiar, is what I call the *modal* conception of plausibility. To assert that a hypothesis is plausible in the (non-probabilistic) modal sense is just to assert that it is a serious possibility, something that we have reason to investigate further and that we *might* be brought to believe. Typically, there is no assertion of *degree*. The point of calling the hypothesis plausible is to single it out from a mass of bare possibilities. This view is implicit in remarks made by the physicist Norman Campbell about Fourier's use of analogy in support of his theory of heat conduction:

> *Some* analogy is essential to it; for it is only this analogy which distinguishes the theory from the multitude of others . . . which might also be proposed to explain the same laws (Campbell, 1957, 142).

A good analogy with a known theory establishes Fourier's theory as a serious possibility, or (as I shall say) as prima facie plausible.

The modal interpretation of plausibility also has nineteenth century antecedents. Herschel, whose philosophy of science rivalled that of Mill, insists that for a proposed hypothesis to merit serious attention, it must postulate a *vera causa* (true cause). For Herschel, this means that the hypothesis must be "analogous to causes that are already known to have produced similar effects in other cases" (see Snyder, 2006, 201)—that is, to causes known to exist in nature.[3] Like Campbell, Herschel uses the requirement of analogy to screen out frivolous hypotheses. Herschel and

---

[3]Of course, Herschel is speaking of empirical hypotheses. The vera causa requirement cannot be applied literally to mathematical conjectures.

Mill thus agree that analogical arguments play an important part in inductive reasoning, but they appear to disagree as to whether the assertion of plausibility is modal or probabilistic.

For mathematical analogies, I suggest that the modal conception of plausibility is the primary one. Of course, there are familiar philosophical difficulties with the assignment of any probability other than 0 or 1 to a mathematical proposition, but *that* is not the basis for my suggestion.[4] The suggestion derives instead from pervasive features of analogical arguments in mathematics, as illustrated by Examples 1 (Rectangles and Boxes) and 2 (Triangles and Tetrahedra).

In the first place, there is a practical point: a probabilistic interpretation of plausibility would serve no purpose in these analogical arguments. A numerical probability value would be a needless distraction, since the point of each analogy is simply to persuade us to take a conjecture seriously. Secondly, most mathematical analogies are not comparative. They are useful even when we can't make, or don't care to make, comparative plausibility judgments. Our two examples are typical: they offer individual analogical arguments without any rival conjectures. But even if there are rival conjectures, supported by rival analogical arguments, we are often not concerned about making comparisons. It is enough to know that we have to take all of these conjectures seriously. Finally, in thinking about inductive reasoning in mathematics, there is *no need for belief updating*. We don't need rules for updating our degree of inductive support for mathematical propositions.

In light of these observations, we may conclude that there is little practical point in modelling analogical arguments in mathematics with probabilities. After all, the main motivations for using numerical probabilities to represent our beliefs are to allow for comparisons (and hence decision-making) and for belief updating.

The major reason for interpreting plausibility as a modal notion, however, is that this interpretation is faithful to how analogical arguments in mathematics are actually assessed. Analogical arguments in mathematics do not proceed in the way suggested by Mill, on which the degree of strength of the argument is an increasing function of the degree of overall similarity between domains. On that view, every similarity confers some probability on the conclusion. That is not what happens with mathematical analogies. They are accepted or rejected not on the basis of overall similarity, but rather based on a pointed assessment of whether or not the proof employed in the source domain is "fit for imitation" (Pólya, 1954, 46) in the target domain.

The next section turns this remark of Pólya's into a simple test for plausibility, but the preliminary idea can be made clear by considering Example 1, Rectangles and Boxes. To see whether we have a good analogy, we need to start with a proof of the result about rectangles. Here is a simple proof.

---

[4]Indeed, the probabilistic conception of plausibility is important, and can be useful if we wish to model the strength of an analogical argument (in mathematics or elsewhere). My contention here is simply that the modal conception of plausibility is both ubiquitous and fundamental in mathematics.

Assumption: Perimeter is constant, i.e., $2x + 2y = c$ for some constant c.
Step 1:   Area xy is maximized when $x(c/2 - x)$ is maximized (From the assumption).
Step 2:   By the First Derivative Test, this occurs when $x = c/4$.
Step 3:   From the preceding steps, $y = c/4$ as well. We have proved that the rectangle with maximum area is a square.

Is this proof "fit for imitation" in the target domain (boxes)? Yes: there are obvious analogs of perimeter and area, and an analog of the First Derivative Test for multi-variable calculus. So we have a plausible argument.

For analogical arguments in mathematics, the determination of plausibility requires checking that the proof in the source domain is "fit for imitation" in the target domain (short of actually constructing the target proof). There must be no obvious disanalogy that would prevent us from adapting the proof. If there is such a critical disanalogy, then the analogical argument fails to establish the plausibility of its conclusion, regardless of the overall similarity between the two domains. This observation supports a modal (and not probabilistic) reading of plausibility.

12.4   Test for Plausibility of Mathematical Analogies

Mathematical analogies pose a significant challenge to the community of scholars that is trying to understand analogical reasoning. Most current theories of analogical reasoning—purely philosophical approaches, such as Hesse (1966), as well as computational theories that take their cue from Gentner (1983) and her "structure-mapping" approach—are based on some version of the idea that analogical arguments should be assessed on the basis of *overall similarity* between source and target domains. Such approaches mesh nicely with a probabilistic conception of plausibility and with Mill's idea that each similarity in an analogical argument contributes a measure of plausibility to the conclusion. Already, we have seen that ordinary strategies for assessing analogical arguments in mathematics are at odds with these ideas.[5] Those strategies are not based on overall similarity, and they fit naturally with a modal (rather than probabilistic) interpretation of plausibility. This section offers a quasi-formal test for plausible analogical arguments in mathematics, based upon the informal observations of the preceding section.

The test rests upon two fundamental principles.[6] First, for any good analogical argument, we require a clearly stated *prior association*: a logical, causal or explanatory relationship, in the source domain, between certain features and the feature Q that is projected to hold in the target domain. Second, there must be a

---

[5]For a very different set of tensions between the structure-mapping approach and mathematical analogies, see (Schlimm, 2008).

[6]In chapter 4 of Bartha (2010), I apply these principles widely to a variety of analogical arguments.

potential for generalization of this association from the source domain to the target, which would support the conclusion of the analogical argument.

In the case of mathematical analogies, the prior association consists of an explicit proof. Schematically, we represent this in the following form:

Definition 1 (Prior Association). $\varphi \vdash Q$

Here, Q is the proposition whose analog is projected to hold in the target domain, and φ is a set of explicit assumptions used in the proof (as unstated background assumptions are not part of the prior association). The prior association is that Q follows from φ via a mathematical derivation. The analogical argument suggests that a similar entailment relationship implies an analogous proposition (Q^*) in the target domain.

The prior association must be mathematically acceptable; the proof should satisfy a competent mathematician. Of course, many proofs will meet this standard. The key idea here is that once a proof is given, some definite set of explicit assumptions $\varphi = \{\varphi_1, \ldots, \varphi_n\}$ becomes the set of *critical factors* for assessing the analogical argument. It is perfectly in order to acknowledge an element of judgment in identifying what these factors are.

In Example 1 (Rectangles and Boxes), the critical factors include the formulas for area and perimeter, the assumption that perimeter is constant, and the First Derivative Test. In Example 2 (Triangles and Tetrahedra), it turns out that there are quite distinct ways of proving that the medians of a triangle intersect in a common point. If we use analytic geometry, we can reason as follows. Represent the vertices A, B, C as ordered pairs in the Cartesian coordinate system. Then the mid-points are

$$ X = \frac{A+B}{2}, \ Y = \frac{B+C}{2}, \text{ and } Z = \frac{A+C}{2}, $$

while the medians are the sets of points $\{(1-t)C + tX : 0 \leq t \leq 1\}$ and so on. The point $(A+B+C)/3$ lies on each median, as can be seen by taking $t = 2/3$. This proves that the medians are concurrent. The critical factors are the following:

1. X, Y, and Z are mid-points (given by the specified formulas).
2. The medians join vertices to mid-points.
3. The 1-parameter algebraic representations for the medians.

Now we turn to the other component of the test for plausibility, assessment of the potential for generalization of the proof to cover the target domain. The important idea here is that good analogies in mathematics generally are a precursor to generalization. Here it is sufficient to think of generalization as the formulation and proof of a result in a setting that comprehends both the source and target domains. To treat analogies as the first step in such a process seems entirely reasonable, since most (perhaps all) successful mathematical analogies do lead to such generalizations. Of course, when we argue by analogy, we cannot appeal to (and are often unaware of) the generalization. But we can apply Pólya's criterion by considering whether the source proof is "fit for imitation" in the target domain. This gives us a simple test for plausibility[7]:

Definition 2 (PfP: Prima facie Plausibility for Mathematical Analogies).

1. *Overlap.* Some fact used in the source proof must belong to the positive analogy, P.
2. *No-critical-difference.* Nothing used in the source proof can correspond to something known to be false in the target domain. That is, no critical assumption belongs to the negative analogy.

If a general theorem holds in an abstract setting that includes the source and target domains as special cases, then the general proof could be adapted, step-by-step, to yield proofs in our two domains. So if, for any fact used in the proof of Q, either there is no meaningful analog in the target T or the analog is false, then the *No-critical-difference* condition fails. There is then no good reason to expect a generalization, and the analogical argument is not plausible.

Let us apply these criteria to Example 2, relative to the proof just sketched. The concepts analogous to 'triangle', 'midpoint' and 'median' are 'tetrahedron', 'centroid', and 'median'. So facts 1 and 2 identified above as relevant have true analogs in solid geometry by the set-up of the problem. Furthermore, the algebraic representation for the centroids and medians of a tetrahedron is similar to that used for a triangle. For instance, if the vertices of the tetrahedron are A, B, C and D, then the centroids are $(A+B+C)/3$, $(A+B+D)/3$, and so on, exhibiting a form and symmetry like the two-dimensional case.[8] Fact 3 thus has an analog, and our analogical argument passes the test. On our model, the conclusion is shown to be prima facie plausible. Note that the test might fail (and indeed does fail) if we employ an alternative proof of the same conclusion, but this is consistent with the view that the particular proof employed is an integral part of the analogical argument.

In summary, for a mathematical analogy to satisfy the test of plausibility, everything used in the source proof must correspond to something not known to be false in the target domain. Otherwise, we have no reason to hope that there is a more general result of which the source and target theorems are special cases.

Note that if the *Overlap* and *No-critical-difference* conditions are satisfied, then there is scope for strengthening an analogical argument in two ways. First, we can make the argument better by moving more of the relevant properties from the neutral to the positive analogy. Second, we can make the nature of the correspondence sharper (as will be done in subsequent sections).

There is a natural worry here about how to handle proofs that are not parsimonious, i.e., proofs that introduce assumptions not strictly needed to derive the

---

[7]The test makes use of the terminology introduced in Sect. 12.2.

[8]Lacking an account of similarity at this point, I simply assert that there is a natural correspondence here. Shortly, we shall describe this as a case of *geometric similarity*.

conclusion. It might turn out that an analogical argument is defeated by (PfP) when one of these idle assumptions has no analog (or a false analog) in the target. We could have 'saved' the analogy (and the plausibility of the conclusion) by moving to a parsimonious proof. This leads to the objection that my account of plausibility is too strict. An alternative approach would define the set φ of critical factors as those that appear in a parsimonious proof.

My first response is to repeat that distinct proofs lead to distinct analogical arguments. If an analogical argument, as presented, is not amenable to generalization, then it ought to be defeated. The burden rests with the advocate proposing the analogy, who has every incentive to formulate as parsimonious a proof as possible. It is perfectly in order for a critic to single out any feature of the proof as given that does not seem to carry over to the target. But there is a more interesting response. The art of analogical reasoning is to 'see' a common pattern in two domains. Often, this means reformulating a proof in a way that removes disanalogies. To this end, it is fruitful to define the critical factors relative to the actual proof, rather than relative to an ideal, parsimonious proof. The approach we have adopted provides clear guidelines about what to do when an analogy fails: either abandon it or refine the proof to eliminate the disanalogy.

A more significant objection is that, so far, we have not placed any constraints on the similarity relationships between domains. We have relied upon intuitive judgment in assuming that correspondences between rectangles and boxes, or between triangles and tetrahedra, are genuine similarities.

It is helpful at this point to introduce terminology developed by Hesse (1966). The terminology makes reference to a tabular representation scheme for analogical arguments such as the one proposed as schema (AA) of Sect. 12.2. The *horizontal relations* in an analogy are relations of similarity (and difference) in the mapping between domains, while the *vertical relations* are those between the objects, relations, and properties within each domain. These notions can be illustrated using Example 1 (Rectangles and Boxes). The correspondence between area and volume is a horizontal relation. The proof that the square has maximum area indicates vertical relations among the concepts of the source domain.

The objection that we are now considering is that the test for plausibility (PfP) places no restrictions upon how we identify the horizontal relations, i.e., corresponding features in the source and target domains. Yet it is obvious that without restrictions on what counts as legitimate similarity, any theory of analogy will run into trouble. The next section presents nontrivial examples that illustrate this problem, and subsequent sections develop models for three important types of similarity, in effect giving us three special varieties of mathematical analogy. These models enrich our basic test and enable us to address the challenges, to which I turn next.

12.5 Three Challenges

This section raises three difficulties for the account of plausible mathematical analogies just proposed. All of them signal the need to supplement that account by incorporating an analysis of the similarities between the source and target domains into our plausibility criteria.

So far, the proposal is that we evaluate each analogical argument, relative to a given proof in the source domain, by assessing the potential for generalizing that proof to cover the target domain. No explicit assumption in the proof can correspond to something known to be false in the target (which would constitute a *critical disanalogy*), and at least one explicit assumption in the proof must correspond to something known to be true in the target (the *Overlap* condition). If these two conditions are met, then the analogical argument counts as prima facie plausible, i.e., there is a solid case for investigating the conjecture. This test is vague as to what counts as an acceptable proof, but the lack of precision is tolerable: it reflects genuine ambiguity in what counts as a good analogical argument. Mathematicians can and do disagree about questions of plausibility, and the disagreements can often be traced to choices about representation. Our theory of analogical reasoning intentionally leaves such matters open by relativizing to the given proof in the source domain.

As stated at the end of Sect. 12.4, however, we cannot be equally sanguine about identifying legitimate correspondences between features of the source and target domains. Without some restrictions, the theory is vulnerable to easy counterexamples. The remedy, to be developed in Sects. 12.6–12.8, is to supply constraints on correspondence, i.e., standards for acceptable—or, as I shall say, *admissible*—similarity between our two domains. My criteria for admissibility will not be geared towards blocking all possible counterexamples. I shall be concerned mainly with three problems that I take to be of particular importance and interest.

12.5.1 Specious Resemblance

The first problem is that an analogical argument may satisfy (PfP) yet be quite implausible because the resemblances between the two domains derive from contrived manipulations. In order to see that this is an interesting problem, let us begin by considering Example 1 once again, which I take to be a good analogical argument. Our first step in evaluating this analogical argument is to supply a proof of Q, which we did in Sect. 12.3. The critical facts in that proof are the formulas for area and perimeter, the assumption that perimeter is constant, and the First Derivative Test. Looking at the target domain, we see that there are analogous formulas for volume and perimeter, and an analogous theorem about maxima for functions of two variables. This information is summarized in Table 12.3.

From (PfP), it follows that we have a plausible analogical argument. Indeed, the conclusion Q^* is true.

Table 12.3 Rectangles and boxes

Source (two-dimensional rectangles)	Target (three-dimensional boxes)
$Area = x \cdot y$	$Volume = x \cdot y \cdot z$
Perimeter $= 2(x + y)$	Perimeter $= 4(x + y + z)$
First derivative test (one-variable):	First derivative test (two-variables):
at a maximum, $Df(x) = 0$	at a maximum, $Df(x, y) = 0$
Q: for fixed perimeter,	Q^*: for fixed perimeter,
area is maximal if $x = y$	volume is maximal if $x = y = z$

Our test for plausibility seems to work well. Yet the analysis takes for granted that volume is the natural analog to area. There are other possibilities.

Example 1A (Variant on Rectangles and Boxes). Let's re-write the formula for the area of a rectangle with sides x and y as

$$AREA(x, y) = 3x^{2-2} + x^{2-1}y^{2-1} - 3\sin^{2-2}y,$$

which is, of course, equal to xy. Now we obtain a new three-dimensional analog:

$$VOLUME(x, y, z) = 3x^{3-2} + x^{3-1}y^{3-1}z^{3-1} - 3\sin^{3-2}z.$$

The 'resemblance' here is that both *AREA* and *VOLUME* are instances of the formula

$$3x_1^{n-2} + x_1^{n-1} \cdot \ldots \cdot x_n^{n-1} - 3\sin^{n-2}x_n,$$

putting $n = 2$ and $n = 3$, respectively. That gives us an analogical argument that of all boxes with a fixed perimeter, the one that maximizes *VOLUME* is a cube. What is more, the argument appears to satisfy the criteria of (PfP), although it is manifestly implausible.

This is a clear case of specious resemblance, resting on a technical trick. Nobody would take the argument seriously. But how can we rule out such tricks in a principled way when we have stressed the importance of allowing flexibility in representing an analogical argument? If the example seems far-fetched, consider the following case.

Example 3 (Euler Characteristic Formula). Euler's formula states that if the number of faces, edges and vertices of a (convex) polyhedron are F, E, and V respectively, then $F + V = E + 2$. For example, the equation holds good for a cube, which has 6 faces, 8 vertices and 12 edges. It also works for tetrahedra, dodecahedra and other solids.[9] There is an even simpler formula for the two-dimensional case: $V = E$. Any polygon has the same number of edges and vertices.

---

[9]See Lakatos (1976) for the history and limitations of this formula.

Pólya (1954, 43) rewrites the two formulas in an ingenious way to exhibit an analogy:

$$V - E + F = 1 \qquad \text{(polygons—2 dimensions)}$$

$$V - E + F - S = 1 \qquad \text{(polyhedra—3 dimensions)}$$

The left-hand side of each equation is an alternating sum beginning with the number of vertices (0-dimensional elements), and proceeding through the number of edges (1-dimensional), faces (2-dimensional) and solids (3-dimensional elements).[10]

This looks like a case of specious resemblance. New terms are introduced and old ones manipulated in order to 'create' an analogy. Yet it turns out to be an illuminating way to re-write the two formulas. Indeed, it suggests analogous Euler formulas for higher dimensions, and these analogous formulas happen to be correct.[11] A 'manufactured' resemblance thus leads to an insightful analogy. If we insist upon a 'standard' representation, we might never find this innovative reformulation of the Euler formula. How are we to distinguish, when evaluating an analogical argument, whether similarities emerge from legitimate reformulation or from specious manipulations?

12.5.2 Separated Domains

Our second problem is how to deal with analogical arguments based on similarity between 'widely separated' domains. An analogical inference between two neighboring dimensions is prima facie stronger than one between dimensions one and three, one and four, and so on. Perhaps 'gappy' analogical arguments don't reveal anything about plausibility, as in the following (slightly artificial) example.

Example 4 (Punctured Balls). A *punctured* set is a set with one point removed, such as the interval $(-1, 1)$ without the point 0 (pictured in Fig. 12.1a). A set is *disconnected* if it can be written as the union of two disjoint non-empty open sets. Roughly, this means it has two or more separate, non-touching pieces. Lastly, the *unit ball* in $\mathbb{R}^n$ is the set of points whose distance to the origin is less than 1 unit. It is easy to show that in one dimension, the punctured unit ball $(-1, 0) \cup (0, 1)$ is disconnected into two components.

What should we say about arguing, by analogy, that the punctured unit ball in $\mathbb{R}^3$ (pictured in Fig. 12.1b) is disconnected? It seems that we should reject the argument: for one thing, one should look first at the two-dimensional case (where the punctured unit ball is not disconnected). Yet (PfP) provides no restrictions that require looking first at 'closer' cases.

---

[10]For a polygon, F (the number of faces) is 1; for a polyhedron, S (the number of solid elements) is 1.

[11]The full generalization involves concepts of algebraic topology; see Munkres (1984).

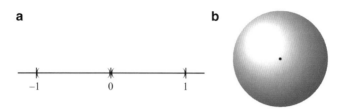

Fig. 12.1 (**a**) Punctured interval. (**b**) Punctured sphere

Our test seems to fail us. This argument appears to satisfy (PfP), yet it is intuitively implausible.

Once again, there is no quick way to fix the problem. Suppose we rule out any analogy between domains that are not 'neighbours' (although as yet, we have no precise concept of neighbouring domains). That will imply something much too strong: analogical arguments that depend upon parity are a priori implausible. But we know that some mathematical properties do depend upon *parity*. For example, polynomial equations of odd degree with real coefficients always have at least one real root, and that is not so if the degree is even. The Hairy Ball Theorem asserts that only in real vector spaces of even dimension is it possible to define a continuous tangent vector field on the surface of the unit ball. More picturesquely, in even dimensions we can comb a hairy ball flat without parting the hair in any way, but we can't do this if the dimension is odd. Sometimes, in thinking about a problem in three dimensions, the one-dimensional case might be a better guide than the two-dimensional case.

12.5.3 Limits

Similarities involving limits play a central part in many interesting mathematical analogies. The problem is that our theory seems to reject most of them because of the No-critical-difference requirement. Unlike the previous two problems, in which our theory gives intuitively implausible arguments a passing grade, the difficulty here is that our theory appears to be too strict.

An excellent illustration of this type of analogy—and the challenge it poses for our theory—is Euler's beautiful exploitation of similarities between polynomials and power series. My presentation is based on (Pólya, 1954).

Example 5 (Polynomials and Power Series). A polynomial of degree n is a function of the form $f(x) = a_0 + a_1 x + \cdots + a_n x^n$. A power series is a function defined by an infinite sum of the form $f(x) = \sum_{n=0}^{\infty} a_n x^n$; for each x, the value $f(x)$ is defined if the partial sums $f(x) = \sum_{n=0}^{M} a_n x^n$ converge to some limit. Each of these partial sums is a polynomial function, so that a power series is the limit of a sequence of polynomial functions. The concept of a limit is essential for the precise characterization of the correspondence between polynomials and power series. Euler uses the analogy between polynomials (the source domain) and power series (the target domain) to support two conjectures and to obtain a striking result, namely, the infinite sum $1 + 1/4 + 1/9 + 1/16 + \cdots = \pi^2/6$, or in concise notation:

$$\sum_{n=1}^{\infty} \frac{1}{n^2} = \frac{\pi^2}{6}.$$

Euler's first conjecture equates an infinite sum and an infinite product. In the source domain (polynomials), it is not difficult to prove the following result:

$$a_0 + a_1 x + \cdots + a_n x^n = a_0 (1 - \frac{x}{\alpha_1}) \ldots (1 - \frac{x}{\alpha_n}).$$

Here, the polynomial $f(x) = a_0 + a_1 x + \cdots + a_n x^n$ has the n roots $\alpha_1, \ldots, \alpha_n$ (counting multiplicities). So we can equate a finite sum (left side) with a finite product (right side). By analogy, if f is defined by a power series and has roots α_i (counting multiplicities), Euler conjectures that f is expressible as the infinite product

$$a_0 \prod_{n=1}^{\infty} (1 - \frac{x}{\alpha_i}).$$

Since $\sin x / x$ has the power series

$$\frac{\sin x}{x} = 1 - \frac{x^2}{3!} + \frac{x^4}{5!} - \cdots$$

and has roots $\pm k\pi$ for $k \neq 0$, Euler's conjecture yields:

$$\frac{\sin x}{x} = (1 - \frac{x}{\pi})(1 + \frac{x}{\pi})(1 - \frac{x}{2\pi})(1 + \frac{x}{2\pi}) \cdots$$
$$= (1 - \frac{x^2}{\pi^2})(1 - \frac{x^2}{4\pi^2}) \cdots$$

Euler's second conjecture is to derive an identity from this first step. In the source domain, begin with the equation just discussed:

$$f(x) = a_0 + a_1 x + \cdots + a_n x^n$$
$$= a_0 (1 - \frac{x}{\alpha_1}) \ldots (1 - \frac{x}{\alpha_n})$$

The coefficient of x in these two expressions must be the same, which gives us the relation:

$$a_1 = -a_0 (\frac{1}{\alpha_1} + \ldots + \frac{1}{\alpha_n}).$$

The expression in parentheses is the sum of the reciprocals of all roots of f. Euler conjectures that the same relationship between the coefficients a_0 and a_1 holds for the infinite case, i.e., for power series. Applying it to $\sin x/x$, he obtains

$$-\frac{1}{3!} = (-1)(\frac{1}{\pi^2} + \frac{1}{4\pi^2} + \ldots),$$

which, after simplification, gives the desired series for $\pi^2/6$.

Euler's argument poses a challenge to our theory for the following reason. In proving the equality of the two representations of a polynomial, as finite sum and finite product, we use facts about the *degree* of the polynomial. But there is no analog for degree, and hence nothing corresponding to these results, for an infinite power series. According to (PfP), this counts as a critical difference. Thus, on our theory, it looks as if Euler's argument is blocked, and consequently establishes nothing about plausibility. Other analogical arguments based on taking limits are blocked in similar ways. That looks bad for our theory, since Euler's argument appears prima facie plausible.

Again, there is no simple way to fix our theory. The No-critical-difference condition leads to trouble here, but it is at the heart of our account of plausible analogical arguments and is not easily modified.

To meet all three of the challenges raised in this section, we need to incorporate constraints on similarity into our evaluation criteria. As a first step, let's make the following simple adjustment.

Definition 3 (PfP*).

1. *Overlap*. Some explicit assumption in the proof must correspond admissibly to a fact known to be true in the target domain. (The positive analogy is non-trivial.)
2. *No-critical-difference*. No explicit assumption in the proof can correspond admissibly to something known to be false in the target domain. (No critical assumption belongs to the negative analogy.)

The only change from (PfP) is that the correspondences have to be admissible, i.e., an acceptable basis for an analogical argument. In Sect. 12.6–12.8, I shall propose models that help us to decide when the similarities in an analogical argument are admissible. These models will permit us to amend our basic test for plausibility and to deal with the three challenges just raised.

12.6   Algebraic Similarity

An algebraic similarity is a mathematical property shared by a cluster of relations, functions and objects in the two domains. For example, addition and multiplication (of real numbers) are similar in that both operations are commutative ($x+y=y+x$ and $x \cdot y = y \cdot x$) and associative (($x+y)+z = x+(y+z)$ and $(x \cdot y) \cdot z = x \cdot (y \cdot z)$). We characterize such similarities as second-order identities, in the following manner.

Definition 4 (Algebraic Similarity: AS). An *algebraic similarity* between relations $R_1, \ldots, R_m$, functions $F_1, \ldots, F_n$, and constants $c_1, \ldots, c_p$ of S, and corresponding $R_1^*, \ldots, R_m^*$, functions $F_1^*, \ldots, F_n^*$, and constants $c_1^*, \ldots, c_p^*$ of T is expressed by a (second-order) propositional function

$$\Psi(\mathfrak{R}_1, \ldots, \mathfrak{R}_m; \mathfrak{F}_1, \ldots, \mathfrak{F}_n; \mathfrak{c}_1, \ldots, \mathfrak{c}_p),$$

such that the two sentences

$$\Psi(R_1, \ldots, R_m; F_1, \ldots, F_n; c_1, \ldots, c_p),$$

and

$$\Psi(R_1^*, \ldots, R_m^*; F_1^*, \ldots, F_n^*; c_1^*, \ldots, c_p^*),$$

that result from substituting the respective relations, functions, and constants for the variables are true in the respective domains S and T.[12]

For example: *commutativity* of both addition $(x + y = y + x)$ and multiplication $(x \cdot y = y \cdot x)$ is an algebraic similarity expressed by the function

$$\Psi : (x)(y)(\mathfrak{F}xy = \mathfrak{F}yx).$$

Here, $\mathfrak{F}$ is a two-place function variable. Ψ is true when either operation ($+$ or $\cdot$) is substituted for $\mathfrak{F}$.

As a second example: *distributivity* of addition and multiplication, $x(y + z) = xy + xz$, is expressed by

$$\Psi : (x)(y)(z)(\mathfrak{F}_1(x, \mathfrak{F}_2(y, z))) = \mathfrak{F}_2(\mathfrak{F}_1(x, y), \mathfrak{F}_1(x, z)).$$

That the distributive property holds in two domains, such as the real numbers and the complex numbers, counts as an algebraic similarity. If we substitute multiplication for $\mathfrak{F}_1$ and addition for $\mathfrak{F}_2$ we obtain a proposition that holds in both domains.

In practice, there is no need to identify algebraic similarities at this level of abstraction. Instead, we simply rely on the obvious resemblance between the corresponding first-order propositions. The following example illustrates how algebraic similarity, in conjunction with (PfP*), can be used to evaluate an analogical argument.

Example 6 (Trigonometric and Hyperbolic Functions). The eighteenth century mathematician John Playfair (1778), in an attempt to justify arithmetical operations

---

[12]The fraktur variables—representing relations, functions and constants—are free in Ψ. The other relation, function and constant symbols denote entities in the source and target domains.

Table 12.4 Unit hyperbola and unit circle

Unit hyperbola	Unit circle
Sectors, area $1/2a$	Arcs, length a (or sectors, area $1/2a$)
$\cosh a = (e^a + e^{-a})/2$	$\cos a = (e^{ia} + e^{-ia})/2$
$\sinh a = (e^a - e^{-a})/2$	$\sin a = (e^{ia} - e^{-ia})/2$
1 (constant)	i (constant)

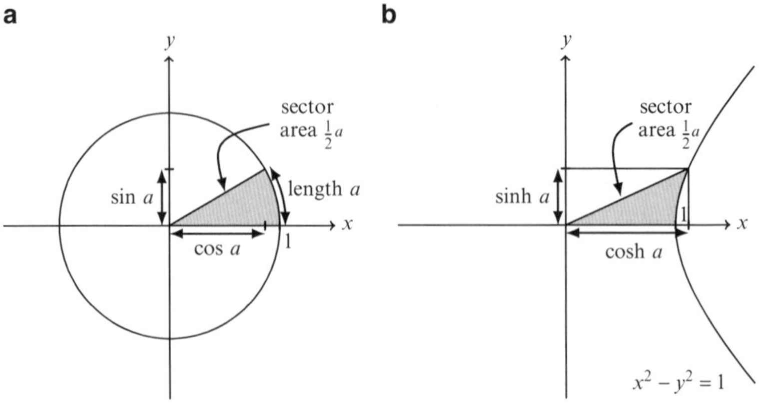

a **b**

Fig. 12.2 (**a**) Unit circle. (**b**) Unit hyperbola

on imaginary numbers, developed an interesting analogy between the unit hyperbola and the unit circle, and in particular between the hyperbolic and trigonometric functions. He began by noting a number of interesting correspondences, summarized in Table 12.4.

Playfair's geometric interpretation of the analogy is illustrated by Fig. 12.2a, b.

It was well known that propositions about real numbers could be "proven" by formal manipulations involving complex numbers, but there was considerable debate about the merits of such proofs in light of philosophical misgivings about the status of imaginary numbers. For example, the double angle formula for cosines can be derived from the above equations as follows:

$$\cos(2a) + 1 = \frac{e^{2ia} + e^{-2ia} + 2}{2}$$
$$= \frac{(e^{ia} + e^{-ia})^2}{2}$$
$$= 2\cos^2 a$$

Playfair provided an interesting defense for this type of derivation. He acknowledged that any such deduction is "unintelligible" per se, but "points out a perfectly precise" valid deduction for the hyperbolic functions, making no use of imaginaries.

This valid deduction can be obtained by direct substitution of 1 for i in the original derivation. For example, the above derivation is readily transformed into a proof of an expansion formula for hyperbolic cosine:

$$\cosh(2a) + 1 = \frac{e^{2a} + e^{-2a} + 2}{2}$$
$$= \frac{(e^a + e^{-a})^2}{2}$$
$$= 2\cosh^2 a$$

In Playfair's view, this identity for hyperbolic functions provides analogical support for the original theorem (and a justification for the formal manipulations with imaginary numbers). The same sort of relationship holds for other trigonometric identities, such as

$$\sin a \cos b = \frac{1}{2}[\sin(a+b) + \sin(a-b)].$$

and

$$\sinh a \cosh b = \frac{1}{2}[\sinh(a+b) + \sinh(a-b)].$$

Now as Playfair noted, the procedure sometimes fails. For instance,

$$\sin a \sin b = \frac{1}{2}[\cos(a-b) - \cos(a+b)]$$

is true, but

$$\sinh a \sinh b = \frac{1}{2}[\cosh(a-b) - \cosh(a+b)]$$

is false. Similarly,

$$\sin^2 a + \cos^2 a = 1$$

is true, while

$$\cosh^2 a + \sinh^2 a = 1$$

is false. What goes wrong in these cases is that the derivations of the true identities make use of the assumption $i^2 = -1$. Since $1^2 \neq -1$, Playfair's substitution procedure fails to yield valid derivations for the two 'identities' that fail.

Let me now connect this example to my theory. I am interested in the analogy that runs *from* the trigonometric functions to the hyperbolic functions (opposite to the direction on which Playfair focuses). This analogy concerns whether a formally correct derivation of a trigonometric identity can be transformed, line by line, into a derivation for the analogous hyperbolic identity. Our theory correctly indicates that the first two hyperbolic identities above are plausible, because every critical assumption belongs to the positive analogy. By contrast, the analogical arguments for the final two (failed) identities are blocked by the *No-critical-difference* condition. They require the critical assumption $i^2 = -1$, which belongs to the negative analogy.

The example provides a good illustration for our amended theory of plausibility, (PfP*). Playfair's analogy rests entirely upon algebraic similarities, reflected in the parallel structure of the relevant functions and assumptions. In this case, our test for plausibility works well, correctly sorting out which identities can be transferred from the trigonometric to the hyperbolic case. That is as much as we can hope for in a theory of analogical arguments.

Algebraic similarity can be integrated into the model in a relatively straightforward manner. That might seem surprising, since the definition of algebraic similarity appears to allow for trivial or specious resemblances. For instance, the operations of adding two numbers and taking the minimum of two numbers are both commutative. This similarity appears slight, yet according to our definition it counts as a legitimate algebraic similarity. We might worry that this leads to a theory that finds specious analogical arguments to be plausible.

Our theory takes care of this worry by offering an independent basis for determining which assumptions are critical, namely, those that occur explicitly in the proof. Thus, if we are extrapolating a result proved using only the commutativity of addition, then $x + y$ and $\mathrm{minimum}(x, y)$ are appropriately similar. Otherwise, this similarity is either irrelevant or insufficient to secure the plausibility of the analogical argument.

Every algebraic similarity is admissible. This type of similarity is quite unproblematic. In order to come to grips with the difficulties noted in Sect. 12.5 and to represent more interesting analogical arguments, we turn to other kinds of mathematical similarity.

12.7 Geometric Similarity

Two functional or relational expressions (henceforth just *expressions*) are geometrically similar if they look the same up to a change in the value of some parameter or parameters. In characterizing them, I limit my attention to cases involving a single parameter.

It helps to begin with a few examples of expressions that exhibit geometric similarity:

- $Area(x,y) = x \cdot y$ and $Volume(x,y,z) = x \cdot y \cdot z$ (from Example 1)
- $Length(x,y) = (x^2 + y^2)^{1/2}$ and $Length(x,y,z) = (x^2 + y^2 + z^2)^{1/2}$ (two and three dimensions)
- $f(x) = x^2$ and $g(x) = x^4$
- $f(x) = x^{n/4}$ and $g(x) = x^{n/2}$

In these examples, the parameter can be a positive integer or a real number. Furthermore, the parameter may indicate the number of arguments, as in the first two examples, or it may function as an individual constant, as in the last two. That gives us three types of geometric similarity (since number of arguments must be an integer).

Definition 5 (Geometric Similarity: GS). Two expressions F and F^* are geometrically similar if one of the following three cases obtains:

(a) $F = \mathfrak{F}(m)$ and $F^* = \mathfrak{F}(n)$, where $\mathfrak{F}(k)$ is an expression parameterized by a positive integer k in some range including m and n.
(b) $F = \mathfrak{F}(\alpha)$ and $F^* = \mathfrak{F}(\beta)$, where $\mathfrak{F}(t)$ is an expression parameterized by a real number t in some range including α and β.
(c) $F = \mathfrak{F}(x_1,\ldots,x_m)$ and $F^* = \mathfrak{F}(x_1,\ldots,x_n)$, where $\mathfrak{F}(x_1,\ldots,x_k)$ is an expression with k arguments, defined for a range of values of k including m and n.

Each definition involves a one-parameter family of relational or functional expressions. The similarity is captured by the common form, $\mathfrak{F}$. The two similar expressions are obtained by supplying distinct values for the parameter. The parameter space may be a set of integers, as in cases (a) and (c), or a set of real numbers, as in case (b). As examples of each case, we have the following pairs:

(a) $f(x) = x^2$ and $f^*(x) = x^4$ (both instances of $\mathfrak{F}(k) = x^k$);
(b) $f(x) = x^{n/4}$ and $f^*(x) = x^{n/4}$ (both instances of $\mathfrak{F}(t) = x^{tn}$);
(c) $Area(x,y) = x \cdot y$ and $Volume(x,y,z) = x \cdot y \cdot z$ (both instances of $\mathfrak{F}(x_1,\ldots,x_k) = x_1 \cdot \ldots \cdot x_k$).

Geometric similarity is highly sensitive to form. It is not preserved under mathematical and logical equivalence. For instance,

$$F(x_1,x_2) = x_1 \cdot x_2$$

and

$$F^*(x_1,x_2,x_3) = x_1 \cdot \sin(x_2 \cdot x_3)$$

are mathematically equivalent to the instances $n = 2$ and $n = 3$ of

$$\mathfrak{F}(x_1,\ldots,x_n) = x_1 \cdot \ldots \cdot x_n + (n-2)[x_1 \cdot \sin(x_2 \cdot \ldots \cdot x_n) - x_1 \cdot \ldots \cdot x_n].$$

But F and F^* are not themselves instances of $\mathfrak{F}$, so they are not geometrically similar. Without this restriction, any two expressions could be shown to be geometrically similar and the relation would be trivial.

Let's look at how geometric similarity functions in analogical arguments. It will quickly become clear that the purely syntactic definition (GS) is inadequate—and for the very reasons that emerged in the challenges of Sect. 12.5. We begin with a set of examples, of which the first few have already been introduced.

Example 1 (Rectangles and Boxes). This analogy rests on geometric similarities. Two- and three-dimensional perimeter are related as instances of *Perimeter*$(n) = 2^{n-1}(x_1 + \ldots + x_n)$, with $n = 2$ and $n = 3$. Similarly, area and volume are instances of generalized volume, *Volume*$(n) = x_1 \cdot \ldots \cdot x_n$. Assuming the geometric similarities are admissible, the analogical argument is plausible. So far, so good.

But trouble arises right away. In Sect. 12.5, we presented a superficially similar but intuitively implausible analogical argument, Example 1A. That example was constructed so that all of the relevant functions satisfy our purely syntactic definition of geometric similarity. This is the problem of specious resemblance.

Example 3 (Euler Formula). The similarities between the different formulas for polygons, polyhedra and so forth can be fully represented in terms of a single parameter (dimension), and again they satisfy our definition of geometric similarity.

Example 4 (Punctured Balls). The analogy between punctured intervals in one dimension and punctured spheres in three dimensions once again satisfies our definition of geometric similarity, yet is not obviously plausible. This is the problem of separated domains.

These examples make it clear that geometric similarity must meet substantive criteria to count as *admissible*, in order to be an acceptable basis for a plausible analogical argument. But precisely what these criteria should be is a difficult question, and one that I shall not entirely resolve. Before taking on this task, I offer one final example. The example is important, in part because it is an analogical argument that fails even though it is (according to my theory) plausible.

Example 7 (Abelian Groups). This example relies on elementary group theory, so I start with some mathematical background. A *group* is a set closed under an operation—multiplication—that satisfies some basic properties: associativity, the existence of an identity, and the existence of inverses. For example, the set of rotations in the plane around the origin is a group. To multiply two rotations is to follow one with another. This operation satisfies all of the group properties: it is associative, the identity element is rotation through $0°$, and the inverse of a rotation of $k°$ is a rotation of $-k°$. A group G is said to be *abelian* or commutative if $ab = ba$ for any two elements a, b of G. For example, the group of two-dimensional rotations is abelian: the net effect of two rotations is the same regardless of which is performed first. Finally, if x is a member of a group G and m is a positive integer, then x^m signifies x multiplied by itself m times. For example, if x is a rotation through $20°$, then x^3 is 3 rotations through $20°$, which is equivalent to rotation through $60°$.

Table 12.5 Abelian groups

Source (S)	Target (T)
For all a, b, $(ab)^2 = a^2b^2$	For all a, b, $(ab)^3 = a^3b^3$
Associative	Associative
Cancellation laws	Cancellation laws
Q: Abelian	$\Rightarrow Q^*$: Abelian

Now to the analogical argument, which is based on the following elementary problem in group theory (Herstein, 1975): if G is a group such that $(ab)^3 = a^3b^3$ for all a, b in G, must G be abelian? Analogical reasoning seems to make a positive answer plausible. For we can prove the analogous result that if $(ab)^2 = a^2b^2$ for all a, b in G, then G is abelian. Here is the proof. If $a, b \in G$, then:

$$abab = aabb \text{ (by assumption and associativity)}$$
$$\Rightarrow \quad bab = abb \text{ (using the left cancellation law, } (ax = ay) \Rightarrow (x = y))$$
$$\Rightarrow \quad ba = ab \text{ (using the right cancellation law, } (xb = yb) \Rightarrow (x = y)).$$

The proof uses only the definition of exponentiation, associativity, and the two cancellation laws. According to my theory, these are the critical facts.

My theory suggests that we have a good analogical argument. The source and target domains are groups satisfying analogous conditions, as shown in Table 12.5. Q is the statement that S is abelian, which (by the above proof) is true. Q^* is the conjecture that T is abelian. There are both algebraic and geometric similarities between S and T. The general properties of groups, associativity and the cancellation laws, count as algebraic similarities. The two identities at the head of each column exhibit geometric similarity. Assuming that all of these similarities are admissible, and noting that everything that is critically relevant in our proof of Q has a true analog in T, our theory tells us that Q^* is plausible. In fact, Q^* is false.[13]

This example is important for three reasons. In the first place, we have an analogical argument that is both putatively plausible and a failure. That establishes the non-triviality of our theory, for plausibility should not always line up with success. Secondly, as a closely related point, the example reminds us that we need to characterize plausibility in a way that makes it at least partly independent of success. Our theory suggests that, short of attempting a proof, this analogical argument counts as plausible because there are no obvious critical disanalogies between the source and target domains. Third and finally, this example reinforces the lesson we have already drawn from Example 1 and its variant Example 1A: we need an account of *admissible* geometric similarity that supplements the purely syntactical definition (GS) given above. Towards that end, I begin by proposing solutions to the first two challenges of Sect. 12.5.

---

[13] A modified result is true: if $(ab)^3 = a^3b^3$ for all a, b in T, and the number of elements in T is not divisible by 3, then T is abelian.

Table 12.6 Specious resemblance: summary of results

Example	Satisfies (GS)?	Legitimate manipulation?	Plausible analogy?	Successful analogy?
Example 1 (boxes)	Yes	Yes	Yes	Yes
Example 1A (boxes: variant)	Yes	No	No	No
Example 3 (Euler formula)	Yes	Maybe	Maybe	Yes
Example 7 (Abelian groups)	Yes	Yes	Yes	No

Table 12.7 Contrived version of area/volume analogy

Two dimensions	Three dimensions
$Area = x \cdot y$	$Volume = x \cdot y \cdot z$
$AREA = 3x^{2-2} + x^{2-1}y^{2-1} - 3\sin^{2-2}y$	$VOLUME = 3x^{3-2} + x^{3-1}y^{3-1}z^{3-1} - 3\sin^{3-2}z$

12.7.1 Specious Resemblance

The problem of specious resemblance is that two expressions with no meaningful similarity can be re-written so as to satisfy the purely syntactic criterion (GS). Yet creative forms of representation are an important part of analogical reasoning in mathematics. We seek a balance that allows us to distinguish between legitimate and specious manipulations. Table 12.6 summarizes facts (columns 1 and 4) and intuitions (columns 2 and 3) about the examples that bear on this problem.

To 'solve' the problem of specious resemblance, we need to account for the facts and intuitions in Table 12.6. We must explain why, for instance, the manipulations of the Euler formula may be legitimate while those in the highly contrived Example 1A are not. My proposed solution is roughly as follows. An admissible geometric similarity must both satisfy the syntactic conditions of definition (GS) and be motivated by the proof that serves as the prior association on which the analogy is based. Thus, in Example 1A, *AREA* (which is mathematically equivalent to *Area*) and *VOLUME* meet criteria (GS) for geometric similarity, as Table 12.7 shows. Yet if we consider the proof (in Sect. 12.3) for the theorem that of all rectangles with fixed perimeter, the square has maximum *Area*, we find no motivation for re-writing *Area* as *AREA*.

I suggest, then, that the way to deal with specious resemblance is with the following requirement of *internal coherence*:

Definition 6 (Internal Coherence: IC). For a geometric similarity to count as admissible, the relevant relations and functions should be expressed using standard representation, unless some justification internal to the domain can be given for a non-standard representation. More specifically, any novel representation should be justified in terms of the proof that is the prior association for the analogical argument.

The real problem lies not with representations that are complex or novel, but rather with those that have no independent motivation. The proposal is that any novel way of representing features of the source or target domain must have a motivation that is independent of the analogy and its purposes, a rationale internal to the domain. More specifically, the novel representation should be linked to the proof that is the basis of the analogical argument. In Example 1A, the only way to motivate the artificial formula *AREA* is to point to the function *VOLUME* and say, "I want the formula for area to look like that (so as to provide analogical support for my conjecture)." The motivation is external to the source domain, which makes the reformulation illegitimate.

There is another way to argue for *internal coherence*. The problem that it is meant to solve can be viewed as a problem about hidden assumptions. The expressions for *AREA* and *VOLUME* lack an important property possessed by the original expressions for *Area* and *Volume*: symmetry between x and y. The formulas for *Area* and *Volume* remain the same if we interchange x and y, but that does not hold for *AREA* and *VOLUME*. Symmetry is exploited in the proof on which the analogy in Example 1A is based, but the manipulations destroy the symmetry. That raises an important general problem: in evaluating an analogical argument, how do we know whether or not there are relevant assumptions that are not explicit? Features such as symmetry may be important in a mathematical proof, and highly relevant to the correct assessment of an analogical argument, even though they are not made explicit. The problem is that our theory seems to make no provision for identifying these tacit yet critical disanalogies.

The problem is interesting because we might well think that demanding *inferential explicitness* in a proof is all that we could possibly want when it comes to mathematical clarity. A proof is inferentially explicit if each step is (or can be) justified as following from earlier steps by the application of an elementary rule or principle. This type of explicitness, familiar to logicians, is adequate so long as we are only interested in assessing the validity of a mathematical argument. When we turn to the inductive side of mathematics, however, and specifically to analogy, we have to acknowledge the importance of tacit features. We may not have noticed them, yet they can be critical to an analogical argument.

We might conjecture that there is an ideal of *pragmatic explicitness*, important for inductive arguments in mathematics, that complements the traditional ideal of inferential explicitness in representing deductive arguments. A proof is pragmatically explicit if it spells out all important hidden assumptions (such as symmetry). It should be clear, however, that this idea is a non-starter. The ideal of pragmatic explicitness is unattainable, or at least something that we can never be sure of having attained. We can agree that a proof is sufficiently explicit to count as valid, but we can never be sure that we have attained pragmatic explicitness.

Accordingly, our theory defines critical relevance using the standard of inferential explicitness. It would be a mistake to resolve the problem of specious resemblance (or hidden assumptions) by insisting that all proofs be pragmatically explicit, even if we could be clear about what that meant. In the first place, any proof will almost always have interesting hidden features. In the second place, the success of an analogical argument often hinges on the fact that some or all of those tacit features are irrelevant, i.e., no obstacle to generalization. To require that we identify every tacit feature about a proof in the source domain would hobble analogical reasoning. It is enough to require that all assumptions used in an inferentially explicit proof be capable of imitation in the target. When an analogical argument meets this standard yet supports a conjecture that happens to be false, that failure brings out the importance of the hidden features but it does not show that the original conjecture was implausible.

Our best hope, then, is the somewhat vague criterion of internal coherence. We have seen that this requirement lets us deal effectively with both Example 1 and Example 1A. In my view, it also succeeds with Example 7 (Abelian Groups). There is no specious resemblance, so the similarity is admissible and the analogical argument counts as plausible, in spite of the fact that it fails.

What about Example 3, the Euler formula? Pólya's manipulations (described in Sect. 12.5) are certainly motivated by the desire to display an analogy between two and three dimensions, but that is acceptable only if they also have a natural motivation independent of the analogical argument. Let us see what our theory says about this example. In the first place, what is the analogical argument? Let us focus on the argument from the two- and three-dimensional Euler formulas to the analogous formula in four dimensions:

$$V \ (vertices) - E \ (edges) + F \ (faces) - S \ (solids) + K \ (\textit{4-dimensional objects}) = 1$$

If we have no more to go on than the formulas for two and three dimensions (see Sect. 12.5), with inductive support but no formal proof, then there is no prior association and, according to our theory, the analogical argument establishes nothing at all about plausibility.

The first person to prove the three-dimensional Euler formula in a rigorous manner was Cauchy. Cauchy's proof, restricted to convex solids, involves first collapsing the solid to a two-dimensional structure of vertices, edges and regions, then adding edges until all regions in the complex are triangular, and finally deleting edges and vertices down to a single triangle, all the while ensuring that each transformation makes no difference to the sum of $V - E + F$. I submit that relative to such a proof, Pólya's way of writing the Euler formula is sufficiently motivated, so that the similarity between the three and four dimensional cases may be considered admissible. Our theory does a fair job of explaining our intuitions about the Euler example.

Of course, we can find plenty of objections to the requirement of internal coherence. With a little ingenuity, one could probably produce an 'internal' justification even for a case as artificial as Example 1A. I acknowledge the problem, but still maintain that the requirement imposes the right kind of constraint on representation.

A second objection is that requiring internal coherence is too conservative. Too many geometric similarities become inadmissible because they depend upon unusual representations. Recall, however, that the person formulating an analogy has considerable latitude in drawing up the proof, and can probably motivate any reasonable change in representation.

In the end, it must be conceded that I have not completely resolved the problem. The requirement of internal coherence is not a precise condition, but only a general guideline about representation. Disagreement about the plausibility of a mathematical analogy may well turn on whether some novel representation can be made to appear legitimate, and the theory I have put forward does not settle all such disputes.

12.7.2 Separated Domains

Our second problem is to deal with geometric similarity between 'widely separated' domains. In Example 4 (punctured balls), the analogical inference that proceeds from the one-dimensional case to the three-dimensional case appears to be implausible, because the natural thing to do is to look first at the two-dimensional case. Yet the argument seems to qualify as plausible according to our theory, which so far provides no restrictions along these lines. Furthermore, as noted in Sect. 12.5, complications arise because there can be legitimate analogies that skip over intermediate domains.

I believe that the best way to deal with this problem is to extend the basic model of Sect. 12.4, which applies only to individual analogical arguments, to handle multiple analogies. Any case of geometric similarity with separated domains is a situation where multiple analogical arguments are available, corresponding to different possible parameter values. In Example 4, for instance, we are considering an analogy between one and three dimensions, but there is clearly a two-dimensional alternative source domain to consider as well. The extension to multiple analogies is developed in detail in (Bartha, 2010)[14]; here, it suffices to present an abbreviated version.

Suppose that we have a set of analogical arguments $\mathscr{A}_1, \mathscr{A}_2, \ldots$ with a common target domain, but possibly supporting different conjectures about the target. Suppose further that each of these individual arguments already meets the criteria of (PfP); in particular, no critical factors belong to the negative analogy for any of these arguments. We want to define a *ranking* or partial ordering denoted by $\prec$ on this set: $\mathscr{A}_1 \prec \mathscr{A}_2$ if $\mathscr{A}_2$ is superior to $\mathscr{A}_1$. There is one natural choice that we may call the *standard ranking*: argument $\mathscr{A}_2$ is superior to argument $\mathscr{A}_1$ if the positive analogy of $\mathscr{A}_2$ includes all critical factors in the positive analogy of $\mathscr{A}_1$. Everything important that $\mathscr{A}_1$ shares with the target domain is also present in $\mathscr{A}_2$. In cases involving geometric similarity, however, there is additional information to take into account, because the domains are indexed by parameters from a set (real numbers or integers) that already has its own ordering. In Example 4, for instance, it is natural to think that if we want an analogy for a conjecture about $\mathbb{R}^3$, then $\mathbb{R}^2$

---

[14]In Section 4.10.

is a better guide than $\mathbb{R}$ because (looking at the dimensional parameter) $1\leq2\leq3$. I propose to refine the standard ranking by making use of the natural ordering on the parameters.

Definition 7 (Ranking Proposal). Suppose $\mathscr{A}_1$ and $\mathscr{A}_2$ are possible analogical arguments for some conjecture about a target domain T. Suppose that the geometrically similar expressions in the three relevant domains are characterized by parameters α_1 for $\mathscr{A}_1$, α_2 for $\mathscr{A}_2$, and α for T. Then $\mathscr{A}_1 \prec \mathscr{A}_2$ ($\mathscr{A}_2$ is strictly superior to $\mathscr{A}_1$) if and only if:

(a) $\mathscr{A}_2$ is strictly superior to $\mathscr{A}_1$ on the standard ranking (based on critical factors); or

(b) $\mathscr{A}_2$ and $\mathscr{A}_1$ are equally good on the standard ranking, but the parameter value α_2 lies between α_1 and α.

On this proposal, parameter-based considerations never override the standard ordering, but they can break ties.[15] Thus, in Example 4, the two-dimensional punctured disk is strictly superior to the punctured interval as an analog for the three-dimensional case.

A second important consideration when thinking about multiple analogies is *Independence*.[16] The condition of *Independence* asserts that while the plausibility of optimal arguments (with respect to the ranking $\prec$) should be assessed independently of competing analogies, sub-optimal arguments may be ignored. Applying this result to Example 4, we reject the analogy between the punctured interval and punctured sphere. It is not that the relevant geometric similarities don't exist; rather, we reject the analogy because of the existence of a competitor that is manifestly superior.

The *Ranking Proposal* can also handle cases where it seems appropriate to skip over certain intermediate parameter values, as in parity-based analogies (discussed in Sect. 12.5). In such cases, there is a critical factor that is present in two separated domains (e.g., $n = 2$ and $n = 4$) but not present in the intermediate domain ($n = 3$). Since (on the *Ranking Proposal*) proximity in parameter value is secondary to overlap in critical factors, our proposal allows for cases where the best analog is not a neighbour but a separated domain.

A complication arises if we recognize the existence of intermediate analogs, but lack relevant knowledge about them. Suppose that we are contemplating a conjecture about three-dimensional geometry. We have a proven result for one dimension, and it meets all of our criteria to support an individually plausible analogical argument about the three-dimensional case, yet we are ignorant about whether the analogous result holds in two dimensions. In this situation, according to the above proposal, ignorance does not block the one-dimensional analogy. We

---

[15]The proposal can be generalized to cases where geometric similarity is defined using more than one parameter, although this requires the introduction of a metric.

[16]Again, see Bartha (2010) , section 4.10, for a full discussion.

still have a plausible analogical argument, at least until such time as we know more about the two-dimensional case.

To conclude this discussion, let us compare geometric and algebraic similarity. Whereas algebraic similarity is a purely syntactic notion, geometric similarity between two domains combines common syntactic structure with substitution of distinct parameters. The result is a more complex form of similarity, as reflected in the examples and our two challenges: the problem of specious resemblance and the problem of separated domains. We had to introduce an additional (and somewhat imprecise) requirement for a geometric similarity to count as admissible: *internal coherence*. Furthermore, we have seen that a satisfactory discussion of geometric similarity sometimes requires consideration of multiple, rather than just individual, analogies. Our discussion of the two forms of similarity shows that we can find alternative ways of modeling similarity—alternatives that let us raise, and partially answer, interesting questions. The next section introduces a third and even more complex form of similarity.

12.8   Asymptotic Similarity

Similarities between the finite and the infinite, between discrete and continuous systems, and other similarities involving limits are at the heart of some of the most interesting mathematical analogies. These asymptotic similarities build on the two previous types, but the use of limits introduces a new level of complexity. As noted in Sect. 12.5, arguments of this type pose a special challenge for our theory. My principal objective in this section is to show how the model can be refined so as to handle this type of analogical argument.

I shall discuss only two examples. The first is Example 5, Euler's ingenious analogy between polynomials and power series. A polynomial of degree n has the form $f(x) = a_0 + a_1 x + \cdots + a_n x^n$; a power series has the form $f(x) = \sum_{n=0}^{\infty} a_n x^n$, and is defined at x if the partial sums $f(x) = \sum_{n=0}^{M} a_n x^n$ converge to some limit. So a power series is the pointwise limit of a sequence of polynomial functions (in fact, the sequence converges uniformly inside some interval). Euler's argument depends on the idea that mathematical operations behave similarly for polynomials and power series.

The second example, to be presented shortly, is an analogy between finite and infinite matrices. A finite matrix with m rows and n columns has $m \cdot n$ elements. We refer to the element in row i and column j as a_{ij}. An infinite matrix is a matrix A of the form

$$
\begin{array}{cc|cc}
a_{11} & a_{12} & a_{13} & \cdots \\
a_{21} & a_{22} & a_{23} & \cdots \\
\hline
a_{31} & a_{32} & a_{33} & \cdots \\
\vdots & \vdots & \vdots & \ddots
\end{array}
$$

which has an element a_{ij} for each pair of positive integers i and j. The similarity between infinite and finite matrices is once again best described in terms of limits. An infinite matrix may be conceived as the limit of a sequence of finite matrices. The above infinite matrix A is the 'limit' of the sequence whose nth member is the square matrix A_n consisting of the elements belonging both to the first n rows and to the first n columns. A_2 is the 2×2 matrix in the upper left corner of the figure.

Because limits and convergence are topological notions, we need to introduce some topological ideas in order to define asymptotic similarity. A topology τ on a set X is a family of subsets (called open sets) that must include X, the empty set, all unions of members of τ, and all finite intersections of members of τ. Specifying a topology is equivalent to specifying which sequences converge: the sequence $x_1, x_2, x_3, \ldots$ converges to the limit x if and only if every open set containing x also contains all but finitely many x_n.

For example, if X is the set of real numbers, the sequence $\frac{1}{2}, \frac{1}{4}, \frac{1}{8}, \ldots$ converges to 0 because the terms of that sequence eventually become as close to 0 as one wishes, provided we go far enough along. More generally, a sequence of real numbers x_n converges to a limit x if for any non-empty open interval $(x - \varepsilon, x + \varepsilon)$ around x, all but finitely many terms of the sequence belong to that interval. This definition of convergence is equivalent to specifying that the open sets (i.e., the topology) for the real numbers consist of all open intervals or unions of open intervals.

We need one final topological notion. Given a topology on X, a subset D is dense in X if for every element x in X, there is a sequence $\langle d_n \rangle$ of elements of D that converges to x. This means that every element in X can be closely approximated by a nearby object in D. For example, the rational numbers are a dense subset of the real numbers. For any real number x, any open interval containing x, no matter how small, contains rational numbers. Equivalently: for any x, there is a sequence $d_1, d_2, \ldots$ of rational numbers that converges to x.

We really need these mathematical notions to have a realistic hope of defining asymptotic similarity. Let me explain, informally, the definition that I shall propose, referring to Example 5. In the first place, when we talk about similarities between polynomials and power series, we don't primarily mean that a particular polynomial is similar to a particular power series. Rather, we mean that the two classes share similar properties: operations such as addition and factorization (if Euler is right) behave similarly for polynomials and power series. That is the notion of asymptotic similarity that we want to capture.

The next thing to notice is that the set of polynomials can be regarded as a dense subset of the set of power series that converges on some closed interval $[-a, a]$. A polynomial of degree n is simply a power series, $f(x) = \sum_{n=0}^{\infty} a_n x^n$, with coefficients $a_{n+1} = a_{n+2} = \ldots = 0$. The appropriate topology here is uniform convergence. A power series f converges on $[-a, a]$ if and only if there is a sequence $\langle P_n \rangle$ of polynomials, with P_n of degree n, such that for each $\varepsilon > 0$ there is an N such that $n \geq N$ implies $|P_n(x) - f(x)| < \varepsilon$ for all $x \in [-a, a]$. The polynomials thus constitute a dense subset.

My suggestion is that asymptotic similarity combines limit operations with the two elementary types of similarity already discussed. An asymptotic similarity

Fig. 12.3
Asymptotic similarity

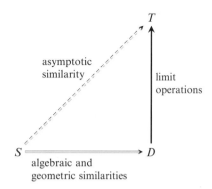

between a source domain S and a target domain T is founded upon algebraic or geometric similarities between S and an appropriate dense subset D of the target, as represented in Fig. 12.3, the diagram above. (In the power series example, S and D are the same, but that need not be the case in general.)

The full analogy (dotted double arrow) is obtained by "lifting" these similarities to T. Frequently, it turns out that conjectures based on such an analogy can be rigorously proven using the limit construction and the 'leading special case' where the conjecture is restricted to D. This pattern of reasoning is familiar to every mathematician.

Using these ideas, we are finally in a position to define asymptotic similarity of relations and functions defined on S and T. The definition is relative to the choice of (appropriate) topology on the target domain, and it makes use of the concept of the restriction of a function or relation to D. If R^* is a relation on the objects of T, then $R^*|_D$ is $R*$ restricted to objects in D; similarly, if F^* is a function on T, then $F^*|_D$ is F^* restricted to D. These notions let us characterize the relations illustrated by the figure above in precise terms.

Definition 8 (Asymptotic Similarity). Relations R and R^* (or functions F and F^*) defined on S and T, respectively, are asymptotically similar if there is an (appropriate) dense subset D of T such that R and $R^*|_D$ (or F and $F^*|_D$) are algebraically or geometrically similar.

There is no special requirement for an asymptotic similarity to count as admissible, other than that the constituent geometric similarities (if any) must be admissible, and that the choice of topology must be appropriate.

Let us see how these ideas apply to Example 5 (polynomials and power series). As already noted, the class of polynomials, $D = S$, is a dense subset of the class of convergent power series. Any relation or function on S is (trivially) identical to one defined on D, which counts as a trivial case of either algebraic or geometric similarity. In keeping with our theory, Euler's elaborate analogical argument (as outlined in Sect. 12.5) should count as plausible.

In order to reach that conclusion, however, we need to make one final change to our test for plausibility. The *No-critical-difference* condition in (PfP*) characterizes plausibility in terms of correspondences between the source and target domains. For analogies involving limits, we must substitute correspondence between S and a dense subset D of T. Here is the final version of our test for plausible mathematical analogies.

Definition 9 (PfP**: Plausibility Test for Analogical Arguments in Mathematics).

1. *Overlap.* Some explicit assumption in the proof must correspond admissibly to a fact known to be true in the target domain (or an appropriate dense subset, in the case of asymptotic similarity). (The positive analogy is non-trivial.)
2. *No-critical-difference.* No explicit assumption in the proof can correspond admissibly to something known to be false in the target domain (or an appropriate dense subset, in the case of asymptotic similarity). (No critical assumption belongs to the negative analogy.)

The modified clauses capture the familiar heuristic principle, "what holds up to the limit is also true at the limit" (Pólya, 1954, 205). To apply this principle, it suffices that any element of T can be approximated by a sequence of elements in D, each of which exhibits no known critical difference from elements in the source domain. What is true for polynomials of arbitrarily high degree is true for power series. With this final revision, Euler's analogical argument counts as plausible.

It might be objected here that mathematics is full of examples where Pólya's maxim fails. We can rearrange the order of terms in any finite series and get the same sum, but that is not generally true for infinite series. A function that is the pointwise limit of a sequence of continuous functions need not itself be continuous. And so on. In spite of the many exceptions, however, the maxim is still an excellent guide to plausible conjectures—occasional failure is not a decisive obstacle. Furthermore, our definition of asymptotic similarity and its incorporation into the test for plausibility leads to an approach that is actually more refined than Pólya's maxim. By combining ordinary analogies with limit operations, we are able to avoid many obvious counterexamples.

We illustrate this point by returning to the analogy between finite and infinite matrices introduced at the start of this section. Here, Pólya's maxim fails, but our test leads to a correct result.

Example 8 (Finite and Infinite Matrices). As noted earlier, there is an analogy between the set S of finite matrices and the class T of infinite matrices, with real number entries. For both domains, let a_{ij} represent the element in row i and column j. Consider the following analogical argument. When we sum up the elements of a finite matrix (i.e., a matrix in S), the order of addition is unimportant. The result is the same whether we first compute the sum for each row, or whether we first compute the sum for each column:

$$\sum_{i=1}^{m}\sum_{j=1}^{n} a_{ij} = \sum_{j=1}^{n}\sum_{i=1}^{m} a_{ij}$$

By analogy, we might conjecture that a similar result holds for T:

$$\sum_{i=1}^{\infty}\sum_{j=1}^{\infty} a_{ij} = \sum_{j=1}^{\infty}\sum_{i=1}^{\infty} a_{ij}$$

What does our theory tell us about this conjecture? Is it plausible?

That equality holds in the finite case is trivial to prove. In fact, it follows from the commutativity and associativity of addition. If we can put S into appropriate correspondence with a dense subset D of T, then by our theory, the conjecture will be plausible. A natural choice for D might be the set of infinite matrices whose entries all become 0 beyond some row and beyond some column (that is, there exist integers M and N such that $a_{ij} = 0$ if $i \geq M$ or $j \geq N$). Plainly S (finite matrices) and D (infinite matrices whose rows/columns are eventually 0) are isomorphic, so that the relevant algebraic similarities obtain. The difficulty, however, is to specify an appropriate topology on T for which D is a dense subset.

We might try to define the distance between two matrices with entries $\{a_{ij}\}$ and $\{b_{ij}\}$ as

$$\sum_{i=1}^{\infty}\sum_{j=1}^{\infty} |a_{ij} - b_{ij}|$$

with the intention of defining convergence (and hence a topology) relative to this distance function.[17] But there is no guarantee that this series will converge, and in general it does not converge. If we can find no plausible definition of convergence with respect to which D is a dense set, then the analogy fails to provide support for its conclusion. In fact, the conclusion is false. Let A be the following infinite matrix:

$$\begin{pmatrix} -1 & 0 & 0 & 0 & \cdots \\ \frac{1}{2} & -1 & 0 & 0 & \cdots \\ \frac{1}{4} & \frac{1}{2} & -1 & 0 & \cdots \\ \frac{1}{8} & \frac{1}{4} & \frac{1}{2} & -1 & \cdots \\ \vdots & \vdots & \vdots & \vdots & \ddots \end{pmatrix}$$

Then

$$\sum_{i=1}^{\infty}\sum_{j=1}^{\infty} a_{ij} = -2, \text{ but } \sum_{j=1}^{\infty}\sum_{i=1}^{\infty} a_{ij} = 0,$$

so that the order of summation affects the result.

---

[17]If this sum of *positive* terms converges to a finite limit, then it is absolutely convergent and so the order of summation is unimportant.

At this point, our theory suggests an interesting strategy. Let T consist not of all infinite matrices, but only of those for which the sum

$$\sum_{i=1}^{\infty} \sum_{j=1}^{\infty} |a_{ij}|$$

is finite. The distance function proposed above is well-defined and induces a topology on T, and further, D is a dense subset of T. So according to our theory, the conjecture about the order of summation becomes plausible for this restricted target domain. Indeed, it is true.

Example 8 is interesting for a number of reasons. Most importantly, it shows that our theory can avoid problematic analogical inferences from the finite to the infinite by forcing us to define precisely how objects in the target domain are obtained as limits from objects in the source domain. If we are unable to articulate this connection, then we should reject the argument.

The example also points to an important limitation of our account: we say nothing about what counts as an 'appropriate' topology. If we adopt the so-called trivial topology on T, according to which the only open sets are the empty set and T itself, then any non-empty subset of T will be dense. But the trivial topology is manifestly not appropriate in most settings. Since I offer no account of what constitutes an appropriate topology, I concede that I have only the beginnings of a theory of asymptotic similarity and limit-based analogical reasoning. Still, the analysis of the two examples in this section suggests that the theory has some promise.

This concludes our discussion of similarity in mathematical analogies, and also wraps up the discussion of criteria for good analogical arguments.

12.9 Plausibility and Analogy in Mathematics

Why should we think that the "plausibility test" developed in the last few sections is justified? Why does an analogical argument that satisfies the conditions of that test (*Overlap* and *No-critical-difference*) actually make its conclusion plausible? After briefly reviewing our understanding of plausibility, I outline what I take to be a promising approach to justification.[18]

In Sect. 12.3, I drew a distinction between modal and probabilistic conceptions of plausibility. This essay has focused on the modal conception, according to which a mathematical conjecture is plausible if there is reason for further investigation. A successful analogical argument, then, is one that yields a practical conclusion: an 'epistemic license' to proceed. The task of this section is to show why passing our

---

[18]In chapter 8 of (Bartha, 2010), I provide a more general (and detailed) argument.

plausibility test justifies such a conclusion. For simplicity, I shall focus on (PfP), the simplest form of the plausibility test.

My justification begins with two broad principles that characterize symmetry-based reasoning[19]:

1. *Positive symmetry principle.*
   Structurally similar problems must receive correspondingly similar solutions.
2. *Negative symmetry principle.*
   An asymmetry can only come from a preceding asymmetry.

These principles have theoretical applications (e.g., in drawing conclusions about asymmetric molecular structure based on asymmetries in experimental results). They also constrain our practical reasoning, as the following (non-mathematical) example shows.

Example 9 (Used Car Purchase[20]). Suppose that two used cars are indistinguishable in make, model, age, general condition, odometer reading and color—in short, in every parameter that you deem relevant to establishing a price. In such a case, it is irrational to value one car at $10,000 and the other at $12,000. The two cars are perfectly analogous, and this symmetry should be reflected when you assign values, or ranges of possible values, to the two cars. You can be imprecise or declare yourself unable to assign a value, but both cars must be treated in the same way.

More generally, if $V = F(X_1, \ldots, X_n)$ describes an evaluation function that depends solely on parameters $X_1, \ldots, X_n$, and if for $1 \leq i \leq n$ we have $X_i(a) = X_i(b)$ for two objects a and b, then a and b are perfectly symmetrical with respect to the parameters relevant to V. In such a case, we must have $V(a) = V(b)$. The two objects agree in all respects relevant to your evaluation, so you should assign them the same value (or range of values). This is meant to be an unproblematic symmetry constraint, derived from the two broad symmetry principles. Remarkably, from this simple principle and the used car example, a series of modifications brings us to a justification for our plausibility test.

The first step is to move to a situation of partial information. Suppose you know only that the two cars are indistinguishable in most of the relevant respects: make, model, age, general condition and odometer reading. You have not yet learned the color, which is the only other determinant of value. Symmetry still constrains you to assign the same range of possible values to the two cars. Suppose now that you learn that one of the cars is blue, and you assign a value of $12,000. A generalized symmetry constraint implies that the range of possible values assigned to the second car must include $12,000, as the second car would have this value if it turned out to be blue.

---

[19] Versions of these principles are identified by van Fraassen (1989).

[20] This example is a modified version of one proposed by Russell (1986) and by Davies and Russell (1987) in their discussions of analogical reasoning.

The second step is to modify the goal. Instead of wanting to assign a dollar value to used cars, your objective now is simply to decide whether any car merits a closer look. Let C denote the set of cars that you take to be serious candidates (or live options) for an eventual purchase. The members of C are cars that have been deemed either worth purchasing or worth further investigation. Suppose that car a already belongs to C, and that you learn that car b is indistinguishable from car a in make, model, age and general condition; you know nothing about the odometer reading or color of car b. You recognize that, due to symmetry, you would have to regard b as a serious candidate if it turned out to match a on these final two parameters. But you have to make a decision about b right now! I suggest that in this situation, symmetry (in the form of partial analogy) justifies a decision to place b in the set C of serious candidates. Roughly speaking: if you were to judge now that b is not a serious candidate, you would fail to respect the requirement that 'belonging to C' must lie in the range of possible values for b. To reject b now is (for all intents and purposes) to rule out obtaining any further information about it.

More generally: suppose $V = F(X_1, \ldots, X_n)$ is your evaluation function over some set of possibilities (for purchase or acceptance or some other type of endorsement) and your concern is to identify a set C or serious candidates. Suppose that a belongs to C, that $X_i(a) = X_i(b)$ for some of the X_i, and that you do not believe that $X_k(a) \neq X_k(b)$ for any k (because you have no information about the values $X_k(b)$). Then b should belong to C as well. I call this the *modal symmetry constraint* because "belonging to C" is a modal notion.[21]

The link between this principle and the test (PfP) for a prima facie plausible analogical argument is not difficult to appreciate. Let the objects a and b be domain-specific conjectures, rather than used cars. Let the parameters $X_1, \ldots, X_n$ be the relevant critical factors. "Belonging to C" becomes "being a serious candidate for further investigation." The requirement that $X_i(a) = X_i(b)$ for some of the X_i is just the *Overlap* condition: there must be a nontrivial positive analogy. The condition that we not believe $X_k(a) \neq X_k(b)$ for any k is precisely the *No-critical-difference* condition: no critical factor can belong to the negative analogy. To a first-order approximation (i.e., neglecting everything that is not a critical factor and neglecting concerns about the nature of the similarity relations), we have shown that an analogical argument that passes the test for prima facie plausibility establishes that it is rational to treat a conjecture as a serious candidate for further investigation. Furthermore, it should be clear that the conclusion does not follow for an analogical argument that fails the test.

Note, finally, that our test is fallible. Analogical arguments can satisfy (PfP) and still fail. Example 7 (Abelian Groups) provides us with a case of a perfectly good analogical argument whose conclusion turns out to be false. Obviously there *are* relevant differences between the source and target, since the theorem is true in one case but false in the other. Indeed, the proof in the source domain makes only one use of the cancellation laws, and that is what leads to the failure of the analogy. That

---

[21] For a detailed defense of this principle, see section 8.2 of Bartha (2010).

does not count as a relevant difference on the present account, and I think that is right. The inference is plausible, even though it fails.

12.10   Conclusion: A Larger Role for Mathematical Analogies

This chapter began by noting that, in addition to its role in discovery and in plausibility arguments, analogy plays a *systematizing* role in mathematics. An analogy can function as a guide in an ongoing research program. Some research may indeed be characterized as a prolonged effort to understand and extend a relationship of analogy between two structures. For instance, in the mid-twentieth century, the analogy between logic and topology was refined by Mostowski, Kleene, Addison and others, culminating in the theory of the arithmetical and analytical hierarchies. A second example is the attempt to develop systematically the analogy between the p-adic numbers and real-closed fields (Macintyre, 1986).

We might speculate that an analogy closely identified with such a research project has a kind of life history. Initially, it rests on a few points of similarity. It guides researchers in formulating new concepts and theorems, and developing more complex similarities. An analogy that is successfully developed in this way culminates in a deductive theory; a failed analogy, weakened by too many points of difference, will not lead to any such breakthrough. This large role appears to contrast with the humble analogies (and analogical arguments) that help mathematicians to solve a particular problem. Alternatively, within this large and dynamic conception of analogical reasoning, individual analogical arguments, such as those that have been our focus in this chapter, may appear as temporal cross-sections, snapshots in an evolving drama.

That leads to two interesting questions. First: should the systematic or unifying function of analogy somehow be taken into account in our model for evaluating individual arguments, and perhaps even lead to rejection or at least to substantial alteration of that model? Second (and more optimistically): might our model for evaluating individual analogical arguments shed any light on the problem of understanding the dynamic evolution of analogical reasoning?

As regards the first question, the simple answer is that we need a 'static' theory of individual analogical arguments that works independently of the big picture. When we consider analogies in terms of their systematic or unifying function, certainly the best possible 'justification' for an analogy is ultimately the successful construction of a comprehensive deductive theory. But in its heuristic role, analogy is an aid to inductive reasoning. A heuristic model of analogical reasoning must recognize this practical orientation. We need guidelines that help us decide what to do next: to decide which conjectures are likely to be worth investigating, and which proof strategies might turn out to be fruitful. Criteria used to evaluate the plausibility of analogical arguments must not depend upon the final mathematical theory—

otherwise they would be, practically speaking, useless. Furthermore, realistic criteria should be fallible: they should pronounce some conjectures plausible even though they turn out to be false, and other conjectures implausible although they turn out to be true. All of these remarks suggest that a model for evaluating individual analogical arguments should be independent of an account that explains their role in a broader research context.

Certainly, we can find reasons in favour of the opposite position, that our heuristic model must take the big picture into account. We can point to examples of analogical arguments that appear plausible not because they conform nicely to the models presented in this chapter (they don't), but rather because they are located within an evolving program of research. We have already seen such an example in the Euler characteristic formula, Example 3. As explained in Sects. 12.5 and 12.7, a creative mathematician can manipulate that formula to find an analogy between three and higher dimensions, and may find such an analogy plausible even though the most basic criterion of our theory (that we start from a rigorous proof in the source domain) is not met.

In order to respond to this argument, let me turn things around by suggesting that the heuristic model proposed here can make a genuine contribution to our understanding of the larger, dynamic role of analogies. Our model (PfP) proposes that an individual analogical argument is plausible if it has a reasonable potential for generalization, and it offers an interpretation of this potential in terms of two basic criteria. If an analogy that looks likely to be fruitful in light of experience somehow fails to meet those criteria, our model does not mandate total rejection of the argument; it leaves open the option of modification. It is at this point that historical considerations exert their influence. We may have a vague sense that the analogy can be refined. We may have found this sort of analogy to be so reliable that we are confident that some version will succeed. In the case of the Euler characteristic formula, experience may suggest to us that a rigorous proof for the three-dimensional case will almost certainly involve removing and adding vertices, edges and faces. In all such cases, broadly speaking, we retain the basic intuition that the analogy exhibits reasonable potential for generalization. If we decide to persist with the analogy, then our model provides clear guidance about how it may be restricted or reformulated in order to meet our criteria of plausibility.

Actual practice illustrates some of the most basic ways in which this can happen. When mathematicians discover a critical difference, they respond in one of three ways: they reformulate the proof (in the source domain) to avoid the critical assumption that has no target analog; they restrict the target domain so that the disanalogy disappears; or they give up on the analogy. The first two responses are natural strategies for restoring the potential for generalization. Several of our examples (Examples 6– 8) illustrate how our theory provides this sort of guidance.

Just as deductive logic and the probability calculus constrain individual arguments in science, we need a basic model for assessing individual analogical arguments. And just as deductive logic and the probability calculus have important roles to play within a dynamic model of mathematical (or scientific) evolution, so too a basic model for assessing individual analogical arguments contributes to a larger understanding of analogical reasoning.

Acknowledgements This chapter is based largely on material in my book (Bartha, 2010), with some additions and clarifications. Sections of that book are reproduced here by kind permission of Oxford University Press. Many issues that are neglected or only briefly mentioned here are discussed in the book. I also wish to acknowledge the helpful contributions of the editors.

References

Bartha, P. (2010). *By parallel reasoning: The construction and evaluation of analogical arguments*. New York: Oxford University Press.
Campbell, N. (1957). *Foundations of science*. New York: Dover.
Davies, T., & Russell, S. (1987). A logical approach to reasoning by analogy. In J. McDermott (Ed.), *IJCAI 87: Proceedings of the tenth international joint conference on artificial intelligence* (pp. 264–270). Los Altos, CA: Morgan Kaufmann.
Descartes, R. (1954 [1637]). *The geometry of René Descartes* (D. E. Smith & M. L. Latham, Trans.). New York: Dover.
Gentner, D. (1983). Structure-mapping: A theoretical framework for analogy. *Cognitive Science, 7*, 155–170.
Hadamard, J. (1949). *An essay on the psychology of invention in the mathematical field*. Princeton, NJ: Princeton University Press.
Herstein, I. (1975). *Topics in algebra* (2nd ed.). New York: Wiley.
Hesse, M. (1966). *Models and analogies in science*. Notre Dame, IN: University of Notre Dame Press.
Hume, D. (1947 [1779]). *Dialogues concerning natural religion*. Indianapolis, IN: Bobbs-Merrill.
Kline, M. (1972). *Mathematical thought from ancient to modern times*. New York: Oxford University Press.
Kokinov, B., Holyoak, K., & Gentner, D. (Eds.) (2009). *Proceedings of the second international conference on analogy (Analogy-2009)*. Sofia: New Bulgarian University Press.
Lakatos, I. (1976). *Proofs and refutations: The logic of mathematical discovery* (edited by J. Worrall & E. Zahar). Cambridge: Cambridge University Press.
Macintyre, A. (1986). Twenty years of p-adic model theory. In J. Paris, A. Wilkie, & G. Wilmers (Eds.), *Logic colloquium 1984* (pp. 121–153). Amsterdam: North-Holland.
Mill, J. S. (1930 [1843]). *A system of logic, ratiocinative and inductive, being a connected view of the principles of evidence and the methods of scientific investigation*. London: Longmans, Green and Co.
Munkres, J. (1984). *Elements of algebraic topology*. Menlo Park, CA: Addison-Wesley.
Playfair, J. (1778). On the arithmetic of impossible quantities. *Philosophical Transactions of The Royal Society, 68*, 318–343.
Poincaré, H. (1952). *Science and hypothesis* (W. J. Greenstreet, Trans.). New York: Dover.
Pólya, G. (1954). *Mathematics and plausible reasoning* (2 Vols.). Princeton, NJ: Princeton University Press.
Russell, S. (1986). *Analogical and inductive reasoning*. PhD thesis, Department of Computer Science, Stanford University.
Schlimm, D. (2008). Two ways of analogy: Extending the study of analogies to mathematical domains. *Philosophy of Science, 75*(2), 178–200.
Snyder, L. (2006). *Reforming philosophy: A Victorian debate on science and society*. Chicago, IL: University of Chicago Press.
Van Fraassen, B. (1989). *Laws and symmetry*. Oxford: Clarendon.

Chapter 13
What Philosophy of Mathematical Practice Can Teach Argumentation Theory About Diagrams and Pictures

Brendan Larvor

There has been a rising tide of interest among argumentation theorists in visual reasoning, most notably in the form of special editions of *Argumentation and Advocacy* in 1996 and 2007. In the hands of the leaders of this development, and particularly Birdsell and Groarke (1996, 2007), the effort has been to assimilate visual reasoning to verbal argumentation. At the same time, there is a more mature but still advancing literature on the use of diagrams in mathematical reasoning (e.g. Dove, 2002; Manders, 1995; Netz, 2003). There have been efforts to bring the two together (see in particular Kulpa, 2009; Sherry, 2009; Inglis and Mejía-Ramos, 2009). In this paper, I wish to use the philosophy of mathematical practice to identify a severe limitation in the attempt to assimilate visual reasoning to verbal reasoning, and by extension to criticise the approach to reasoning that treats all reasoning as if it were verbal reasoning.

13.1 Introduction

The commitment to treating all argumentation as verbal is most firmly and clearly expressed in the summary documents of the pragma-dialectical school led by van Eemeren and (until his death) Grootendorst. This approach to argumentation treats argument as composed of speech acts.[1] This treatment has manifold advantages.

---

[1]"Functionalisation of the research object in pragma-dialectics is achieved by regarding the verbal expressions used in argumentative discourse and texts as *speech acts...*" (Van Eemeren and Grootendorst, 2004, 54).

B. Larvor (✉)
Department of Philosophy, School of Humanities, University of Hertfordshire, de Havilland Campus, Hatfield, Hertfordshire, AL10 9AB, UK
e-mail: phlqbpl@herts.ac.uk

A. Aberdein and I.J. Dove (eds.), *The Argument of Mathematics*, Logic, Epistemology, and the Unity of Science 30, DOI 10.1007/978-94-007-6534-4_13,
© Springer Science+Business Media Dordrecht 2013

If the paradigmatic case of argumentation is someone making a speech, it is easy
and natural to ask who is speaking, to whom, to what end, on what occasion and
to consider the cultural, political and historical context in which this takes place.
In particular, it is a natural step to consider the nature of the relation between
speaker and audience on the point at issue. Thus, this approach is attractive to
anyone who wishes to construe the exchange of reasons as an activity that passes
between embodied human beings who meet to decide matters of human interest
on spatio-temporally specific occasions (in contrast to the abstractions of formal
logic). However, this speech-act approach has a significant implication for anyone
who aspires to consider the full range of argumentative devices available to humans.
We are not restricted to spoken words. We can make gestures, carry out physical
demonstrations of natural phenomena and build models (either physical or virtual).
We can sometimes make a point by wearing (or not wearing) a significant garment.[2]
In order to cope with this teeming variety of argumentative vehicles, the pragma-
dialectical approach has to assimilate all these alternatives to its paradigmatic case
of speech. Van Eemeren and Grootendorst make this clear in their summative work,
"In principle, argumentation is a *verbal* activity, which takes place by means of
language use…" (2004, 54, italics in original). They add in a footnote, "In practice,
argumentation can also be partly, or even wholly, non-verbal…this is not adverse
to our pragma-dialectical approach as long as the (constellation of propositions
constituting the) argumentation is externalizable." (2004, 2, n. 2). That is to say,
arguments are made up of propositions, so if a gesture or the wearing (or absence) of
a garment plays a significant role in a piece of argumentation, it can only be because
it expresses a proposition that might as well (from a logical point of view) have been
expressed verbally. Therefore, a theorist who wishes to bring this approach to bear
on visual argument has two tasks. The first is to show that visual elements (pictures,
graphs, cartoons, *etc.*) can express propositional content. The second is to explain
why, this being the case, arguers choose to express some of their propositional
content visually.

The assumption at issue is this: the analysis of an argument that explains how it
works (or why it does not work) as argument must decompose it into propositions
and identify operations on those propositions that lead from premises to conclusion.
This assumption dates back at least as far as Aristotle, but what is remarkable is the
breadth of the consensus. Even a self-conscious radical such as John Dewey held
that, "The conceptions that represent possibilities of solutions must…if inquiry is
controlled, be propositionally formulated;…" (Dewey, 1938, 394–5). A minimal
version of this claim is unarguable: the output of an argument must be either a
proposition about what is the case or a proposal about what to do. However, most
approaches to argumentation (including the pragma-dialectical approach and the
argument-schemes system developed by Walton et al.) go further than this. They
assume that the *body* of an argument must decompose into propositions, that the
process of argument must exhibit a general logic (that is, a logic that might analyse

---

[2]See Lunsford and Ruszkiewicz (2001) and Gilbert (1997) for attempts to do justice to this variety.

argumentation on any and every subject matter) and that the inferential power of the argument is best explained by reference to this decomposition and general logic. Thus, if there is such a thing as visual argumentation, the pictures and diagrams must present propositions, as spoken language does, but do so in a way that offers some advantage over speech. The contribution of the philosophy of mathematical practice here lies in the challenge it issues to this consensus.

In a series of papers (some in collaboration with David Birdsell), Leo Groarke sets about the two tasks that this consensus places before anyone who believes that there are visual arguments. In his 2007 paper with Birdsell, he attempts to counter criticisms of the very idea of visual argumentation (offered by, among others, Fleming (1996), Johnson (2005), and subsequently by Patterson (2010)). It is clear from this debate that Groarke and his critics share the consensus that arguments are best understood as sequences of (possibly implicit or indirectly expressed) propositions. To take one of many expressions of this commitment, in his paper on music in arguments Groarke concludes that, "concatenations of music, visuals and texts can often be understood as argument in a straightforward way that uses the principles of communication to distil the implicitly propositional content they contain" (2003, 421). The principles of communication he has in mind are drawn from, "an account of argumentative communication along pragma-dialectical lines" (2003, 421). Note the emphasis on communication; Groarke's strategy is to show that images (and perhaps also music) can convey propositions. If he achieves that, then (given the consensus about the nature of arguments) he has completed the task of showing that there can be visual (and perhaps also musical) arguments. The claim of the present paper is that his emphasis on communication in this strategy obscures the particular advantages of pictures and diagrams for making arguments. Groarke is the principal critical target, but the aim is to urge him to march more boldly along a road on which he is already travelling.

Groarke's 2007 paper with Birdsell is the editorial introduction to the special edition of *Argumentation and Advocacy* on visual argumentation. The articles in that issue deal with: Aristotle's classical enthymeme and the visual (Smith, 2007); visual argumentation in Scandinavian political advertising (Kjeldsen, 2007); prison tattooing (McNaughton, 2007); UNICEF Belgium's Smurf public service announcement (Hatfield et al., 2007); folk/traditional art (Roberts, 2007); flag waving (Pineda and Sowards, 2007); and functional brain imaging (Gibbons, 2007). This list is indicative of the sorts of images that theorists of visual argumentation have most often cited as examples and case-studies. In Groarke's own work, the examples that he analyses are often political cartoons and advertisements. Critics of the idea of visual argumentation charge that pictures alone cannot function as arguments because they cannot express propositions precisely, except with the support of very detailed contextualisation and often with supporting text (as is the case with many of the cartoons and advertisements). In reply, Birdsell and Groarke insist that some images convey propositional content better than speech. They divide argumentative images into five types: flags (that attract attention to an argument), demonstrations (that communicate propositional content directly), metaphors, symbols and archetypes. (Confusingly, the flags of nations are not, usually, argument-flags in Birdsell and Groarke's sense.) They then step through these five types, arguing in each case that images of this sort can contribute to arguments because they can express or invoke propositional content.

According to Birdsell and Groarke, the images that they call 'visual demonstration' are inherently propositional:

> An image is a visual demonstration when it is used *to convey information* which can best be presented visually. Demonstrations of this sort are the most effective way to explain or describe the irregular shape of a piece of land; the strategy of a hunting wolf pack; the appearance of Victorian houses in Pacific Heights, San Francisco; or the differences in contour of the skulls of different peoples... Visual demonstrations also can *present abstract information that is not conveyed easily in words*. A graph or chart, for example, may present a whole catalogue of relationships, such as correlations between GNP and military spending in various countries over the past century, that cannot be captured easily (or precisely) in a verbal description (Birdsell and Groarke, 2007, 105, emphasis added).

Since we are preparing a contrast with the treatment of images in the philosophy of mathematical practice, this is the most relevant of Birdsell and Groarke's five categories of image (which is not to say that argument flags, metaphors, symbols and archetypes have no place in mathematics). Notice that the role they give to images is to convey information. The point of the contrast with mathematical practice is this: Birdsell and Groarke's broadly pragma-dialectical analysis of these examples does not reveal them to be examples of visual reasoning. Rather, it construes them as examples of the visual presentation of information. What we can learn from mathematics is that making an argument requires other activities in addition to the presentation of information. To see this contrast, let us turn to its other side.

13.2 Mathematics

One of the most important events in the history of mathematics was the emergence of the symbolic notation that is now familiar to everyone who studies mathematics in school. This change took place in the first quarter of the seventeenth century. Late in the sixteenth century, mathematicians were still writing algebraic procedures out in ordinary prose. This mathematical prose had ready abbreviations for 'the thing sought', 'square' and so on, and these were eventually replaced by symbols. The crucial breakthrough, however, was the introduction of (notation that does the work of) brackets.[3] Then, the same quantity could be written in two ways (that is to say, brackets can be multiplied out, or conversely terms can be gathered). At first, the manipulation of these new symbols was regarded as a useful calculating aid, but such manipulations quickly (that is, in a couple of decades) gained the status of

---

[3]The earliest algebraic notations did not use brackets quite as we have them now but this does not affect the present point.

proof-procedures. Mathematicians had invented syntactic argumentation.[4] The point to note is that, while a proof consisting of a sequence of equations is (*ipso facto*) a sequence of propositions, recognising this order does not give us the argument. Moreover, the actions that constitute the process of argument are not actions that one could perform on just any content, nor even on the same content differently represented. To see this, consider the following procedure from a proof by Pólya:

Define the real numbers $c_1, c_2, c_3, \ldots, c_n, \ldots$ by

$$c_1 c_2 c_3 \ldots c_n = (n+1)^n$$

Then trivially:

$$\sum_1^\infty (a_1 a_2 \ldots a_n)^{1/n} = \sum_1^\infty \frac{(a_1 c_1 a_2 c_2 a_3 c_3 \ldots a_n c_n)^{1/n}}{n+1}$$

This is the first step in a proof that consists entirely of re-arrangements and replacements of terms (Pólya, 1954, II, 147). Given the definition of the c_i, this first step is trivial—provided we have the algebraic notation available to us, for this is a mathematical operation made evident and compelling by the notation. We could write this out in mathematical prose, but it would only be comprehensible to someone who had the skill of mentally converting prose into algebraic notation. One way to see the importance of the notation is to consider why Pólya used the recursive definition of the c_i. Formally, he might just as well have defined c_n as $(n+1)^n / n^{n-1}$, but then our 'trivial' step would not have been as obvious. Even setting up the left hand side would baffle many people who can easily understand the operation as given ("Think of an infinite set of real numbers, and put them in some order. No, it doesn't matter what order. Now consider the sequence of initial segments of that order. Got that? Then form the sum of the geometric means of those initial segments. Now hold that thought while we define another infinite set of real numbers..."). Suppose, *per impossibile*, we have thus concocted the prose equivalent to the expression on the left hand side and defined the c_i without making any use of algebraic notation. Making the step to the whole equation verbally would make yet greater demands on the same skill. Evidently, arguments of this sort are not 'an in-principle *verbal* activity' that can in practice take other vehicles than speech. Rather, arguments like this are in-principle *inscribed*. Syntactic reasoning of this sort is normally and naturally written precisely because it exploits the notation, though it can be spoken aloud among expert practitioners. So, either this kind of reasoning is not argumentation, or the pragma-dialectical approach is mistaken in thinking that argument is always and everywhere best construed as composed of speech-acts. Since the first option is contrary to the spirit of this volume, we should choose the second.

---

[4]For a brief account of this development, see Larvor (2005). For exhaustive detail, see Serfati (2005).

What are the consequences of this decision? It is clear from their discussion that Van Eemeren and Grootendorst were driven to stipulate that argument is in-principle verbal by a worry that non-verbal arguments (such as gestures or absences of garments) would not be 'externalisable', that is, would be so embedded and embodied that they would escape logical analysis. This may not be a problem with mathematical syntactic argumentation, which is by nature external to the body. This is generally the case with mathematics (which, when not inscribed, uses pebbles, pieces of string, fingers or some other easily manipulated representation). However, mathematical argument is not wholly external to the human body. As we shall see, the argumentative use of mathematical representations can depend on understanding, on aspect-dawning, on seeing-as. The consequences of this for the philosophy of argumentation depend on whether this appeal to seeing-as is a unique feature of mathematical argument or whether it is also a feature of the legal and political argumentation that the pragma-dialectical movement tends to take as paradigmatic. This paper will not deal with that question, except to note that most argumentation theorists retain Aristotle's division of the subject into logos, ethos and pathos, and insist on the relevance of all three.

So far, we may not seem to have said anything to trouble Groarke. His claim is that some pictures can convey their information more clearly and effectively than speech. The fact that algebraic notation can do this too may even be grist to his mill. It is the next point that illustrates the limitations of his approach. For, though the proof from which this procedure comes is a series of propositions (and definitions), the progress from one line to the next is not a series of logical operations. Of course, these operations must conform to general logic, but they achieve that indirectly by conforming to local, domain-specific rules (in this case: maintain the value of a quotient by multiplying numerator and denominator by the same factor). Consequently, the exhibition of the proof as a series of propositions does not give us the argument. To see this, imagine the proof written out in mathematical prose but without any symbols. We would have the decomposition into a sequence of propositions (supposing that the propositional content is fully expressed by the prose version) but we would have neither the argument nor its rigour, because the argument proceeds by manipulating, gathering and replacing algebraic terms, in conformity with domain-specific rules. The objects of our activity are not propositions but rather the algebraic expressions that express them. Returning to Groarke and Birdsell's notion of a 'visual demonstration', it is true that algebraic notation can "present abstract information that is not conveyed easily in words". However, the real power of the notation lies not in the clarity with which it presents information. Rather, an algebraic expression represents information in a way that invites manipulations that other representations of the same information do not provide for. We can do things to algebraic expressions that we cannot do to other modes of representation of the same content. These manipulations can be steps in the process of an argument. To anticipate the discussion to come: Groarke and Birdsell restrict themselves to analysing the things we can say with pictures. They do not consider the things we can do with and to diagrams.

13.3 Diagrams

Algebraic notation is quasi-pictorial. In limited ways, it mirrors diagrammatically the structures it expresses (think of matrices, for example, or the way in which an equation evokes weighing scales in balance). There are plenty of intermediate cases between algebraic notation and the diagrams of classical Euclidean geometry (such as Beth trees, Frege's two dimensional logic notation, Feynman diagrams and the arrow-diagrams of category theory). The case of ordinary school algebra suffices to make some orienting points about the limits of the pragma-dialectical approach when brought to bear on mathematics. Now it is time to consider visual reasoning proper.

Take, for example, the familiar proof of the inner angle theorem. This proceeds by constructing an angle which is both equal to the sum of the inner angles of our triangle, and evidently equal to two right angles.

It starts with a triangle:

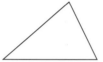

Notice that this triangle does not assert anything. Like the left-hand expression at the start of Pólya's proof, it expresses no proposition (nor is it a definition or a rule of inference). Its linguistic equivalent would be "Consider any triangle". In fact, this image does not express this instruction particularly well, because it is not *any* triangle, it is a specific triangle (insofar as it is really triangular). In particular, it is a poor representation of triangles with obtuse angles on their baselines. Its merit does not lie in the clarity or vividness with which it expresses some propositional (or imperative) content. Rather, its virtue is that we can do things to it. In this proof, we first extend the baseline:

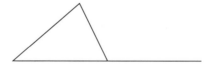

It will be important that we think of this as *extending the baseline*, rather than as adding a new line that meets the triangle at one of its corners. When explaining this proof to someone else, it is best to draw the new line in just this way, placing the chalk or pen somewhere on the baseline and dramatically drawing over it to the right, through the corner of the triangle. It will not matter how long this line is, and in a live performance of the proof, one might mark this by allowing the line to taper off or fade into dots. Next we add another new line, through the same corner of the triangle and parallel to the side opposite that corner:

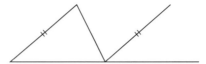

In a live performance, we would not have three separate figures as here; rather we would have one diagram, of a triangle, to which we have added new elements (our two lines and the marks to indicate parallels). This is the significant difference between these diagrams and (most of) Groarke and Birdsell's 'visual demonstrations'. These diagrams do not merely convey information; like algebraic expressions, they are objects of manipulation. There are essential steps in this proof that do not consist of producing a new proposition from a stock of axioms and theorems using a general logic, but rather work by adding to the diagram. Note that, so far, we have not encountered an explicit proposition in this proof (though we will shortly).

Now we complete the proof by using selective attention. First, we ignore one of the triangle's sides and think of the extended baseline as a line lying across a pair of parallel lines. An instructor might convey this by rubbing out the line that we need to ignore, or by using colour, or most likely with gestures, re-drawing (either literally or in the air) the two parallel lines simultaneously and then indicating the extended baseline with a sweeping gesture.[5] The proof works only for someone who can see this, who can shift their gestalt of the diagram so that the irrelevant line disappears and the remaining lines appear as a line across a pair of parallels. To help beginners, an instructor might make gestures that extend the baseline to the left and the parallel lines below the baseline. A previously proved theorem shows us that the angles it creates with the parallels are equal:

Now we perform a similar act of selective attention. This time, we ignore the baseline (and as before, one might indicate this to observers by erasing it, or by dramatically re-drawing the other lines, going to some trouble because the proof depends on this gestalt shift). This time, we want to see the remaining side as a line lying across two parallels. This may require some extra gestures because the lines in view do not look like one line crossing two parallels. They look like a zigzag. An instructor might bring this zigzag closer to the relevant visual stereotype by redrawing it in the air (or perhaps even literally re-drawing it) so that the lines extend either side of their intersections. As before, a prior theorem shows us that two angles are equal:

_____

[5]The emphasis on gesture in this section takes inspiration from (Marghetis and Núñez, 2010) and (Hacking, 2010).

Now all we have to do is gather our information and remind ourselves of the triangle that we started with:

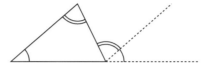

Now we see why the extended baseline has to be a single line rather than two lines pivoting at the right-hand foot of the triangle. We know that the 'angle' formed by the original baseline and its extension is 180°, that is, two right angles, because it is an angle on a straight line. This 'angle' is split up into three angles, which we now know to be equal to the inner angles of the triangle. (Helpfully, it is easier to see the angle formed above the right-hand foot of the triangle as an angle precisely because it is split into three angles; the angle formed below the same foot by the baseline and its extension is also 180°, but somehow seems less like an angle.)

As in the case of algebra, we could render this proof into prose and deliver it as speech, but it would be pretty well impossible to follow. Anyone who could follow it would do so by creating and manipulating mental images, that is, by re-creating and acting on the diagram in imagination. Some of the actions in the proof, such as quoting previously established theorems, are inferential steps licensed and governed by general logic. Other inferential actions in this proof (such as drawing lines and gestalt-shifting) are only possible with images. Here, we have visual reasoning properly so called. Or, if we wish to emphasise the social, intersubjective aspect of proving, we can imagine a viewer or student who interjects with questions. Then, we might rather call it visual argumentation. Visual argumentation need not be wholly visual. Visual arguments typically come with a (spoken or written) commentary to explain the process of argument and guide the visual gestalt-shifts that do the persuasive work. (The 'proofs' in Roger Nelsen's *Proofs without Words* (1993, 2000) are in fact diagrams that challenge the reader to supply the proofs in which they figure.) What qualifies an argument as a piece of visual reasoning is that it includes actions on or in something visual that would not be possible in another medium. The proofs of classical geometry are visual in this sense; they include actions for which no speech-acts could substitute, because they are actions on visual objects. Thus, this emphasis on inferential action supplies an answer to the critics who objected (against Birdsell and Groarke) that pictures cannot be arguments because they usually come with text. We now see that this is what one would expect, because a picture or diagram is not an argument; it is rather something on or with which an argument might be performed. It is the essentially visual character of (some of) the actions in the performance that makes the argument a visual argument.

13.4 Icons

These observations about mathematical practice are not original (though the use of them to criticise the pragma-dialectical view is). The role of notation in shaping and facilitating mathematical thought has been evident to mathematicians since early modernity and was of particular interest to Leibniz (see Larvor, 2010b). The emphasis on inferential action in ancient Greek geometry is a particular feature of Manders's (1995) analyses of Euclidean argument. More broadly, the philosophy of mathematical practice makes a point of treating mathematics as an activity rather than as a body of knowledge (see Mancosu, 2008; Larvor, 2010a). As yet, this field is relatively under-theorised, but there is one book-length treatment that suggests a means to give philosophical articulation to the thought that some arguments are essentially visual. Emily Grosholz, in her (2007) *Representation and Productive Ambiguity in Mathematics and the Sciences*, employs a distinction between 'iconic' and 'symbolic' signs that originates in C.S. Peirce. In fact, Peirce distinguishes *tokens* (general terms), *indices* (proper names, pronouns and indexicals) and *icons*. These last he characterises thus:

> I call a sign which stands for something merely because it resembles it, an *icon*. Icons are so completely substituted for their objects as hardly to be distinguished from them (Peirce, 1885, §362).

Whether a sign functions as an icon depends on whether the reader (or hearer) takes it as referring to a particular or a general condition, and this has a bearing on our case:

> A diagram, indeed, so far as it has a general signification, is not a pure icon; but in the middle part of our reasonings we forget that abstractness in great measure, and the diagram is for us the very thing... At that moment we are contemplating an *icon*. (Peirce, 1885, §362).

Peirce's qualification is essential. In the course of using a diagram, 'in the middle part of our reasonings', we may forget its generality 'in great measure', but if we forgot it altogether, we would fail to make case distinctions and check that our proof works for everything in its scope. (For example, does the proof given above work for triangles with obtuse angles on their baselines?) However, with this qualification, Peirce's observation captures something right about the experience of reasoning in a diagram. In the middle of the proof, and especially while performing the gestalt shifts, the fact that this is a general triangle, a token standing for all triangles, falls away, to be recollected later in the course of proof-checking.

Peirce claims that all three kinds of sign are necessary. Tokens are indispensible even to rudimentary predication because they are the only general signs, and without indices, we could never say anything about any particular item. He continues:

> With these two kinds of signs [tokens and indices] alone any proposition can be expressed; but it cannot be reasoned upon, for reasoning consists in the observation that where certain relations subsist certain others are found, and it accordingly requires the exhibition of the relations reasoned within an icon. ...all deductive reasoning... involves an element of

observation; namely, deduction consists in constructing an icon or diagram the relations of whose parts shall present a complete analogy with those of the parts of the object of reasoning, of experimenting upon this image in the imagination, and of observing the result so as to discover unnoticed and hidden relations among the parts (Peirce, 1885, §363).

For the present purpose, we need not endorse Peirce's strong claim that all deductive reasoning is diagrammatic.[6] Rather, our interest is in his characterisation of diagrammatic reasoning. An icon or diagram must be structurally similar to its object (in Peirce's words, it must 'present a complete analogy'). This structural similarity allows us to 'experiment' and thereby discover previously unknown relations among the parts of the object. Peirce speaks of experimenting 'in the imagination' but this need not preclude the physical manipulations of a diagram produced on a screen, drawn on a chalkboard or traced in the sand. The gestalt-shifting phases of our sample proof remind us that the imagination can play on images present to sight as well as those inwardly present to the mind.

This is, at most, the beginning of an account of diagrammatic reasoning. In any given case, we want to know which 'experiments' on the image are permitted. Which 'analogies' are reliable? If, in the midst of reasoning on a diagram, we must treat it as an icon, that is, see it as 'the thing itself', how do we guard against the errors that arise from conflating the sign with its object? The answers to such questions typically lie in the detail of particular diagrammatic practices. In the case of Euclidean geometry, we have detailed answers in the works of philosophically-minded historians such as Manders and Netz. Manders is developing a theory of mathematical proving as a 'domesticated action', that is, action subject to socially embedded norms. Dove argues for the possibility of giving satisfactory answers to these questions in mathematics generally, including the case of calculus.

If we return to Birdsell and Groarke's examples of 'visual demonstration', we see that the usefulness of these images for argument depends on the 'completeness' of the structural similarity with their objects and the range of 'experiments' permitted. We might use a picture of an irregularly shaped piece of land to argue that dividing farmland into large rectangular fields inevitably leaves some areas uncultivated (we might experiment by drawing rectangles on it). Or, we could use the same image to make a point about the efficiency of an image-compression algorithm (using a rather different set of permitted manipulations). Graphs and charts purport to present the structure of their objects, and Peirce's claim that in the midst of diagrammatic reasoning we take the icon for the thing itself suggests a diagnosis of the ease with which statistical graphs can mislead. The structural similarity is obvious in the case of a diagram showing the strategy of a hunting wolf pack; the permitted manipulations will depend on the intended conclusion (for example, a point about

---

[6]Clearly, the hard case for this claim is the spoken argument in ordinary language that the pragma-dialectical school takes as paradigmatic of argument generally. On the other hand, note that the analysis of ordinary prose arguments by argumentation theorists (from Toulmin onwards) often results in a diagram with labelled parts ('warrant', 'backing', *etc.*). In connection with the diagrammatic quality of algebraic notation, taking the sign for the thing itself is a feature of syntactic reasoning.

the effect of diminishing habitats might require that manipulations of the diagram preserve scale). In the case of the appearance of Victorian houses in Pacific Heights, San Francisco, the relevant structural similarity and the permitted manipulations will depend on the point at issue. A point about the proportions and sculptural decoration of these houses might become vivid in black-and-white versions of the images.

While Peirce's remarks about our use of icons do not amount to a comprehensive account of deductive reasoning, they do facilitate more penetrating analyses of the sort of cases that Birdsell and Groarke cite than the pragma-dialectical approach. Peirce achieves this because he takes diagrammatic reasoning to be sui generis rather than trying to assimilate it to the case of verbal reasoning. I introduced Peirce through Emily Grosholz, and I now wish to mention the use that she makes of Peirce's icons. Grosholz argues that mathematics and science can break through conceptual blocks because mathematicians and scientists are able to use the same sign in two different ways in the course of a single argument. This systematic ambiguity does not degenerate into equivocation because the scientists and mathematicians are able to control it, and this control is possible because the ambiguities are between token-use and iconic-use. Earlier, I argued that syntactic (algebraic) and diagrammatic (geometric) inferential actions cannot be understood if we treat them as speech acts. Such actions exploit features of mathematical representation that are not present in ordinary speech. This claim might be thought vulnerable to the objection that, in principle, an ideal mathematician (that is, a cognitive agent with no 'merely medical' limitations) could understand prose versions of syntactic and diagrammatic proofs without resort to syntactic and diagrammatic means. The short answer to this objection is to give it short shrift. In both philosophy of mathematical practice and argumentation theory the object is to understand the arguments made by finite, fleshy human beings. Grosholz's readings of mathematics and science supply a slightly longer and less bellicose answer. If she is right, then the arguments that she analyses would be incomprehensible if we tried to treat them as speech acts because they depend on a movement between token-use and iconic-use of signs that is invisible if the signs are treated as proxies for spoken words.

13.5   Conclusion

Birdsell and Groarke say that, "...the different modes of visual meaning allow us to assert that something or other is the case." (Birdsell and Groarke, 2007, 106). This is true; we sometimes use images to make assertions. But an assertion is not an argument and if we wish to understand visual argumentation, we have to pay attention to the special cognitive advantages that pictures and diagrams offer. As we saw in the case of geometry, we can manipulate diagrams in ways that allow us to employ our capacity for visual-spatial understanding. Since we are descended from tree-swingers, these are among our most powerful natural endowments. Beyond mathematics, other pictures invite us to perform experiments in imagination. A good example from Birdsell and Groarke is the use of photographs in arguments about whether a plane really did strike the Pentagon on September 11 2001 (Birdsell and Groarke, 2007, 109–110). The text that accompanies these photographs attempts to guide the play of imagination, just as the instructor's words and gestures guide the gestalt shifts necessary to understand the geometric proof.

Peirce's notion of the iconic use of diagrams supplies some materials for a theory of visual argumentation. Iconic use requires a resemblance, and in particular a structural similarity, between the icon and its object. In the moment of iconic use, we act on the icon as if it were the thing itself. This suggests three critical questions to put to cases of visual argumentation:

1. What is the structural resemblance between the icon and its object?
2. What manipulations of the icon and imaginative experiments on it are permitted?
3. What protection do we have against the error of taking the icon for the thing itself permanently (rather than just in the midst of reasoning when this conflation is essential to iconic use)?

In a given case of visual argumentation, we should seek answers to questions 2 and 3 (and a good part of the answer to 1) not in the images but rather in the practices in which they figure. The proofs in Euclidean geometry are reliable because the practice of geometry has suitable answers to these three questions implicit in it. That is how we are able to perform good reasoning on bad drawings. Other kinds of visual argumentation are rigorous only insofar as they too figure in practices that specify structural analogies, control the actions performed on icons and include timely reminders that the icon is not its object.

Acknowledgements I am grateful to the members of the Open University philosophy department for the opportunity they gave me to test this paper on them and to Valeria Giardino for the inspiration of her (2010).

References

Birdsell, D., & Groarke, L. (1996). Toward a theory of visual argument. *Argumentation and Advocacy*, *33*(1), 1–10.

Birdsell, D., & Groarke, L. (2007). Outlines of a theory of visual argument. *Argumentation and Advocacy*, *43*(3–4), 103–113.

Dewey, J. (1938). *Logic: The theory of inquiry*. New York: Holt, Rinehart & Winston.

Dove, I. J. (2002). Can pictures prove? *Logique & Analyse*, *179–180*, 309–340.

Fleming, D. (1996). Can pictures be arguments? *Argumentation and Advocacy*, *33*(1), 11–22.

Giardino, V. (2010). Interpretation is an action: Understanding diagrams by manipulating them. In A. Pease, M. Guhe & A. Smaill (Eds.), *Proceedings of AISB 2010 symposium on mathematical practice and cognition* (pp. 18–20). Leicester: AISB.

Gibbons, M. G. (2007). Seeing the mind in the matter: Functional brain imaging as framed visual argument. *Argumentation and Advocacy*, *43*(3–4), 175–188.

Gilbert, M. (1997). *Coalescent argumentation*. Mahwah, NJ: Lawrence Erlbaum Associates.

Groarke, L. (2003). Are musical arguments possible? In F. H. van Eemeren, R. Grootendorst, J. A. Blair & C. A. Willard (Eds.), *Proceedings of the fifth conference of the international society for the study of argumentation*. Amsterdam: Sic Sat.

Grosholz, E. R. (2007). *Representation and productive ambiguity in mathematics and the sciences.* Oxford: Oxford University Press.

Hacking, I. (2010). Proof, truth, hands, and mind. Howison Lecture in Philosophy, University of California, Berkeley. Online at http://www.uctv.tv/search-details.aspx?showID=20382

Hatfield, K. L., Hinck, A., & Birkholt, M. J. (2007). Seeing the visual in argumentation: A rhetorical analysis of UNICEF Belgium's Smurf public service announcement. *Argumentation and Advocacy, 43*(3–4), 144–151.

Inglis, M., & Mejía-Ramos, J. P. (2009). On the persuasiveness of visual arguments in mathematics. *Foundations of Science, 14*(1–2), 97–110.

Johnson, R. (2005). Why "visual arguments" aren't arguments. In H. V. Hansen, C. Tindale, J. A. Blair & R. H. Johnson (Eds.), *Informal logic at 25.* Windsor, ON: University of Windsor.

Kjeldsen, J. E. (2007). Visual argumentation in Scandinavian political advertising: A cognitive, contextual and reception oriented approach. *Argumentation and Advocacy, 43*(3–4), 124–132.

Kulpa, Z. (2009). Main problems of diagrammatic reasoning. Part I: The generalization problem. *Foundations of Science, 14*(1–2), 75–96.

Larvor, B. (2005). Proof in C17 algebra. *Philosophia Scientiae, 9.* (Reprinted in *Perspectives on mathematical practices: Bringing together philosophy of mathematics, sociology of mathematics, and mathematics education*, by B. Van Kerkhove & J. P. Van Bendegem (Eds.), 2007, Dordrecht: Springer.)

Larvor, B. (2010a). Review of *The philosophy of mathematical practice*, P. Mancosu (Ed.), Oxford: Oxford University Press, 2008. *Philosophia Mathematica, 18*(3), 350–360.

Larvor, B. (2010b). Syntactic analogies and impossible extensions. In B. Löwe & T. Müller (Eds.), *PhiMSAMP. philosophy of mathematics: Sociological aspects and mathematical practice* (pp. 97–208). London: College Publications.

Lunsford, A. A., & Ruszkiewicz, J. J. (2001). *Everything's an argument* (2nd ed.). New York: Bedford/St. Martins.

Mancosu, P. (Ed.). (2008). *The philosophy of mathematical practice.* Oxford: Oxford University Press.

Manders, K. (2008 [1995]). The Euclidean diagram. In P. Mancosu (Ed.), *The philosophy of mathematical practice* (pp. 80–133). Oxford: Oxford University Press.

Marghetis, T., & Núñez, R. (2010). Dynamic construals, static formalisms: Evidence from co-speech gesture during mathematical proving. In A. Pease, M. Guhe & A. Smaill (Eds.), *Proceedings of AISB 2010 symposium on mathematical practice and cognition* (pp. 23–29). Leicester: AISB.

McNaughton, M. J. (2007). Hard cases: Prison tattooing as visual argument. *Argumentation and Advocacy, 43*(3–4), 133–143.

Nelsen, R. B. (1993). *Proofs without words.* Washington, DC: Mathematical Association of America.

Nelsen, R. B. (2000). *Proofs without words II: More exercises in visual thinking.* Washington, DC: Mathematical Association of America.

Netz, R. (2003). *The shaping of deduction in Greek mathematics: A study in cognitive history.* Cambridge: Cambridge University Press.

Patterson, S. W. (2010). "A picture held us captive": The later Wittgenstein on visual argumentation. *Cogency, 2*(2), 105–134.

Peirce, C. S. (1885). On the algebra of logic: A contribution to the philosophy of notation. *The American Journal of Mathematics, 7*(2), 180–202. (Reprinted in *Collected papers of Charles Sanders Peirce* (Vol. 3), §§359–403, by C. Hartshorne & P. Weiss (Eds.), Cambridge, MA: Harvard University Press.)

Pineda, R. D., & Sowards, S. K. (2007). Flag waving as visual argument: 2006 immigration demonstrations and cultural citizenship. *Argumentation and Advocacy, 43*(3–4), 164–174.

Pólya, G. (1954). *Mathematics and plausible reasoning* (Vols. 2). Princeton, NJ: Princeton University Press.

Roberts, K. G. (2007). Visual argument in intercultural contexts: Perspectives on folk/traditional art. *Argumentation and Advocacy, 43*(3–4), 152–163.

Serfati, M. (2005).   *La Révolution Symbolique: La Constitution de l'Ecriture Symbolique Mathématique*. Paris: Éditions Petra. Preface by Jacques Bouverasse.

Sherry, D. (2009).   The role of diagrams in mathematical arguments. *Foundations of Science*, *14*(1–2), 59–74.

Smith, V. J. (2007).   Aristotle's classical enthymeme and the visual argumentation of the twenty-first century. *Argumentation and Advocacy*, *43*(3–4), 114–123.

Van Eemeren, F. H., & Grootendorst, R. (2004).   *A systematic theory of argumentation: The pragma-dialectical approach*. Cambridge: Cambridge University Press.

Part IV
An Argumentational Turn in the Philosophy of Mathematics

Chapter 14
Mathematics as the Art of Abstraction

Richard L. Epstein

14.1 Introduction

I'd like to tell you a story, a story of how I understand mathematics.

There are so many stories already: the platonist's, the formalist's, the constructivist's, the structuralist's, the humanist's, ... But each of these fail to answer satisfactorily at least one of the following questions:

- How do we create and know mathematics?
- How does mathematics compare to our other intellectual activities, particularly science?
- What is mathematical intuition?
- What is a proof in mathematics?
- Is a good proof in mathematics also a good explanation?
- What is mathematical truth, and are mathematical truths necessary?
- How is it that mathematics is useful in our daily lives and in science?

Any story of mathematics should answer all of these. But a good story should also:

- Be consistent with how we actually do mathematics.
- Be useful to mathematicians, leading to new and interesting work in mathematics.

I hope my story is a good one.

R.L. Epstein (✉)
Advanced Reasoning Forum, P.O. Box 635, Socorro, NM 87801, USA
e-mail: rle@AdvancedReasoningForum.org

A. Aberdein and I.J. Dove (eds.), *The Argument of Mathematics*, Logic, Epistemology, and the Unity of Science 30, DOI 10.1007/978-94-007-6534-4_14,
© Springer Science+Business Media Dordrecht 2013

14.2   Analogies and Abstraction in Science

One of the fundamental ways of reasoning about what passes in our lives is reasoning by analogy.

A comparison becomes *reasoning by analogy* when a claim is being argued for: on one side of the comparison we draw a conclusion, so on the other side we can draw a similar one.

This situation or thing is just like that; since we can draw this conclusion about the first, we are justified in drawing a similar conclusion about the second.

Such reasoning is not good until we can say in each particular case what we mean by 'is just like'and 'similar'. No two things, no two situations, no two experiences are exactly the same. We pay attention to some similarities and ignore the differences. If the differences don't matter, or rather, if they don't matter too much, then we are justified in drawing similar conclusions. The point of an analogy is to force us to be explicit about that justification, setting out, if we can, some general claim under which the two sides fall and from which the conclusions follow. Such reasoning is pervasive in our daily lives.

Science sometimes proceeds by analogy. But scientists always proceed by abstracting: choosing some aspect(s) of experience to pay attention to and claiming, perhaps implicitly, that all other aspects of experience in these kinds of situations don't matter. What we pay attention to gives us the constraints for saying whether a claim is true or false.

A scientific theory is true in the context of what we pay attention to, but is false in that it does not take into account all of the world. The hypotheses of scientific theories act as conditions for where the theory can be applied. When we 'falsify' a scientific theory we do not show that it is false. What we show is that it is not applicable to the exper ience described in the falsifying experiment. Then we try to describe carefully what kind of experience that may be, adding more conditions to our theory in the form of further premises. Thus, we can continue to use Newton's theory of kinetics even though, it is said, it is false; we use it so long as the objects we are investigating are not too large nor too small and are not going too fast; we use it where it is applicable.

The abstractions that comprise science are not false, nor are they true. They are schematic claims until we say what we are paying attention to.

14.3   Mathematics as the Art of Abstraction

Mathematics abstracts from experience, too, only much more than any science.

14.3.1   Counting and Addition

We first have numbers as adjectives: one dog, two cats, eighteen drops of water, fourteen sonatas, forty-seven ideas about mathematics. Numbers are labels.

With practice, repeating and learning the rules for naming and writing new numerals, the counting numbers become a measuring stick we carry in our head. We count off objects to find how many, as we use a tape to measure off lengths of objects. There is a definite length we cannot go beyond in measuring with our tape, but there is no definite limit we perceive in how far we can go with counting.

Counting to find how many, however, is not learned independently of using counting for addition. Consider how we learn $3 + 5 = 8$. We take something like pebbles, or pieces of candy, or dots on a paper and count:

$$
\begin{array}{ccccccc}
\bullet & \bullet & \bullet & \quad & \bullet & \bullet & \bullet & \bullet & \bullet \\
1 & 2 & 3 & & 1 & 2 & 3 & 4 & 5 \\
1 & 2 & 3 & & 4 & 5 & 6 & 7 & 8
\end{array}
$$

We show that when we put the two sequences of counting together into one sequence we get the result that the last item in the new sequence is assigned '8.' Then, after a lot of practice, it seems to us that the same results of counting would apply to any other objects like these; the two ways of counting aren't idiosyncratic to just these dots or pieces of candy. We learn to ignore what doesn't matter—we abstract—and get a theory of addition.

We've been doing this so long, we learn it so early, it is so much a part of our culture that we don't see this as a model. Surely '$1 + 2 = 3$' is true.

But 1 drop of water + 2 drops of water $\neq$ 3 drops of water. '$1 + 2 = 3$' is not a truth about the world; it is one of the claims that is needed to apply in a situation in order for arithmetic to be applicable there. We can't use arithmetic for drops of water when we put those together.

Our abstraction from counting is applicable or not. Our 'truths' of arithmetic say what follows from our abstraction and hence when our model is applicable, as in any reasoning by analogy or abstraction. Arithmetic is an application of measuring, and we must measure correctly. We have to learn (as children) how to apply the model.[1]

It seems to us, once we get the hang of it, that counting-addition is univocal. But that only says we have made it very clear how to apply it. There is no reason to think ahead of time that we will never encounter another situation in which it seems by all we have done so far that our counting-addition is wrong. If we do, we will surely restrict counting-addition not to apply to such a case, saying that what we thought were objects are not things, for only things can be counted.[2] That is, '$1 + 2 = 3$' cannot be falsified not because it is a necessary truth, but because we preserve it to be true by applying it only in cases in which it is true.

---

[1] See Appendix 1 for a further discussion of this.

[2] We recognize this in English with the distinction between count and non-count (mass) terms. See (Epstein, 1994) for a discussion of the concept of 'thing' in our reasoning.

Then we take numbers as nouns. We reify our abstracting: the end of the process of abstraction—paying attention to only some of our experience—begins to be treated as a thing, an abstract thing. We abstract from counting and addition to get the natural numbers: $1, 2, 3, \ldots$.

To say that '$1 + 2 = 3$' is true is to say that numbers—not as adjectives or as arising from counting, or as abstracted from those uses—are actual things about which we reason. That is not incompatible with what I've said, but it gives us no insight into how we create and do mathematics and why mathematics is applicable.[3]

14.3.2 The Integers

When we learn how to count and add we also learn how to subtract. We abstract from the process of taking away objects ('Hey, when you took six pieces of candy that left me with only five') to get a theory of subtraction. It may sound odd to call addition and subtraction 'theories' when they are just part of what we do every day. But they are theories just as much as Newton's in that they abstract from our experience.

When we do subtraction along with addition, we find that our calculations—that is, the working out of claims without reference to the things to which they might apply—go a lot more clearly and smoothly if we have some 'things' called zero and negative numbers. That's how and why those were first introduced. They flesh out our abstractions. They make the calculations easier. They don't seem to apply to anything.

And hence we feel uneasy about them. If they aren't abstractions from our experience, how can we trust that the calculations in which we use them give results that are applicable? When negative numbers were first introduced in the sixteenth century they were suspect. How can we understand the equality of ratios? How can $1 : -5 = -4 : 20$ when in the first ratio 1 is 'larger' than -5, while in the second -4 is 'smaller' than 20? Objections and questions about the legitimacy of their use continued until the nineteenth century, when they were given a visual/physical interpretation (see Mancosu, 1996).

---

[3]Benjamin Lee Whorf says:

> Our tongue makes no distinction between numbers counted on discrete entities and numbers that are simply 'counting itself.' Habitual thought then assumes that in the latter the numbers are just as much counted on 'something' as in the former. This is objectification. (Whorf, 1941, 140)

In a similar vein, Aristotle says:

> Whether if soul did not exist time would exist or not, is a question that may fairly be asked; for if there cannot be some one to count there cannot be anything that can be counted, so that evidently there cannot be number; for number is either what has been, or what can be, counted. (Aristotle, 1930, Book IV, 223a)

As in any scientific theory, if we introduce new 'entities' that do not arise by abstraction, there will be, and should be, objections about their use in the theory. When we can see a path of abstraction to them, we begin to have confidence in our theory.

14.3.3 Irrationals

Irrationals were not introduced like negative integers. Irrationals were always part of the abstraction of space we call geometry. Or rather, points on a line were always part of that abstraction, and some of those, it was discovered, couldn't be measured by ratios of integers. The measuring of those points then became viewed as things: irrational numbers.

14.3.4 Complex Numbers

The square root of -1 was introduced into algebraic calculations because it facilitated calculations and gave new results that could be checked by older methods. Such calculations were challenged by many mathematicians as being fantasy, as having no physical counterpart, as not reliable. Yet mathematicians continued to make those calculations in their work because no contradictions arose when they were used correctly, that is, according to the rules that were eventually determined for them.

Eventually complex numbers became accepted because they were given a visual/physical representation as points on a plane. That clear path of abstraction made us feel confident that their use was legitimate.

14.3.5 Plane Geometry

Euclidean plane geometry speaks of points and lines: a point is location without dimension, a line is extension without breadth. No such objects exist in our experience. But Euclidean geometry is remarkably useful in measuring and calculating distances and positions in our daily lives.

Points are abstractions of very small dots made by a pencil or other implement. Lines are abstractions of physical lines, either drawn or sighted. So long as the differences don't matter, that is, so long as the size of the points and the lines are very small relative to what is being measured or plotted, whatever conclusions drawn will be true. Defining a line as extension without breadth is an instruction to use the theory only when we can ignore the breadth of the line.

No one asks (anymore) whether the axioms of Euclidean geometry are true. Rather, when the differences don't matter, we can calculate and predict using Euclidean geometry. When the differences do matter, as in calculating paths of airplanes circling the globe, Euclidean plane geometry does not apply, and another model, geometry for spherical surfaces, is invoked.

Euclidean geometry is a mathematical theory, which, taken as mathematics, would appear to have no application since the objects of which it speaks do not exist in our experience. But taken as a model it has applications in the usual sense, arguing by analogy where the differences don't matter.

14.3.6   Group Theory

Consider a square.

If we rotate it any multiple of 90° in either direction it lands exactly on the place where it was before. If we flip it over its horizontal or vertical or diagonal axis, it ends up where it was before. Any one of these operations followed by another leaves the square in the same place, and so is the same result as one of the original operations.

To better visualize what we're doing, imagine the square to have labels at the vertices and track which vertex goes to which vertex in these operations.

There is one operation that leaves the square exactly as it is: do nothing. For each of these operations there is one that undoes it. For example: the diagonal flip that takes vertex d to vertex a is undone by doing that same flip again; rotating the square 90° takes vertex a to vertex b, and then rotating 270° in the same direction returns the square to the original configuration.

Already we have a substantial abstraction. No rotation or flip leaves the square drawn above in the same place because we can't draw a square with such exact precision (nor can a machine), and were we (or a machine) to cut it out and move it, it wouldn't be exactly where it was before. We are imagining that the square is so perfectly drawn, abstracting from what we have in hand, choosing to ignore the imperfections in the drawing and the movement. If we like, we can then talk about abstract, 'perfect' objects, platonic squares, of which the picture above is only a suggestion, but that seems to be only a reification of our process of abstracting.

Now consider several objects lined up in a row. We permute, that is exchange, places of some of them, again leaving them lined up in a row. Any permutation followed by another is a permutation, a way to get the objects lined up in a row again. There is one permutation that leaves everything unchanged: do nothing. For any permutation we can undo it by reversing the replacements.

Consider, too, the integers and addition. There is exactly one integer which, when we add it to anything, leaves that thing unchanged, namely, 0. And given any integer there is another which when added to it yields back 0, namely the negative of that integer.

Mathematicians noted similarities among these and many more examples. Abstracting from them, in the sense of paying attention to only some aspects of the examples and ignoring all others, they arrived at the definition of 'group.' A typical definition is:

A *group* is a non-empty set G with a binary operation $\circ$ such that:

i. For all $a, b \in G$, $a \circ b \in G$.
ii. There is an $e \in G$, called the *identity*, such that for every $a \in G$, $a \circ e = e \circ a = a$.
iii. For every $a \in G$, there is an $a^{-1} \in G$, called the *inverse* of a, such that $a \circ a^{-1} = a^{-1} \circ a = e$.
iv. The operation is *associative*, that is, for every $a, b, c \in G$, $(a \circ b) \circ c = a \circ (b \circ c)$.

We can prove claims about groups, such as that there can be only one identity in a group. Such a proof does not show that 'There is only one identity in a group' is true—not, that is, unless we reify our abstraction into abstract things called groups. But such a proof does give us a true claim:

'There is only one identity' follows from the assumptions of group theory.

If the assumptions of group theory apply to some thing/situation/part of our experience/process, then the claim about the identity of the group must apply, too. If that claim fails to apply, then we know that group theory is not applicable in that case, not that group theory is false.

14.3.7 Ring Theory

We have the integers with addition and multiplication. We have the real numbers with addition and multiplication. We have the rationals with addition and multiplication. These and many more examples led to the notion of a ring in mathematics. We can abstract from our abstractions.

But don't we mean by 'abstract' to ignore and pay attention to certain aspects of our experience?

The practice of doing mathematics is part of our experience, too. Mathematical abstractions are part of our intersubjective mental life, like laws and sonatas. Our intersubjective mental activities are also suitable subjects for abstraction. Mathematics is a human activity.

We can abstract from our abstractions, going further and further in ignoring aspects of the ordinary experience of our daily lives. Some of us have great pleasure in considering and reasoning in such highly abstract subjects. But the pleasure is merely aesthetic until applications of the abstractions are found, relating abstract subjects such as algebra and plane geometry as René Descartes did.

Perhaps you've heard the phrase 'the unreasonable effectiveness of mathematics' used as shorthand for saying that it is very odd that highly abstract theories of mathematics developed solely within the context of other abstractions can so often be applicable (see Wigner, 1960). Such a view is mistaken. As mathematicians know, it is rare for a mathematical theory of abstractions of abstractions of abstractions to be applicable to our daily experience, as opposed to just the experience of doing mathematics of a few mathematicians. And when such applications can be found, why should it be so shocking? Yes, the theory was derived from abstracting from abstracting, perhaps many levels. An application is the result of going back along that path of abstraction. And that can only be if what the theory pays attention to and what it ignores is aptly chosen. We can make a theory that ignores almost everything and pays attention to bizarre or abstruse or little-used parts of our experience. Sometimes, by chance, those aspects that are considered turn out to be important in other parts of our experience. In those few cases, the 'pure' mathematics is applicable to more of our ordinary lives. We have hit upon a good analogy.[4]

The great mathematicians, those who have some insight into what claims follow in some mathematical theory or who create a theory joining parts of mathematics never before considered similar, who make abstractions of abstractions well, are said to have great intuition, mathematical intuition. That intuition is no different from the intuition that leads a wise person to draw an analogy between dogs and humans in arguing for the humane treatment of animals, or the intuition of the wise person who first 'sees' that light can be understood as waves. In my own experience I find that the intuition I had in seeing the general outlines of this paper before I began writing it, and the intuition I had in proving a new theorem about degrees of unsolvability, and the intuition I had in writing one of my plays are the same mental activity, differentiated only by the subject matter. I see a general picture, I see a few of the details, I begin with that 'vision' or 'insight' ahead of me, and I fill in the details, often arriving at something quite different from what I first imagined.

We do not understand how such intuition works. But mathematical intuition is not something different in kind; it is only different in its subject matter.

---

[4]Reuben Hersh (2006) discusses this issue from a similar perspective.

14.3.8 Transfinite Ordinals

We have our theory of counting with the natural numbers. We can count forever and never reach the end.

But we can imagine that there is an end, even if we are not 'able' to reach it, just as there is an end to every counting we do in our lives. By analogy we postulate an end to the sequence of natural numbers and call it ω (omega). Then we can continue our counting: $\omega + 1, \omega + 2, \omega + 3, \ldots$. And since we've done it once, we'll do it again: assume an end to that counting, $\omega + \omega$. And then we can continue such counting forever.

This sounds like pure fantasy. What's to tell us that this will 'work' in the sense of never leading to contradictions? And why bother? What's the use?

We can describe such ways of counting in a constructive manner as arrays of natural numbers: one sequence is ω; two sequences, with every number in the first coming before every number in the second is $\omega + \omega; \ldots$. When we see a picture of this, we can see a path of abstraction. Moreover, such ways of counting can be seen to correspond to more and more complicated forms of proof by induction.[5]

But when someone considers extending the counting beyond what we can constructively describe, indeed into an infinite that is beyond what could be 'counted' in any sense, we have more serious doubts. Why is this acceptable? Why is this theory based on analogy and postulating new kinds of doings consistent?

14.4 Mathematical Proof

As I mentioned in the discussion of group theory and will make clearer now:

A mathematical theorem does not show that a claim is true.
It shows that the claim follows from the assumptions of the theory.

When we give a proof in Euclidean plane geometry of the claim that there cannot be two distinct lines through a point parallel to another line not through that point we are not showing that claim is true. We've seen that it doesn't even make sense to say it is true, but only true of something, in the sense of applicable. Our proof is good if it is a good argument to establish that the claim about parallel lines follows from the axioms of Euclidean plane geometry.[6]

---

[5]See the presentation by Carnielli and me in (Epstein and Carnielli, 1989)

[6]Compare the presentation of Euclidean geometry in (Epstein, 2006). Albert Einstein says:

> Geometry sets out from certain conceptions such as 'plane,' 'point,' and 'straight line,' with which we are able to associate more or less definite ideas, and from certain simple propositions (axioms) which, in virtue of these ideas, we are inclined to accept as 'true.' Then, on the basis of a logical process, the justification of which we feel ourselves compelled to admit, all remaining propositions are shown to follow from those axioms,

Mathematicians do not say, 'There cannot be two distinct lines through a point parallel to another line not through that point.' They say, 'In Euclidean geometry, there cannot be two distinct lines through a point parallel to another line not through that point.' Mathematicians do not say 'There is a unique identity.' They say, 'In any group the identity is unique,' invoking and establishing thereby that the claim is 'true' in the theory of groups.

The truths of mathematics are truths about inferences.
Mathematics is about what follows from what in our abstractions.

Yet many mathematicians and philosophers say not only that mathematical claims are true, they are necessarily true: there is no possible way such a claim could be false.

There is indeed a necessity in our mathematics, but it is not the necessity of a claim such as '$2 + 2 = 4$.' The necessity is that the claim must follow from the assumptions of the theory. There is no way that the axioms of Euclidean geometry could be true and the claim about parallel lines false. There is no way that the assumptions of group theory could be true and the claim that the identity of a group is unique be false. We demand of a mathematical proof that it establish that the inference from the assumptions of the theory to the claim to be proved is a valid inference: it is impossible for the premises to be true and the conclusion false (at the same time and in the same way).

This requirement on mathematical proofs goes back to before Euclid. It does not rely on any analysis of the forms of claims. It invokes only the notions of possibility and truth. What the notion of validity requires, and what it guarantees when it holds of an inference, is that if the mathematical claims that are the premises are true in any application, then the conclusion will be true in that application, too. The following picture summarizes this schematically:

---

i.e., they are proven. A proposition is then correct ('true') when it has been derived in the recognised manner from the axioms. The question of the 'truth' of the individual geometrical propositions is thus reduced to one of the 'truth' of the axioms. Now it has long been known that the last question is not only unanswerable by the methods of geometry, but that it is in itself entirely without meaning. We cannot ask whether it is true that only one straight line goes through two points. We can only say that Euclidean geometry deals with things called 'straight lines,' to each of which is ascribed the property of being uniquely determined by two points situated on it. The concept 'true' does not tally with the assertions of pure geometry, because by the word 'true' we are eventually in the habit of designating always the correspondence with a 'real' object; geometry, however, is not concerned with the relation of ideas involved in it to objects of experience, but only with the logical connection of these ideas among themselves.

It is not difficult to understand why, in spite of this, we feel constrained to call the propositions of geometry 'true.' Geometrical ideas correspond to more or less exact objects in nature, and these last are undoubtedly the exclusive cause of the genesis of these ideas. (Einstein, 1915, 1–2).

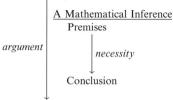

A Mathematical Proof
Assumptions about how to reason and communicate.

argument

A Mathematical Inference
Premises

necessity

Conclusion

The mathematical inference is valid.

A mathematical inference is a statement of a theorem, for example:

The axioms of Euclidean plane geometry.
<u>Therefore</u> the parallel lines postulate.

The conclusion of a mathematical proof is that the mathematical inference is valid, though mathematicians are rarely so careful as to say that explicitly. Rather, they just show the inference is valid. The mathematical proof as a whole must be a good argument for that.

14.5   Mathematical Proofs Are Arguments

Reuben Hersh, a mathematician reflecting on his and others' work, characterizes mathematical proofs:

> Mathematical discovery rests on a validation called 'proof,' the analogue of experiment in physical science. A proof is a conclusive argument that a proposed result follows from accepted theory. 'Follows' means the argument convinces qualified, skeptical mathematicians. Here I am giving an overtly social definition of 'proof' (Hersh, 1997, 6).[7]

Hersh, as many others, has conflated the two arrows in the picture above. The mathematical proof must be a good argument, and that can be loosely described as one that convinces qualified, skeptical mathematicians. But the 'follows from' that must be established is that the mathematical inference is valid.

We can say a great deal about what constitutes a good argument. To begin:

> An *argument* is an attempt to convince someone, possibly yourself, that a particular claim, called 'the conclusion' is true. The rest of the argument is a collection of claims called 'premises,' which are given as the reason for believing the conclusion is true.

---

[7]Though I criticize Hersh here, there is much good in his book, and reading it stimulated me to write this paper.

The following are necessary conditions for an argument to be good:

- The premises are plausible.
- The premises are more plausible than the conclusion.
- The conclusion follows from the premises.

Plausibility is a measure of how much reason we have to believe that a claim is true. In mathematical proofs it is not the assumptions of the mathematical inference that must be plausible. After all, how can we say that the axioms of Euclidean plane geometry are plausible when they aren't even true or false? Some of them aren't even simpler than the claim about parallel lines.[8]

The plausibility conditions apply to the mathematical proof as a whole: the assumptions about the nature of reasoning and abstractions and how we communicate must be plausible and more plausible than the conclusion that the mathematical claim follows from the assumptions of the theory.

Many times in the history of mathematics the assumptions of mathematical proofs—the part labeled 'argument'—have been questioned. At one time it seemed (or still seems) very dubious that we could reason using negative numbers, or the square root of -1, or that we could invoke infinities, or use the method of proof by contradiction, or use the law of double negation, or invoke the axiom of choice. In these cases, though no one doubted that the mathematical inference was indeed valid if given those assumptions, they doubted that the proof, the argument using those assumptions, really established that the inference was valid. They found the premises of the argument—not of the inference—dubious.

The premises of mathematical proofs, the claims about reasoning and communication, the assumptions about mathematics that we use in our everyday work in mathematics are usually left unsaid; they are treated briefly in undergraduate texts and even more briefly in undergraduate mathematics courses. They become of interest to mathematicians only when paradoxes or wrong proofs arise in some area of mathematics. Uncovering and making those assumptions explicit so that we can debate them is what logicians and philosophers of mathematics do. Logicians and philosophers then join in the debates with working mathematicians, for it is the working mathematicians who will finally decide if some method or assumption is acceptable in the canon of mathematics.

Two examples about what assumptions are currently acceptable in mathematical proofs illustrate these points.

Some mathematicians have given proofs that certain very large numbers are prime by setting out probabilistic analyses. They show that within a very small possibility of being wrong, a particular number is prime. But they then want to say more: this suffices to show that the number is prime. After all, they say, if it is certainty we want from our proofs, many mathematical proofs that are very long and have many steps left to the reader are less certain than the small probability of error of their straightforward probabilistic proofs.

---

[8]See the axiomatization due to Lesław Szczerba of Euclidean plane geometry in (Epstein, 2006). Szczerba assures me that no one has a simpler independent axiomatization.

But it is not certainty that is at issue. The issue is whether the inference in a mathematical proof must be shown to be valid, not whether the argument, the mathematical proof, is convincing. If we accept probabilistic inferences in mathematics, then mathematics is no different from any science, for all sciences use strong arguments and not just valid ones to show that a claim follows from the assumptions of their theories. At present there is little division in the mathematical community about this issue: almost everyone agrees that a mathematical proof must show that the inference is valid.

The other new method of proof utilizes computers to evaluate many complicated cases that could not be done by hand, eliminating each as a possible counterexample, and concluding a claim such as the four-color theorem. Mathematicians are hesitant to accept such proofs because they rely on our trusting that the computer software is right and that the computer itself is functioning correctly. How can that be part of mathematics? In response, it is again said that very long proofs that leave many steps to the reader and are accepted on the word of one or two referees are much more dubitable than such computer proofs.

Here the issue is not whether a mathematical proof should establish that an inference is valid, for the computer proof is claimed to do just that. The issue is about what counts as a good argument in mathematics.

Mathematical arguments, just like arguments in our daily lives, leave much unsaid. And of what is said, much is only hints or sketches, with lots explicitly left to the reader. We accept such arguments because they are a form of communication. We can see how to understand and evaluate an argument made by a person, filling in the gaps when needed. When we cannot fill in the gaps, when questions cannot be answered, we reject the argument as a mathematical proof. In contrast, a proof by computer can only be followed step by step in the hopes that we can see how each step is used in the proof, and that is impossible when there are so many steps that the prover had to have recourse to a computer. We cannot imagine the intention of the computer as a guide for how to repair a proof, for computers do not have intentions.

We can, however, try to verify that the program run by the computer does what it is intended to do. We can perform tests on a few inputs where the outputs are already known, we can examine how the program is written, and declare that the program is correct. It is possible that the program might not be, but the chances of that, it is believed, are small enough considering the cost of more extensive checking. But that, still, leaves us only with accepting or rejecting a proof by a computer, not understanding it as one would a human communication. The debate on the acceptability of computer programs in mathematical proofs continues to divide the mathematical community.[9]

---

[9]See (Davies, 2005) for fuller discussions of these and more examples of methods of proof that are currently in debate in the mathematical community.

14.6   A Comparison of Two Proofs of a Simple Claim in Arithmetic

Claim. $1+2+\ldots+n = \frac{1}{2}n \times (n+1)$

Proof. $1 = \frac{1}{2}1 \times (1+1)$. This is called the *basis of the induction*. Suppose that $1+2+\ldots+n = \frac{1}{2}n \times (n+1)$. This is called the *induction hypothesis*. Then:

$$1+2+\ldots+n+(n+1) = [\frac{1}{2}n \times (n+1)] + (n+1).$$

$$\text{So } 1+2+\ldots+n+(n+1) = \frac{1}{2}(n^2+n) + \frac{1}{2}(2n+2),$$

$$\text{so } 1+2+\ldots+n+(n+1) = \frac{1}{2}(n^2+3n+2),$$

$$\text{so } 1+2+\ldots+n+(n+1) = \frac{1}{2}(n+1) \times (n+2).$$

$$\text{That is, } 1+2+\ldots+n+(n+1) = \frac{1}{2}(n+1) \times ((n+1)+1). \qquad \text{Q.E.D.}$$

Some proof like this is what all mathematicians have encountered in learning mathematics. We are told that anything less rigorous than this doesn't count as a proof. We are told it establishes that '$1+2+\ldots+n = \frac{1}{2}n \times (n+1)$' is true.

But the proof does not establish that '$1+2+\ldots+n = \frac{1}{2}n \times (n+1)$' is true. What it shows is that the claim follows if we accept the method of proof by induction. But isn't that true? True of what—our abstraction? No. The method of proof by induction is a condition of our theory: it is part of what establishes the subject of arithmetic. That method is not obvious to students until they have been drilled on it; it is very hard to use. Any claim that follows from the method of proof by induction on the natural numbers must be applicable to any instantiation of the theory. If such a claim fails, then that's not what we are talking about.

This is what the ultra-constructivists who deny induction miss (see, for example, Van Dantzig, 1956). Their program is not about what the natural numbers are, nor the right way to reason about them, nor disproofs of induction. They are going back to counting and deciding not to make the abstraction from our abilities to say that we can 'count' forever.[10] They propose a more 'realistic' theory of counting and arithmetic, where a theory can be said to be more realistic than another if it abstracts less from the same experience(s).

---

[10]Or in the case of David Isles (1981, 2004) that arithmetical functions such as multiplication or exponentiation do not obviously lead to places on the list of numbers we can count to by 1's.

Claim. $1 + 2 + \ldots + n = \frac{1}{2} n \times (n+1)$

Proof. ● ○ ○ ○ ○ ○ ○

  ● ● ○ ○ ○ ○ ○

  ● ● ● ○ ○ ○ ○

  ● ● ● ● ○ ○ ○

  ● ● ● ● ● ○ ○

  ● ● ● ● ● ● ○   Q.E.D.

Is this a proof? When I saw it I felt for the first time that I understood and could believe '$1 + 2 + \ldots + n = \frac{1}{2} n \times (n+1)$.' Before, I only knew from memory that there is a proof using a manipulation of symbols that I could reconstruct fairly easily. Each time I did that algebraic proof I saw indeed that the claim followed by induction. But here I could see that it was true, really true.[11]

True? The picture convinced me, without any recourse to induction, that the equality will apply to any things to which our basic model of counting and arithmetic apply. I see from the picture that any things I can count up to some number, call it n, can be put, along with other such things, in an array that justifies the equality by counting again. And a 6 by 7 array was enough to convince me.[12]

But how do we know that we'll always get the same result, regardless of how large we expand the diagram? We know in the same way that we know '$3 + 5 = 8$' will be true for any selection of objects to which we wish to apply our methods of counting and addition. We can give a proof of '$3 + 5 = 8$' from axioms for arithmetic, but that is less convincing and assumes a great deal more as a theory of addition.

The first proof shows that the equality follows from the assumptions of our more general theory of counting and arithmetic in which we also accept proof by induction; the assumptions about reasoning, though not about the nature of our abstractions, are relatively explicit. This second proof by picture leads us to

---

[11]The picture comes from (Nelsen, 1993, 69). Martin Gardner, in (Gardner, 1973) from which Nelsen takes this example, says:

> The first n consecutive positive integers can be depicted by dots in triangular formation. Two such triangles fit together to form a rectangular array containing $n(n+1)$ dots. Because each triangle is half of the rectangle, we see at once that the formula for the number of dots in each triangle is $n(n+1)/2$. This simple proof goes back to the ancient Greeks (Gardner, 1973, 114).

[12]A colleague said that the picture is convincing for $n = 6$ but not for any larger numbers. But then why not say it is good only for circles colored and laid out in this manner? See (Sherry, 2009) for a fuller discussion of this point.

believe that the equality applies to anything to which our more basic theory of counting and arithmetic apply without invoking induction; the assumptions about the modeling are relatively clear, but the methods of proof are not. In neither case are we showing that a claim is true; we are showing that it follows, that it is part of our theory. The truths of mathematics are truths about inferences and applications of theories.

14.7 A Proof of Pythagoras' Theorem, and Progress in Mathematics

Consider another proof by diagram, this time of Pythagoras' theorem.

First, we can represent the product of two numbers a and b by a rectangle with sides of length a and b. Then we have the identity:

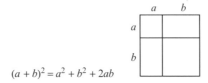

$$(a + b)^2 = a^2 + b^2 + 2ab$$

Now we note the area of a right triangle with sides a and b is $\frac{1}{2}ab$:

Then we can get Pythagoras' theorem: the square of the length of the hypotenuse in a right triangle is the sum of the squares of the lengths of the legs. All we need do is calculate the area of the large square below by the two different methods:

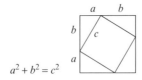

$$a^2 + b^2 = c^2$$

It is clear to me once I have understood the diagrams that it is impossible to come up with a triangle that would not satisfy the identity.[13]

---

[13]We can easily see that the method does not depend on these specific triangles and rectangles. Indeed, in ancient times often just one example, such as using '5' and '8' here (which is the ratio

Yet some say that these diagrams do not constitute a proof. Rather, it is only in Euclid that we find a real proof of Pythagoras' theorem. That view, I believe, is a reflection of the desire to make our implicit assumptions explicit. Geometric assumptions necessary for proving Pythagoras' theorem are explicitly set out in Euclid. Or at least they seemed to be for over two millennia. But now the same criticism of Euclid can be given as is given of the diagrams: too much was left as unstated assumptions. The first time we had a real proof of Pythagoras' theorem, it is said, was when Hilbert formalized Euclidean geometry.

What is a proof depends on what we tolerate as implicit assumptions. If progress in mathematics means (at least in part) making more explicit the assumptions by which we prove results, then it is not clear that Hilbert's system is progress over the diagrams because the assumptions he has made explicit seem to go far beyond what was in our use of the diagrams. What progress there is in mathematics depends more on the assumptions that are being made explicit having greater generality and applicability. That is, progress in mathematics in this sense depends on greater generality through abstraction.[14]

This is not, however, the exclusive kind of progress in mathematics. Solving a difficult and long-known problem counts as progress, too. But universally it is acknowledged that such a solution is not of much interest, that is, is not really considered progress, if it does not employ new methods that allow for the solution of other problems or a generalization of the original problem. That is what is frustrating about evaluating the use of computers in the proof of the four-color theorem: nothing new is used in the proof, no new generalities or ideas, just the brute force of the computer working through many, many cases.

14.8 Formal Proofs

Some mathematicians think an objective standard can be given for what counts as a proof in mathematics. They say an argument counts as a mathematical proof only if it could be formalized, by which they mean within a system of formal logic (see Fallis, 1998).

---

as close as the printer or computer monitor can manage), was given and the reader was expected to see that the method of proof or calculation was quite general. See (Bashmakova and Smirnova, 2000).

The issue of when we are justified in understanding specific parts of a diagram as variable is examined in (Kulpa, 2009). The counterpart in reasoning with formulas is given a precise answer in classical predicate logic by the theorem on constants: if a name appears in a theorem but does not appear in the axioms, then it can be replaced by a variable that is quantified universally (Epstein, 2006, Lemma 10.b). Kulpa's paper also surveys recent work on formalizing reasoning with diagrams.

[14]See (Breger, 2000) for a discussion of this point and more on the nature of abstraction in mathematics.

In (Epstein, 2006) I provide a derivation in first-order logic of the claim 'The identity is unique' from the axioms of group theory. But that is not a mathematical proof. I give the mathematical proof first, and then argue that the formalization is apt.

A proof in a fully formal system of logic that a claim follows from some axioms is not a proof in mathematics. It is evidence that can be used in a mathematical proof: Why should I believe that this claim follows from these others? Because— and we point to the formal proof. It is the pointing that is crucial. It relies on many assumptions, most particularly that the formal system chosen for the formalization is an apt model of reasoning and a good one for this mathematical proof, and that the steps that have been added—for there are always steps that have to be filled in—are appropriate.[15] We cannot fully formalize the argument that constitutes the mathematical proof without leading to an infinite regress.

We cannot replace proofs in mathematics with formal proofs, though we can use formal proofs as evidence in mathematical proofs. Formalizing mathematical proofs can lead to uncovering or clarifying assumptions behind such informal proofs and seeing how or whether such assumptions are needed.

14.9   Mathematics as Pure Intuition

At the other extreme, some say that mathematics is entirely subjective. The noted mathematician R. L. Wilder says,

> What is the role of proof? It seems to be only a testing process that we apply to these suggestions of our intuition.
>
> Obviously, we don't possess, and probably will never possess, any standard of proof that is independent of time, the thing to be proved, or the person or school of thought using it (Wilder, 1944).

The celebrated mathematician G. H. Hardy seems to concur:

> There is strictly speaking no such thing as a mathematical proof; we can, in the last analysis, do nothing but point; ... proofs are what Littlewood and I call gas, rhetorical flourishes designed to affect psychology, pictures on the board in the lecture, devices to stimulate the imagination of pupils (Hardy, 1928, 18).

And the mathematician and philosopher L. E. J. Brouwer says,

> In the construction of [all mathematical sets of units which are entitled to that name] neither the ordinary language nor any symbolic language can have any other rôle than that of

---

[15]As I show with examples in (Epstein, 2006), first-order logic is rarely appropriate because much mathematics requires second-order assumptions. Second-order logic and set-theory give no unique standard because there are many systems of those that differ too much. Further, all those systems are based on a metaphysics that denies a mathematics of process as distinct from things (see Epstein, 2010). Moreover, it is quite common to establish a theorem in a formalized theory by semantic means rather than with a syntactic proof, as can be seen in (Epstein, 2006).

serving as a nonmathematical auxiliary, to assist the mathematical memory or to enable different individuals to build up the same set (Brouwer, 1912, 81).

To do mathematics we must use our intuition. But that does not mean that mathematics is subjective. It is intersubjective, like law, like drama, like etiquette. To appeal to each individual mathematician's intuition leaves us no criteria at all for judging whether we have a proof, just as appealing to only a judge's intuition gives no standard for what is legal. And we know at least one clear standard for mathematical proofs: they must establish that an inference is valid.

The view of mathematics as subjective introspection leaves it difficult for us to explain how we learn mathematics, how we judge whether what we have is a proof, what mathematical truth is, ... All the questions we began this paper with remain unanswered. Though intuitionists who followed Brouwer have created a full and rich mathematical theory, they did so by denying his quote above: they use and work with proofs just as all other mathematicians do.

14.10   Set Theory and the Existence of Infinities

Given any few small objects we can collect them together. Given any things we can describe, we can make up a description of all of them at once. So proceeding by analogy, mathematicians in the nineteenth century assumed that we can 'collect' any things whatsoever into a new entity called the 'set' of those things.

We can describe a beautiful theory that way that has many applications. But it leads to contradictions, such as the set of all sets that are not elements of themselves. Not every beautiful way of postulating new things in analogy with old ones is good.

But the utility of such a theory is so desirable that mathematicians worked to rescue it, modifying the analogy to say that only certain ways of collecting are acceptable. And thus we have modern set theory. And we hope that it is consistent, for it is a very useful high-level abstraction in which we can codify and apply to many areas of mathematics.

The assumptions of our new set theory countenance 'collecting' all natural numbers into one set. We can also 'collect' all points on the line into one set—as if a line that is finite but forever extendible were actually extended and completed as an infinite thing.

That makes a lot of mathematicians uneasy, from the ancients, through the seventeenth century, and continuing today. Intuitionists and constructivists deny that such abstractions are legitimate in mathematics. They can see no path of abstraction that leads to such a fantastical analogy. Those who use set theory say it should be accepted because it is fruitful and (so far, it seems) consistent.

The utility and consistency of a mathematical theory, some argue, are sufficient for us to investigate it. Indeed, not even utility but just a sense that the theory is beautiful has been enough in the past century for mathematicians to publish papers on new theories.

But, as we've learned from Kurt Gödel, it is rare that we can prove a theory to be consistent (see Epstein and Carnielli, 1989). So we try to relate it to other theories we know arise from a path of abstraction and for which we have inductive evidence of consistency: no contradiction has arisen in the many years that many mathematicians have worked in that theory.

Some go farther. They say that consistency of a mathematical theory is all that is needed for us to conclude that the things of which it speaks exist.

> In particular, in introducing new numbers, mathematics is only obliged to give definitions of them, by which such a definiteness and, circumstances permitting, such a relation to the older numbers are conferred upon them that in given cases they can definitely be distinguished from one another. As soon as a number satisfies all these conditions, it can and must be regarded as existent and real in mathematics. (Georg Cantor 1883, 182).[16]

> A mathematical entity exists, provided its definition implies no contradiction. (Henri Poincaré 1921, 61).

> If the arbitrarily given axioms do not contradict one another with all their consequences, then they are true and the things defined by the axioms exist. This is for me the criterion of truth and existence. (David Hilbert, Letter to Frege of December 29, 1899).[17]

The infinities beyond infinities of set theory exist, they say. And they often say so without the qualifier 'if our set theory is consistent.' They have gotten used to working in the theory and thinking of these objects, so how could they not exist? If the theory is contradictory, they feel that they can modify it once again to retain their world of abstract objects. They do not consider that if they modify their theory, the objects of which it speaks might not be the ones they had been thinking of earlier.

But we do not need that mathematical objects postulated by our theories exist in order for our theories to be used and to be useful. We only need that the theory is consistent, for then we can act *as if* they exist: it is not logically impossible for them to exist. And possibilities are all we need in order to reason about mathematical inferences in our mathematical proofs: an inference is valid if there is no *possible way* for the premises to be true and conclusion false.

Some find it remarkable that our theories are consistent and cite that as evidence that mathematical claims are indeed true or false (Putnam, 1975, 73). But the process of abstraction, ignoring some of our experience and focusing on just part of it, is not likely to lead to an inconsistency. It is only when we postulate something in addition to our experience that we risk inconsistency. The great difficulty in analyzing processes of abstraction is distinguishing between those cases where we only ignore certain aspects of our experience, as with addition and multiplication of counting numbers, and those cases where we postulate some additional ability or capacity of ourselves or some extension of our experience, as in set theories that allow for infinite collections.[18]

---

[16]The translation here comes from (Dauben, 1979, 128–129).

[17]Frege (1980, 39–40). Frege strongly disputes Hilbert's view (op. cit., 43–47).

[18]This issue in historical context is discussed in (Detlefsen, 2005).

14.11 Mathematical Proofs as Explanations

It is often said that a good mathematical proof does more than just show that a mathematical claim is true, it provides a good explanation of why the claim is true. We want to know not just that '$1 + 2 + \ldots + n = \frac{1}{2}n \times (n + 1)$' is true, but why it is true (see, for example, Mancosu, 1996, 2000).

An explanation in this sense is characterized as follows.

> An *inferential* explanation is a collection of claims that can be understood as 'E because of $A, B, C, \ldots$'. The claims $A, B, C, \ldots$ are called the *explanation* and E is the claim being explained. The explanation is meant to answer the question 'Why is E true?'

For an inferential explanation to be good, the inference from $A, B, C, \ldots$ to E must be valid or strong. And the claim being explained must be highly plausible: we do not explain anything we do not already believe.

To view a proof of a mathematical claim as an explanation raises the same problems as viewing a mathematical proof as an argument for that claim: we must accept that mathematical claims are true or false, not just true or false in application, and we have to come up with a way to understand how one mathematical claim is more or less plausible than another. It is this latter that has particularly stymied attempts at analyzing mathematical proofs as explanations (see Mancosu, 2000).

Further, for an explanation to be good, at least one of the claims doing the explaining must be no more plausible than the claim being explained. Otherwise, we would have an argument for the conclusion. Thus we would have to have that some inferences to '$1 + 2 + \ldots + n = \frac{1}{2}n \times (n + 1)$' are to be judged as arguments for that claim, establishing the truth of it, and some are to be judged as explanations, telling us why the claim is true. There would not be a uniform standard by which to judge mathematical proofs.

None of these problems arise if we go back to the schematic diagram of mathematical proofs presented above (Sect. 14.4). The mathematical inference to the mathematical claim is neither an argument nor is it an explanation; it is a pure inference, to be judged solely as to whether it is valid or not. The mathematical proof is an argument that the inference is valid. Part of the feeling that some proofs are better at showing 'why a claim is true' has to do with what criteria we have for such an argument to be good.

We have seen necessary conditions for an argument to be good (Sect. 14.5). We can also say that one argument is better than another if its premises are more plausible and it is more clearly valid or strong. For some kinds of arguments, such as generalizations and analogies, more can be said about what constitutes a good or better argument. But consideration of what conditions are needed for one argument to be better than another in mathematics has been obscured by seeing it in terms of explanations. For example, Mark Steiner says,

> Now I see no reason, except dogmatism, not to accept this story at face value: The embedding of the reals in the complex plane yields explanatory proofs of otherwise

unexplained facts about the real numbers. The explanatory power of such proofs depends on our investing the complex numbers with properties they were never perceived as having before: length and direction (Steiner, 2000, 137).

There is more than dogmatism as a motive to reject Steiner's story, for it depends on our assuming that mathematical claims have truth-values. Yet if mathematical claims have truth-values, we cannot 'invest' the complex numbers with properties. Either they have those properties or they don't. In any case, mathematicians don't claim that the complex numbers have those properties: we represent them using those quantities.

To see better what Steiner might be getting at, consider the following quotation from Philip Kitcher:

> And as in other sciences, explanation can be extended by absorbing one theory within another. It is customary to praise scientific theories for their explanatory power when they forge connections between phenomena which were previously regarded as unrelated. Within mathematics the same is true and it has become usual to defend the 'abstract' approach to mathematics by appealing to the connections which are revealed by studying familiar disciplines as instantiations of general algebraic structures (Kitcher, 1975, 259–260).

It is not clear what Kitcher means by 'explanation' and 'explanatory power,' for we are not explaining anything in his examples. Rather, we are setting out further analogies, connecting our abstractions to show that they have instantiations we didn't previously see, and showing that some of our abstractions can be abstracted further to relate them by analogy to other abstractions. As always, good analogies help us 'see' the relationships in the sides of the analogy. It's not different from marijuana compared to alcohol, or humans compared to dogs: we don't explain anything with such an analogy, but we do see common aspects and reason to similar claims based on those aspects when the differences don't matter.

We need such further analogies or instantiations of our abstractions because our abstractions have become too abstract to reason about well, or because they are so abstract we are not sure they are related to anything in experience beyond what they have been abstracted from, or because we are postulating new entities that need to be shown to have an instantiation in something less abstract, or because we have run out of ways to conceive of further progress in the area and have need of some other way to visualize the subject. All these help us understand our abstractions better. But they are not explanations.

Consider what Ernest Nagel says about a particular mathematical claim that needs explaining:

> Why is the sum of any number of consecutive odd integers beginning with 1 always a perfect square (for example, $1 + 3 + 5 + 7 = 16 = 4^2$)? Here the 'fact' to be explained (called the *explicandum*) will be assumed to be a claimant for the familiar though not transparently clear label of 'necessary truth,' in the sense that its denial is self-contradictory. A relevant answer to the question is therefore a demonstration which establishes not only the universal truth but also the necessity of the explicandum. The explanation will accomplish this if the steps of the demonstration conform to the formal requirements of logical proof and if, furthermore, the premises of the demonstration are themselves in some sense necessary.

> The premises will presumably be the postulates of arithmetic; and their necessary character will be assured if, for example, they can be construed as true in virtue of the meanings associated with the expressions occurring in their formulation (Nagel, 1961, 16).

The fact to be explained cannot be 'the sum of any number of consecutive odd integers beginning with 1 is always a perfect square' because that is not obviously true. It doesn't need to be explained but demonstrated. The following proof of that claim does all one could hope for in making clear why the claim is 'true.'

Claim. $1 + 3 + 5 + ... + ((2n) + 1) = (n + 1)^2$

Proof.

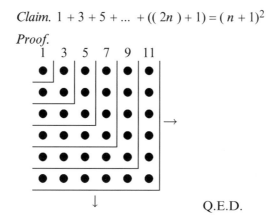

Q.E.D.

As Martin Gardner says,

> Think of the pattern as extending any desired distance to the right and down. Each reversed L-shaped strip contains the odd number of circles indicated at the top. It is obvious that each additional strip, that is, each new odd number in the series $1 + 3 + 5 + ...$, enlarges the square by one unit on a side, and that the total number of dots in each square bounded by the nth odd number is n^2 (Gardner, 1973, 114, from which this diagram is taken).

The picture convinces us that the claim is true in any application of the assumptions of our theory of addition and multiplication. To say that it is thus true by virtue of the meaning of the expressions requires us to pack a very great deal into what is understood by 'meaning.' When a mathematical theory can be shown not to apply to some situation/experience/thing/process, such as addition to drops of water, we do not say the theory is false, but that this is a bad application. Has the meaning of the expressions changed? Or are we only clarifying what we implicitly assumed were the meanings? The appropriateness of an application seems hard to assimilate to implicit meanings of expressions in a theory.

Finally, consider what Michael Resnik and David Kushner say about proofs as explanations:

> We can account for what is probably the most basic intuition behind the idea that there must be explanatory proofs as such, namely that all proofs convince us that the theorem proved is true but only some leave us wondering why it is true. We have this intuition, we submit, because we have observed that many proofs are perfectly satisfactory as proofs but present so little information concerning the underlying structure treated by the theorem

that they leave many of our why questions unanswered. In reflecting on this, we tend to
conflate these unanswered why questions under the one form of words 'why is this true?'
and thus derive the mistaken idea that there is an objective distinction between explanatory
and non-explanatory proofs (Resnik and Kushner, 1987, 154).

The distinction between explanatory proofs and non-explanatory ones is no different
for arguments in daily life, such as newspaper editorials. Sometimes we can follow
the steps of the argument very well but remain unconvinced. We are not being
irrational: we just don't 'see' the connections that make the argument valid or strong.
Similarly, we sometimes need to 'see' the connections in mathematics. It requires
us not only to see the deductive connection but the relation to what we already know
or to something familiar, perhaps through a higher-level abstraction that creates an
analogy. Asking 'Why is this true?' leads to the mistaken idea of looking for the
grounds of the truth of a mathematical claim. Often what is explanatory replaces
symbolic manipulation (e.g., an induction proof) with something more concrete in
our experience (e.g., a picture).

14.12   The Utility of This Story

A good story of mathematics should lead to new and interesting work in
mathematics.

By considering the path of abstracting in a mathematical subject rather than
focusing only on the final abstraction, we can see where we have chosen to ignore
certain aspects of our experience. Then, when the development of the subject
becomes stuck, when we cannot adapt our abstractions to accommodate problems
that resist solution, we can look to what we have ignored and see if it is possible to
take more into account. This is the usual method of scientists; it works equally well
for mathematicians.[19]

---

[19]This is how I developed the general framework for semantics in (Epstein, 1990). Rather than
viewing propositional logics as about abstract things called 'propositions,' I saw them as ways
to reason using ordinary language and abstractions of that. Rather than looking for the 'right'
logic that captures exactly the properties of such abstract things, I saw that what we pay attention
to in our abstracting, what aspects of ordinary language claims we deem important, determines
the appropriateness of the logic we choose. As we vary the aspect we deem important, we vary
the logic. That variation, I saw, can be described by devising an abstraction of our propositional
logics—an abstraction of our abstractions. The general structures that arise are then worth
investigating, not only for their own interest but for the relations among logics they illuminate
and for the assumptions about how to reason well they uncover. I did not deny the abstract, for
how could I show that there are no such things as abstract propositions? Rather, I focused on the
process of abstracting, and that gave rise to new mathematics that is grounded in experience. See
(Epstein, 1988) for a summary.
    This is also what I do in (Epstein, 2006). I present formal logic as an abstraction of reasoning.
By considering propositions as actual utterances or inscriptions rather than making the abstraction
to treat different utterances as the same thing, I give a formal logic that resolves the liar paradox.

By considering how we develop and use mathematics, we are no longer left grasping at abstractions hoping they will somehow reveal to our intellects new solutions, abstractions, or analogies through their ineffable nature. By remembering that mathematics is a human activity, we can solve more.

We can encourage this view with our teaching. By introducing the process of abstracting rather than focusing only on the final abstraction we can help students grasp new concepts and make use of them, not only as a subject to be learned but as a tool for modeling further.[20]

14.13   Grounding Our Stories of Mathematics

The view of mathematics that I present is grounded in mathematics as a human activity. Any story of mathematics has to account for that. The assumptions about the world that I invoke, such as that there are people who do mathematics who communicate and who have intuitions, are not controversial. In that sense, I have offered a story that has minimal metaphysics.

I model what I see mathematicians doing. I do not try to account for all. Abstract things arise—or come to our attention if you are a platonist—via our abstracting. All mathematics is applied mathematics; we've just forgotten that it's applied because of its familiarity. Mathematics is abstraction, proofs, and applications, repeated over and over, limited only by our experience and imagination.

But the question can still arise: Why these abstractions?

We can ask for ultimate explanations. Perhaps we have intersubjective work in mathematics because there are platonic objects that ground our insights; or because mental constructions and subjective thoughts are somehow shared by all thinking creatures; or because . . . . These ultimate explanations are the basis of other views of mathematics. They are beyond testing, except for whether they answer the questions that a good story of mathematics should answer.

We can view various philosophies of mathematics with their accompanying metaphysics as providing us with some ultimate account of why we do mathematics as I have described. Or we can see such stories of abstract objects and mental constructions as psychological props that allow us to reason better about our abstractions. No matter; such stories are compatible with what I have said here.

As for me, I am content to do mathematics and to reflect on how we communicate when we do that. Come, let us reason together.

By considering how we assign truth-values to atomic propositions rather than assuming that they come with truth-values, I develop a simple modification of classical mathematical logic that deals with names that do not refer.

[20]Walter Carnielli and I (Epstein and Carnielli, 1989) use the historical development of the theory of computable functions to recapitulate the path of abstraction in that subject. This is the 'genetic' method of teaching, long favored by George Pólya, as discussed in (Pólya, 1963).

Acknowledgements A previous version of this chapter was published as 'On Mathematics' in Richard L. Epstein, *Logic as the Art of Reasoning Well*, Advanced Reasoning Forum, 2008, 411–441. I am grateful to Fred Kroon, Charlie Silver, Carlo Cellucci, Jeremy Avigad, David Isles, Reuben Hersh, Ian Grant, Paul Livingston, and Andrew Aberdein for their comments on earlier versions.

Appendix 1: Mathematics as an Innate Ability

The discussion of counting and arithmetic above reflects my understanding of experiments done by Jean Piaget with children as described in *The Child's Conception of Number*. There he shows how hard it is for children to learn to count and how counting and addition and subtraction are learned together.

George Lakoff and Rafael E. Núñez dispute that children have to learn mathematics:

> The very idea that babies have mathematical capacities is startling. Mathematics is usually thought of as something inherently difficult that has to be taught with homework and exercises. Yet we come into life prepared to do at least some rudimentary form of arithmetic. Recent research has shown that babies have the following numerical abilities:
>
> 1. At three to four days, a baby can discriminate between collections of two and three items [reference supplied]. Under certain conditions, infants can even distinguish three items from four [reference supplied].
> 2. By four and a half months a baby 'can tell' that one plus one is two and that two minus one is one [reference supplied].
> 3. A little later, infants 'can tell' that two plus one is three and that three minus one is two [reference supplied]. [Experiments are described in which babies stare at slides showing various numbers of objects, the length of the stare indicating to the researcher that the baby is discriminating between different configurations.]
>
> The ability to do the simplest arithmetic was established using similar habituation techniques (Lakoff and Núñez, 2000, 15–16).

Lakoff and Núñez describe it correctly in (1): Babies can discriminate between collections of objects. When they say that babies can discriminate between numbers and do simple arithmetic, consider that most any mammal and many birds can distinguish between collections of two versus three objects. That does not indicate any ability to do mathematics. That does not show that such creatures or that babies have any concept of number. Abstracting the notion of a number as distinct from a number of objects is the first crucial step in mathematics, and there is no evidence that other mammals, birds, or babies can make that step. The familiarity of working with number notations apart from things that are numbered has become so familiar to the authors that they cannot even see they have begged the question in their description of what the experiments showed. See (Bryant and Nuñes, 2002) for a review of Lakoff and Núñez's view. For a survey and critique of work on the idea of mathematics as innate knowledge, see (De Cruz and Smedt, 2010).

Appendix 2: Comparisons to Other Views of Mathematics

One of the virtues of the story I have told is its seeming familiarity. But it is new. In this appendix I'll compare it to two well-known views of mathematics that seem similar.[21]

Deductivism

It is not a new idea that any substantive claim of mathematics is really an inference with the assumptions of the theory as premises and that claim as conclusion, or is a conditional with the assumptions conjoined as antecedent and the substantive claim the consequent. Long ago Gottfried Wilhelm Leibniz said:

> As for 'eternal truths:', it must be understood fundamentally they are all conditional; they say, in effect: given so and so, such and such is the case. For instance, when I say: *Any figure which has three sides will also have three angles*, I am saying nothing more than that given that there is a figure with three sides, that same figure will have three angles (Leibniz, 1765, Bk IV, Ch. xi, §14).

P. H. Nidditch says:

> [I]t has become more and more widely accepted during the past hundred years, with the result that it is now the orthodox doctrine, that to say of a mathematical proposition *p* that it is true is merely to say that *p* is true in some mathematical system S, and that this in turn is merely to say that *p* is a theorem in S. ... This view of the nature of mathematical truth ... was first put forward with full explicitness and clarity by the Scottish philosopher Dugald Stewart. 'Whereas in all other sciences,' he says, 'the propositions which we attempt to establish express fact, real or supposed—in mathematics, the propositions which we demonstrate only assert a connection between certain suppositions and certain consequences. Our reasonings, therefore, in mathematics, are directed to an object essentially different from what we have in view, in any other employment of our intellectual faculties—not to ascertain *truths* with respect to actual existence, but to trace the logical filiation of consequences from our assumed hypotheses.' (Nidditch, 1960, 287).

Charles Sanders Peirce says,

> The most abstract of all the sciences is mathematics. That this is so, has been made manifest in our day; because all mathematicians now see clearly that mathematics is only busied about *purely hypothetical questions*. ... Mathematics does not undertake to ascertain any matter of fact whatever, but merely posits hypotheses, and traces out their consequences (Peirce, 1931, 23; 109).

And Bertrand Russell says:

> Pure mathematics consists entirely of assertions to the effect that, if such and such a proposition is true of anything, then such and such another proposition is true of that thing.

---

[21] For a comprehensive discussion of current views of mathematics see the excellent survey in the introduction to (Ferreirós and Gray, 2006).

It is essential not to discuss whether the first proposition is really true, and not to mention what the anything is, of which it is supposed to be true. Both these points would belong to applied mathematics (Russell, 1986, 75).

Maxime Bôcher says:

The nominalism of the present day mathematician consists in treating the objects of his investigation and the relations between them as mere symbols. He then states his propositions, in effect, in the following form: If there exists any objects in the physical or mental world with relations among themselves which satisfy the conditions which I have laid down for my symbols, then such and such facts will be true concerning them (Bôcher, 1904).

This sounds very much like what I said above:

The truths of mathematics are truths about inferences.
Mathematics is about what follows from what in our abstractions.

But there are major differences.

First, deductivists nowadays typically understand the inferences of mathematics to be either in or justified by formal logic. Here is what Hilary Putnam says:

There is another way of doing mathematics, however, or at any rate, of viewing it. This way, which is probably much older than the modern way, has suffered from never being explicitly described and defended. It is to take the standpoint that mathematics has *no* objects of its own at all. You can prove theorems about anything you want—rainy days, or marks on paper, or graphs, or lines or spheres—but the mathematician, on this view, makes no existence assertions at all. What he asserts is that certain things are *possible* and certain things are *impossible*—in a strong and uniquely mathematical sense of 'possible' and 'impossible'. In short, mathematics is essentially modal rather than existential, on this view, which I have elsewhere termed 'mathematics as modal logic'.

Let me say a few things about this standpoint here.

(1) This standpoint is not intended to satisfy the nominalist. The nominalist, good man that he is, cannot accept modal notions any more than he can accept the existence of sets. We leave the nominalist to satisfy himself (Putnam, 1975, 70).

Putnam understands the modal nature of mathematics as justified, legitimated, or somehow essentially explicated by formal modal logic.[22] But there is nothing in our experience that can count as a possible way the world could be in which Euclidean plane geometry is true. What counts as a possibility, and what justifies our use of inferences with claims that are neither true nor false, is an application of the theory where what we count as true or false is constrained by our agreements, explicit or not, as to what we will pay attention to in our reasoning.

Putnam's comments about the nominalist show that his conception is based on some more ample metaphysics than mine, for there is nothing in the view I present

---

[22] In (Putnam, 1967) he says that his view of mathematics as modal logic is equivalent to taking mathematics as based on set theory. In (Putnam, 1975, 72) he says:

The main question we must speak to is simply, what is the point? Given that one can either take modal notions as primitive and regard talk of mathematical existence as derived, or the other way around, what is the advantage to taking the modal notions as the basic ones?

here that prevents a nominalist from accepting it—at least in those cases where the abstraction is from sufficiently clear experience. Reasoning commits us to some notion of possibility, but that notion need not be unacceptable to a nominalist (see Epstein, 1999, 2012).

By not seeing the nature of possibility in terms of abstractions based on the same methodology as used in the sciences, deductivists are led to a kind of mystery about how mathematics can be applied. As Alan Musgrave says:

> Russell sought a way to bring geometry into the sphere of logic. And he found it in what I shall call the *If-thenist manoeuvre*: the *axioms* of the various geometries do not follow from the logical axioms (how *could* they, for they are mutually inconsistent?), nor do geometrical *theorems*; but the *conditional statements linking axioms to theorems* do follow from logical axioms. Hence, geometry, *viewed as a body of conditional statements*, is derivable from logic after all. … Russell argued that the discovery of non-Euclidean geometries forced us to distinguish *pure geometry*, a branch of pure mathematics whose assertions are all conditional, from *applied geometry*, a branch of empirical science (Musgrave, 1977, 109–110).

> If-thenism has nothing to say about un-axiomatised or pre- axiomatised mathematics, in which creative mathematicians work. Therefore, *even if* its account of axiomatised mathematics is acceptable, as an account of mathematics as a whole it is seriously defective (Musgrave, 1977, 119).

David Hilbert tried to deal with this problem:

> It is surely obvious that every theory is only a scaffolding or schema of concepts together with their necessary relations to one another, and that the basic elements can be thought of in any way one likes. If in speaking of my points I think of some system of things, e.g. the system: love, law, chimney-sweep … and then assume all my axioms as relations between these things, then my propositions, e.g. Pythagoras' theorem, are also valid for these things. In other words: any theory can always be applied to infinitely many systems of basic elements.
>   [T]he application of a theory to the world of appearances always requires a certain measure of good will and tactfulness; e.g., that we substitute the smallest possible bodies for points and the longest possible ones, e.g., light rays, for lines. We also must not be too exact in testing the propositions, for these are only theoretical propositions (Letter to Frege, December 29, 1899 in Frege, 1980, 40–41).

But it isn't good will and tact. The issue is the nature of abstraction and application.

A version of deductivism described by Alan Musgrave seems closer to what I have presented:

> I think, for example, that the sophisticated *evolutionary* Platonism of Popper need not trouble an If-thenist. Popper tries to combine a Platonistic view of the *objectivity* of human knowledge with the Darwinian view that human knowledge is an *evolutionary product*. Thus he insists that the natural numbers are a human creation (part and parcel of the creation of descriptive languages with devices for counting things), but that once created they become *autonomous* so that objective discoveries can be made about them and their properties are not at the mercy of human whim (Popper, 1972, 158–161). An If-thenist could agree with much of this. We create, first of all, languages in which to express certain *empirical* claims: 'Two apples placed in the same bowl as two other apples give you four apples'; 'Two drops of water placed together give you one bigger drop of water'; etc. Then we come to treat numbers and their addition in a more abstract way (so that the second statement just given

does not count as an empirical refutation of '1 + 1 = 2'). This is, at bottom, to create a more or less explicit collection of 'axioms' for the natural number sequence. And then we find that, once these are granted, we must also grant other statements about numbers like 'There are infinitely many prime numbers'. We *discover*, in other words, that our axioms have certain unintended logical consequences. The objectivity of mathematics is guaranteed by the fact that *what follows from what* is an objective question, and we need not postulate a realm of 'abstract mathematical entities' to ensure it (Musgrave, 1977, 123, footnote).

The objectivity of mathematics comes from the objectivity of the inference relation *relative to our assumed metaphysics*. In any case, neither Popper nor Musgrave developed these ideas into an analysis of mathematical proof and applications of mathematics, nor did they relate this to the methodology of the sciences.

Mathematics as an Empirical Science

After writing the body of this text I discovered (Sawyer, 1943) in which Sawyer shows more clearly and thoroughly than I the development of mathematical theories by a process of abstraction. That part of my views, at least, has been commonplace among mathematicians for a long time. David Sherry (2009) presents an analysis of the use of diagrams in mathematics on that basis that is not only compatible with but supports the view I have presented.

John Stuart Mill is supposed by many to claim that mathematics is an induction from experience. But when he uses the words 'induction' and 'generalization' in discussions of mathematics he seems to mean what I mean by 'abstraction.' Read that way his views are very similar to mine:

> We can reason about a line as if it had no breadth; because we have a power, which is the foundation of all the control we can exercise over the operations of our mind; the power, when a perception is present to the senses, or a conception to our intellects, of *attending* to a part only of that perception or conception, instead of the whole.
>
> Since, then, neither in nature, nor in the human mind, do there exist any objects exactly corresponding to the definitions of geometry, while yet that science can not be supposed to be conversant about nonentities; nothing remains but to consider geometry as conversant with such lines, angles, and figures, as really exist; and the definitions, as they are called, must be regarded as some of our first and most obvious generalizations concerning these natural objects. The correctness of these generalizations, *as* generalizations, is without a flaw: the equality of all the radii of a circle is true of all circles, so far as it is true of any one; but it is not exactly true of any circle; it is only nearly true; so nearly that no error of any importance in practice will be incurred by feigning it to be exactly true. When we have occasion to extend these inductions, or their consequences, to cases in which the error would be appreciable—to lines of perceptible breadth or thickness, parallels which deviate sensibly from equidistance, and the like—we correct our conclusions, by combining them with a fresh set of propositions relating to the aberration; just as we also take in propositions relating to the physical or chemical properties of the material, if those properties happen to introduce any modification into the result.

When, therefore, it is affirmed that the conclusions of geometry are necessary truths, the necessity consists only in this, that they correctly follow from the suppositions from which they are deduced (Mill, 1874, Bk II, Chp. V, §1).

Whether Mill's views really are similar to or even compatible with mine will have to await further study.[23]

References

Aristotle (1930). Physics (R. P. Hardie & R. K. Gaye, Trans.). In W. D. Ross (Ed.), *The works of Aristotle*. Oxford: Clarendon.

Bashmakova, I. G, & Smirnova, G. S. (2000). Geometry: The first universal language of mathematics. In E. Grosholz & H. Breger (Eds.), *The growth of mathematical knowledge* (pp. 331–340). Dordrecht: Kluwer.

Bôcher, M. (1904). The fundamental conceptions and methods of mathematics. *Bulletin of the American Mathematical Society, 11*, 115–135.

Breger, H. (2000). Tacit knowledge and mathematical progress. In E. Grosholz & H. Breger (Eds.), *The growth of mathematical knowledge* (pp. 221–230). Dordrecht: Kluwer.

Brouwer, L. E. J. ([1983] 1912). Intuitionism and formalism (Trans. by A. Dresden of 'Intuitionisme en formalisme', Inaugural Address at the University of Amsterdam). In P. Benacerraf & H. Putnam (Eds.), *Philosophy of mathematics: Selected readings* (pp. 77–89). Cambridge: Cambridge University Press.

Bryant, P., & Nuñes, T. (2002). Children's understanding of mathematics. In U. Goswami (Ed.), *Blackwell handbook of childhood cognitive development* (pp. 412–439). Oxford: Blackwell.

Cantor, G. (1883). *Grundlagen einer allgemeinen Mannigfaltigkeitslehre*. Leipzig: Teubner.

Dauben, J. W. (1979). *Georg Cantor: His mathematics and philosophy of the infinite*. Cambridge, MA: Harvard University Press.

Davies, E. B. (2005). Whither mathematics? *Notices of the American Mathematical Society, 52*, 1350–1356.

De Cruz, H., & Smedt, J. D. (2010). The innateness hypothesis and mathematical concepts. *Topoi, 29*, 3–13.

Detlefsen, M. (2005). Formalism. In S. Shapiro (Ed.), *Oxford handbook of philosophy of mathematics and logic* (pp. 236–317). Oxford: Oxford University Press.

Einstein, A. ([1961] 1915). *Relativity: The special and general theory* (R. Lawson, Trans.). New York: Crown.

Epstein, R. L. (1988). A general framework for semantics for propositional logics. Text of invited address to the VII Latin American Symposium on Mathematical Logic. *Contemporary Mathematics, 69*, 149–168.

Epstein, R. L. (1990). *Propositional logics*. Dordrecht: Kluwer. (2nd ed., Oxford: Oxford University Press, 1995. 2nd ed. with corrections, Belmont, CA: Wadsworth, 2000)

Epstein, R. L. (1994). *Predicate logic*. Oxford, UK: Oxford University Press. (Reprinted, 2000, Belmont, CA: Wadsworth)

Epstein, R. L. (1999). The metaphysical basis of logic. *Manuscrito, 22*(2), 133–148.

Epstein, R. L. (2006). *Classical mathematical logic*. Princeton, NJ: Princeton University Press.

Epstein, R. L. (2010). The internal structure of predicates and names with an analysis of reasoning about process. Typescript available at www.AdvancedReasoningForum.org

---

[23]John Skorupski (2005) discusses Mill's views and suggests a program for empiricists that is similar to what I have done here.

Epstein, R. L. (2012). Valid inferences. In J.-Y. Béziau & M. E. Coniglio (Eds.), *Logic without frontiers: Festschrift for Walter Alexandre Carnielli on the occasion of his 60th birthday* (pp. 105–112). London: College Publications.

Epstein, R. L., & Carnielli, W. A. ([2008] 1989). *Computability: Computable functions, logic, and the foundations of mathematics*. Belmont, CA: Wadsworth & Brooks/Cole. (3rd ed., ARF, 2008)

Fallis, D. (1998). Review of Hersh, 1993 et al. *Journal of Symbolic Logic, 63*, 1196–1200.

Ferreirós, J., & Gray, J. J. (Eds.). (2006). *The architecture of modern mathematics*. Oxford: Oxford University Press.

Frege, G. (1980). *Gottlob Frege: The philosophical and mathematical correspondence*. Edited by G. Gabriel, H. Hermes, F. Kambartel, C. Thiel & A. Veraart. Chicago, IL: University of Chicago Press.

Gardner, M. (1973). Mathematical games. *Scientific American, 229*(October), 114–118.

Hardy, G. H. (1928). Mathematical proof. *Mind, 38*, 11–25.

Hersh, R. (1997). *What is mathematics, really?* London: Jonathan Cape.

Hersh, R. (2006). Inner vision, outer truth. In R. Hersh (Ed.), *18 unconventional essays about the nature of mathematics* (pp. 320–326). New York: Springer.

Isles, D. (1981). Remarks on the notion of standard non-isomorphic natural number series. In F. Richman (Ed.), *Constructive mathematics, proceedings of the New Mexico State University Conference*, Vol. 873 of *Lecture notes in mathematics* (pp. 111–134). Berlin: Springer.

Isles, D. (2004). Questioning articles of faith: A re-creation of the history and theology of arithmetic. *Bulletin of Advanced Reasoning and Knowledge, 2*, 51–59.

Kitcher, P. (1975). Bolzano's ideal of algebraic analysis. *Studies in History and Philosophy of Science, 6*, 229–269.

Kulpa, Z. (2009). Main problems of diagrammatic reasoning. Part I: The generalization problem. *Foundations of Science, 14*(1–2), 75–96.

Lakoff, G., & Núñez, R. E. (2000). *Where mathematics comes from: How the embodied mind brings mathematics into being*. New York: Basic Books.

Leibniz, G. W. ([1981] 1765). *New essays on human understanding* (Trans. and edited by Peter Remnant and Jonathan Bennett). Cambridge: Cambridge University Press.

Mancosu, P. (1996). *Philosophy of mathematics and mathematical practice in the seventeenth century*. Oxford: Oxford University Press.

Mancosu, P. (2000). On mathematical explanation. In E. Grosholz & H. Breger (Eds.), *The growth of mathematical knowledge* (pp. 103–119). Dordrecht: Kluwer.

Mill, J. S. (1874). *A system of logic, ratiocinative and inductive, being a connected view of the principles of evidence and the methods of scientific investigation* (8th ed.). New York: Harper & Brothers.

Musgrave, A. (1977). Logicism revisted. *British Journal for the Philosophy of Science, 28*, 90–127.

Nagel, E. (1961). *The structure of science*. New York: Harcourt, Brace & World. (Reprinted by Hackett Publishing Company, Indianapolis, IN, 1979)

Nelsen, R. B. (1993). *Proofs without words*. Washington, DC: Mathematical Association of America.

Nidditch, P. H. (1960). *Elementary logic of science and mathematics*. Glencoe, IL: Free Press.

Peirce, C. S. (1931). *Collected papers* (Vol. I). Cambridge, MA: Harvard University Press.

Poincaré, H. (1921). *The foundations of science* (G. B. Halstead, Trans.). New York: The Science Press.

Pólya, G. ([1977] 1963). *Mathematical methods in science*, Vol. IX of *Studies in mathematics*. Revised edition edited by Leon Bowden. Washington, DC: Mathematical Association of America.

Popper, K. R. (1972). *Objective knowledge*. London: Oxford University Press.

Putnam, H. (1967). Mathematics without foundations. *Journal of Philosophy, 64*(1), 5–22. (Reprinted in Putnam, *Mathematics, matter, and method: Philosophical papers*, (Vol. 1, pp. 43–59). Cambridge: Cambridge University Press).

Putnam, H. (1975). What is mathematical truth? In *Mathematics, matter and method: Philosophical papers* (Vol. 1, pp. 60–78). Cambridge: Cambridge University Press.

Resnik, M., & Kushner, D. (1987). Explanation, independence, and realism in mathematics. *British Journal for the Philosophy of Science, 38*, 141–158.

Russell, B. ([1901] 1986). Mathematics and the metaphysicians. In *Mysticism and logic* (pp. 75–95). London: Unwin.

Sawyer, W. W. (1943). *Mathematician's delight*. Harmondsworth: Penguin.

Sherry, D. (2009). The role of diagrams in mathematical arguments. *Foundations of Science, 14*(1–2), 59–74.

Skorupski, J. (2005). Later empiricism and logical positivism. In S. Shapiro (Ed.), *Oxford handbook of philosophy of mathematics and logic* (pp. 51–80). Oxford: Oxford University Press.

Steiner, M. (2000). Penrose and platonism. In E. Grosholz & H. Breger (Eds.), *The growth of mathematical knowledge* (pp. 133–141). Dordrecht: Kluwer.

Van Dantzig, D. (1956). Is $10^{10^{10}}$ a finite number? *Dialectica, 9*, 273–277.

Whorf, B. L. (1941). The relation of habitual thought to language. In L. Spier (Ed.), *Language, culture, and personality: Essays in memory of Edward Sapir* (pp. 75–93). Menasha, WI: Sapir Memorial Publication Fund. (Reprinted in Carroll J. B. (Ed.). (1956). *Language, thought, and reality: Selected writings of Benjamin Lee Whorf* (pp. 134–159). Cambridge, MA: MIT Press)

Wigner, E. (1960). The unreasonable effectiveness of mathematics in the natural sciences. *Communications on Pure and Applied Mathematics, 13*(1), 1–14.

Wilder, R. L. (1944). The nature of mathematical proof. *The American Mathematical Monthly, 51*(6), 309–323.

Chapter 15
Towards a Theory of Mathematical Argument

Ian J. Dove

15.1 Introduction

In this paper, I assume, perhaps controversially, that translation into a language of formal logic is not the method by which mathematicians assess mathematical reasoning. Instead, I argue that the actual practice of analyzing, evaluating and critiquing mathematical reasoning resembles, and perhaps equates with, the practice of informal logic or argumentation theory. It doesn't matter whether the reasoning is a full-fledged mathematical proof or merely some non-deductive mathematical justification: in either case, the methodology of assessment overlaps to a large extent with argument assessment in non-mathematical contexts through informal logical techniques.

One cannot hope to recapitulate all of informal logic in a single paper. Still, to claim that there is a role for informal logic in mathematics requires some stage setting. As a gloss on the practice, consider Maurice Finocchiaro's definition of informal logic.

> [Define] informal logic as the formulation, testing, systematization, and application of concepts and principles for the interpretation, evaluation, and practice of argumentation or reasoning (Finocchiaro, 1996, 93).[1]

---

[1]Page numbers are to the anthologized version (Finocchiaro, 2005).

I.J. Dove (✉)
Department of Philosophy, University of Nevada, Las Vegas, 4505 Maryland Parkway, Box 455028, Las Vegas, NV 89154-5028, USA
e-mail: ian.dove@unlv.edu

A. Aberdein and I.J. Dove (eds.), *The Argument of Mathematics*, Logic, Epistemology, and the Unity of Science 30, DOI 10.1007/978-94-007-6534-4_15,
© Springer Science+Business Media Dordrecht 2013

For present purposes, focus on the application of concepts and principles for the interpretation[2] and evaluation of argumentation and reasoning. There was, perhaps, a time when one may have thought that the only principle needed for the analysis of arguments was (formal) logical paraphrase and that the only concept needed for argument evaluation was (deductive) validity. But it is against this mythical hegemony of (formal) logic as *the* tool of argument appraisal that informal logic becomes theoretically (and pedagogically) important. The early history of informal logic[3] should be traced at least to Stephen Toulmin's *The Uses of Argument* (1958). In the preface to the updated edition (Toulmin, 2003), Toulmin clearly targets formal logic.

> When I wrote it, my aim was strictly philosophical: to criticize the assumption, made by most Anglo-American academic philosophers, that any significant argument can be put in formal terms (Toulmin, 2003, vii).

What is important to take away from this early history isn't a particular set of techniques; instead, informal logic is best understood as a reaction against formal logic[4] as *the* theory of argument assessment. Much of the work that comprises the first part of Finocchiaro's definition—formulation, testing and systematization of concepts and principles of argument appraisal—is an ongoing theoretical enterprise, in which Finnochiaro and others are actively engaged.

Regarding the application of the concepts and principles of this enterprise to arguments or reasoning *in the wild*, there is only a small literature regarding informal approaches to reasoning in mathematics.[5] However, in the next two sections I make room for informal logic in two mathematical settings. Before turning to those, I must briefly consider the competing conceptions of argument within informal logic.

It may be surprising to anyone outside of informal logic that there is any debate about the proper definition of argument. Those of us weaned on formal logic may think this debate is a non-starter because there is a perfectly acceptable definition of argument which is synonymous with the definition of derivation available in any textbook on formal logic: an argument is a sequence of statements/sentences/propositions/formulas such that each is either a premise or the consequence

---

[2]What Finocchiaro calls 'interpretation' is often called 'analysis'. I prefer the latter term and generally use it instead.

[3]I unceremoniously identify informal logic and argumentation theory and argumentation studies as a single entity—really a single kind of response to practical difficulties in theorizing about and teaching principles of argument appraisal/assessment. That the phrase *informal logic* is more often used by philosophers or that the phrase *argumentation theory* is more often used by researchers in communication studies is beside the point. The goal is the same: to produce a workable theory of argument appraisal.

[4]Where 'formal logic' is understood as a theory of entailment which may include both formal semantics and formal proof theory.

[5]This literature includes: (Aberdein, 2007, 2005; Finocchiaro, 2003b; Krabbe, 1997), for example. Poincaré may be an interesting forerunner to this approach, cf. (Detlefsen, 1992, 1993).

of (some set of) previous lines, and the last of which is the conclusion.[6] Such a definition is required by and antecedent to the techniques of formal paraphrase that constitute the analytic enterprise. However, a close look at arguments as they occur in philosophy journals, say, reveals that there may be more to the argumentative estate. For example, in most good papers, an author considers possible objections to his/her arguments as part of the usual manner of defense. It may be possible to recast objections and replies within the strict premise/conclusion structure—one may grant this provisionally. Yet, whether an argument, properly so-called, ought to contain dialectical elements could be a matter of definition. Even granting the strict premise/conclusion conception, there is room for debate about what counts as a premise.

Besides questions of whether all arguments must contain dialectical elements, other considerations include: whether purpose defines arguments; whether rhetorical elements must be considered in the definition of argument; whether an argument ought to be defined in terms of its structure; etc. The following list provides a sample of attempts to include/exclude these elements.

> An argument, in the logician's sense, is any group of propositions of which one is claimed to follow from the others, which are regarded as providing support or grounds for the truth of that one (Copi and Cohen, 1994, 5, quoted in Finocchiaro, 2003a, 295).

> The simplest possible argument consists of a single premise, which is asserted as true, and a single conclusion, which is asserted as following from the premises, and hence also to be true. The function of the argument is to persuade you that since the premise is true, you must also accept the conclusion (Scriven, 1976, 55–6, quoted in Finocchiaro, 2003a, 297).

> An argument is a type of discourse or text—the distillate of the practice of argumentation—in which the arguer seeks to persuade the Other(s) of the truth of a thesis by producing reasons that support it. In addition to this illative core, an argument possesses a dialectical tier in which the arguer discharges his dialectical obligations (Johnson, 2000, 168, quoted in Finocchiaro, 2003a, 297).

> *Argumentation* is a verbal, social and rational activity aimed at convincing a reasonable critic of the acceptability of a standpoint [read: conclusion] by putting forward a constellation of propositions justifying or refuting the proposition expressed in the standpoint (Van Eemeren and Grootendorst, 2003, 1).

> Now arguments are produced for a variety of purposes. Not every argument is set out in formal defense of an outright assertion. But this particular function of arguments will claim most of our attention [. . .]: we shall be interested in justificatory arguments brought forward in support of assertions, in the structures they may be expected to have, the merits they can claim and the ways in which we set about grading assessing and criticizing them. It could, I think, be argued that this was in fact the primary function of arguments, and that the other uses, the other functions which arguments have for us, are in a sense secondary and parasitic upon this primary justificatory use (Toulmin, 2003, 12).

---

[6]Some such definition is often prefatory to formal logic textbooks. The openness of 'statements/sentences/. . . ' is simply to cover the expected variations in such definitions. See for example, (Mates, 1972, 5) versus (Allen and Hand, 2001, 1).

As I am here interested in making room for informal logic in mathematical justification I will let the last conception do most of the work. This paper will be interested in 'justifactory arguments brought forward in support of assertions.' I will construe support broadly to include both deductive and non-deductive justifications.

15.2 The Role of (Informal) Logic in Assessing Proofs

Ralph Johnson (2000) has a widely read and rightly respected theory of argument appraisal. Against its many virtues, however, there is one vice: he unnecessarily eschews, except in one case as a negative illustration, mathematical examples. This decision, I think, can be traced to the conception of proof he accepts.

> If we are looking for conclusive arguments, we can do no better than take the paradigm of mathematical proofs; for example, the proof that there is no greatest prime number. Here, a proof is a sequence of steps, each of which is either an axiom (or an otherwise incontestable step) or a valid derivation from previously accepted lines (Johnson, 2000, 231).

Suppose Johnson is correct that a proof really is a (finite) sequence of statements such that each statement is either an axiom (or accepted claim) or the logical consequence of an earlier statement. Would this preclude the use of informal logic in mathematics?[7] One difficulty with this notion of proof is that it contains an unstated qualifier—any proof on this account is a successful proof. There is a treasure trove of failed proofs in mathematics. And, I assert, the methods by which these failures were discovered employed the techniques of informal logic.[8] This is not to say that the techniques of informal logical appraisal are only applicable in finding errors. It is a general technique of appraising reasoning. Consider the following *proof* (Maxwell, 1959, 10–12) that any given triangle, *ABC*, is isosceles.

Proof. It is required to prove that *AB* is necessarily equal to *AC*.

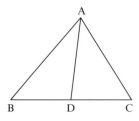

---

[7]Johnson nowhere, as far as I can tell, dismisses the use of informal logic in mathematics. Instead, his concern is to distinguish proof from argument-proper.

[8]If this assertion is too strong or too hasty, qualify it to read: The methods for assessing mathematical proofs resemble the methods for assessing non-mathematical arguments.

If the internal bisector of angle A[9] meets BC in D, then by the angle bisection theorem,

$$\frac{DB}{AB} = \frac{DC}{AC}$$

Now, $\angle ADB = \angle ACD + \angle CAD = C + \frac{1}{2}A$, so that, by the sine rule applied to the triangle ADB,

$$\frac{DB}{AB} = \frac{\sin BAD}{\sin ADB}$$

$$= \frac{\sin \frac{1}{2}A}{\sin\left(C + \frac{1}{2}A\right)}$$

Further, $\angle ADC = \angle ABD + \angle BAD = B + \frac{1}{2}A$ so that

$$\frac{DC}{AC} = \frac{\sin \frac{1}{2}A}{\sin\left(B + \frac{1}{2}A\right)}$$

Hence,

$$\frac{\sin \frac{1}{2}A}{\sin\left(C + \frac{1}{2}A\right)} = \frac{\sin \frac{1}{2}A}{\sin\left(B + \frac{1}{2}A\right)}$$

Moreover, $\sin \frac{1}{2}A$ is not zero, since the angle A is not zero, and so

$$\sin\left(C + \frac{1}{2}A\right) = \sin\left(B + \frac{1}{2}A\right),$$

which is the same as

$$C + \frac{1}{2}A = B + \frac{1}{2}A$$

which means $C = B$. The triangle is therefore isosceles.   □

The conclusion of this reasoning is absurd for the given triangle. But what has gone wrong? Leave aside any problems one thinks accrue when reasoning contains diagrams, for, in this case, the problem isn't the diagram at all. Instead, suppose this *proof* occurs as an example in a paper on mathematical method, as it does here. There is, of course, no imperative to find the mistake. One could, I suppose, continue living as if nothing terrible had happened. But I would hope that a rational being confronted by this proof would want to discover the error—even if that being weren't mathematically inclined. Moreover, I like to think that if one attempts to discover the error in the proof and fails, this would cause some tension in one's web of belief. Leaving these hopes and wishes to one side: the important aspect of this thought experiment is to speculate about how someone would go about appraising this argument.

---

[9]The angles in the original triangle, ABC, are labeled simply A, B or C respectively. The other angles are labeled more fully, e.g. $\angle ADC$, to avoid ambiguity.

On the one hand, if you know that the conclusion is false, then there is a sense in which you know there must be a mistake before you begin your investigation of the reasoning. On the other hand, perhaps, you don't know where the reasoning fails, or, if the conclusion weren't obviously absurd, you might feel less compelled to critique the reasoning. However, believing that a conclusion is false isn't usually enough to warrant rejection. There are many odd results, both inside and outside of mathematics. Anecdotal evidence, gathered by the author, suggests that mathematically inclined critics initially point to the diagram as the faulty source. One explanation for this may be that there is a well known faulty argument for the same conclusion that is often claimed to rest on a mistaken construction. In the present case, however, the problem isn't with the diagram or the associated auxiliary constructions. Those critics without (recent) mathematical inclinations suppose that the fault lies with either the appeal to the angle bisection theorem or the sine rule applied to the triangle. Again, neither of these is a mistake here. The error, as Maxwell points out (Maxwell, 1959, 12) is that equality of sine does not mean equality of angle. For example, $\sin \pi = \sin 2\pi$ but $\pi \neq 2\pi$. So one cannot/shouldn't infer from $\sin\left(C + \frac{1}{2}A\right) = \sin\left(B + \frac{1}{2}A\right)$ that $\left(C + \frac{1}{2}A\right) = \left(B + \frac{1}{2}A\right)$.

The mistake in this case is, in some respects, uninteresting. One treats a function symbol as if it were manipulable in the same way as variables or constants.[10] Still, the identification of the mistake, I trust, is interesting insofar as the process one uses to discover the mistake is analogous to the process one might use to criticize an unpalatable argument in non-mathematical settings. Thus, even for cases of proofs, there could be a role for the informal logician—notice that the mistake wasn't discovered by appealing to some formalization of the statements in a suitable language. Rather, this appraisal takes place in natural language.

15.3 Assessing Mathematical Evidence

Mathematicians regularly use evidentiary or non-deductive methods such as computer-assisted proofs, probabilistic sieves, partial proofs and abduction.[11] And mathematicians seem to be pretty good at assessing these uses. Although some mathematical mistakes may remain temporarily unnoticed, claims that are supported by less than deductive means are often ripe for reconsideration— they may even receive closer scrutiny than their deductively justified cousins. In this section I consider how some of these means of support are assessed by

---

[10] One way to understand the mistake is that the argument treats the 'f' of $f(x)$ to be capable of manipulation. Suppose $f(x) = x^2$. Then for any n, $f(n) = f(-n)$. Then, dividing both sides by f, one gets $n = -n$. Since sin is a function symbol, it isn't itself manipulable. Instead, the function is manipulable only when it is given with an attending argument.

[11] James Franklin (1987, revised version reprinted in this collection) gives a brief sample of some of the acceptable non-deductive methods.

mathematicians. I compare the mathematical method with techniques from informal logic to show they are similar if not identical. This leads me to conclude that the logic of assessment underwriting this practice is informal logic.

15.3.1 Computers and Probabilistic DNA Evidence

In a series of papers, Don Fallis (1996, 1997, 2003) argues that probabilistic evidence ought to be given more credence in mathematics. A surprising use of probabilistic evidence in graph theory concerns the discovery of Hamiltonian Paths among sets of points by encoding the points with unique DNA *addresses* such that a Hamiltonian Path exists only if the sequences of DNA addresses, when mixed in an appropriate solution, combine to form a double helix. The method is reliable and repeatable. That is, the method could result in false-rejections—cases where no double helix is found though there does exist a Hamiltonian Path—though not false positives (Fallis, 1996, 1997). Moreover, even if there were a false positive, there is an ancillary method by which these results can be checked to determine whether the DNA constitutes a real Hamiltonian Path. Fallis argues that there aren't good epistemological reasons to reject such evidence as proofs. As that argument is outside the scope of this paper, bracket it. Instead, I'm interested in how a mathematician would appraise this evidence.

Informal logic, as a theory of argument appraisal, distinguishes two endeavors: analyzing and evaluating arguments. To see this process at work in a non-mathematical, evidentiary case, consider the following.

> During excavations of the Bronze Age levels at El Mirador Cave, a hole containing human remains was found. Tapaphonic analysis revealed the existence of cutmarks, human toothmarks, cooking damage, and deliberate breakage in most of the remains recovered, suggesting a clear case of gastronomic cannibalism (Cáceres et al., 2007, 899).[12]

There are many different informal techniques one could use to analyze this bit of reasoning: as a Toulmin Diagram (Toulmin, 2003, 87ff.), as a numbered step argument or as an argumentation scheme. Let's use a scheme. Schemes present a standard format or template for argument types. To assess an argument's structure one reconstructs the argument using the structure of the chosen scheme. This makes it easier to identify the explicit argumentative elements as well as any material that is left tacit in the reasoning. Along with the canonical form, the scheme also provides evaluative critical questions. In this case, the scheme known as *Argument from Sign*[13] seems appropriate. It has two kinds of premises. One is called the *specific premise*: Some finding, *A*, is true in this situation. Next there is a *general*

---

[12]In what follows I've left out the qualifier 'gastronomic' because, for the purposes of providing an example, it is unnecessary. Gastronomic Cannibalism is distinguished by physical anthropologists from Ritualistic Cannibalism solely in terms of purpose (Cáceres et al., 2007, 899).

[13]For a textbook treatment of this scheme, see (Walton, 2006, 112–4).

premise: Such a finding is indicative of some object, event or action, *B*. This leads to a conclusion: *B* obtains, occurs or happens (in this situation). Reconstructing the above argument as an Argument from Sign, one gets as a specific premise: '[Most of the human remains recovered at El Mirador Cave contained] cutmarks, human toothmarks, cooking damage and deliberate breakage.[14] The general premise is: '[the existence of cutmarks, toothmarks, cooking damage and deliberate breakage] suggests cannibalism.' The conclusion, then, is, '[The Bronze Age inhabitants of El Mirador Cave practiced (in at least one instance) cannibalism.]'[15]

To evaluate an argument using a scheme, one answers the critical questions associated with the scheme. The critical questions typically fall under two broad categories: (a) acceptability of the premises, and (b) amount of support the premises, if true, would confer upon the conclusion. In the case of Argument from Sign, the two critical questions are: (1) How strongly is the sign correlated with the result? and (2) Are there counter-signs that indicate a different result or which undermine the acceptability of the sign in this case? The answers to these questions are not always readily available. However, in this case, later in the article the authors answer question (1) affirmatively (Cáceres et al., 2007, 905) and question (2) in the negative (Cáceres et al., 2007, 912–3). This means that the argument, though defeasible, strongly supports its conclusion.

I'll apply this same appraisal technique to the DNA arguments considered (Fallis, 1996, 1997). Take a graph, *G*, with directed edges and for which the starting node, v_{in}, and ending node, v_{out}, are specified. A *Hamiltonian Path* starts from v_{in}, ends at v_{out}, and travels across each edge of *G* exactly once. The *Hamiltonian Path Problem* is to determine for an arbitrary directed graph whether there exists a Hamiltonian Path through the graph. This problem has been proven to be NP-complete (Karp, 1972, 85ff.), which means that non-deterministic algorithms are likely to be the only kinds of solutions available. Leonard Adleman provides such an algorithm whose implementation involves the use of DNA.

> *Step 1*: Generate random paths through the graph.
> *Step 2*: Keep only those paths that begin with v_{in} and end with v_{out}.
> *Step 3*: If the graph has *n* vertices, then keep only those paths that enter exactly *n* vertices.
> *Step 4*: Keep only those paths that enter all of the vertices of the graph at least once.
> *Step 5*: If any paths remain, say 'Yes'; otherwise say 'No.' (Adleman, 1994, 1021–2).

Adleman implemented this algorithm by encoding a graph's vertices as DNA *addresses*. Then, given DNA ligation, the process by which single strands attach to form double strands, mixing a sufficient quantity of the addresses will produce random paths through the graph (Step 1). Moreover, given accepted techniques for distinguishing DNA sequences, Adleman could complete steps 2–4. Step 5 depends upon whether there are any double helices in the mix.

---

[14]The use of square brackets, '[' and ']', indicates that this material is either paraphrased from the original material or added to the material.

[15]The conclusion is clearly indicated by the phrase 'suggests a clear case of gastronomic cannibalism'.

Fallis, recall, wants to argue that from the standpoint of epistemology, the DNA implementation of this algorithm, with its explicit error measure, entails as much certainty as that of any sufficiently long mathematical proof (Fallis, 1996, 166). The details of this fascinating argument are tangential to the present project. Instead, consider a particular (hypothetical) case of Adleman's DNA implementation of the algorithm analyzed and evaluated as an Argument from Sign. That is, suppose that Adelman implements the algorithm for a particular graph G such that the result in step 5 was 'No.' In such a case the specific premise would be: 'The DNA implementation of the algorithm gave 'No' as the answer regarding graph G.' The general premise would be: 'A 'No' answer for graph G is a sign/indication that there does not exist a Hamiltonian Path for G.' And the conclusion would be: 'Graph G doesn't have a Hamiltonian Path.' The evaluation would follow precisely the evaluation of the argument regarding cannibalism at El Mirador Cave above.

A related mathematical method involves the use of computers in proofs. The most famous and oft discussed example was the use of a computer in proving the Four-Color Theorem.[16] Again, whether the appeal to computers ought to be considered a legitimate case of *proving* is tangential to this discussion. Instead, I'm interested in how mathematicians would assess any particular case of appealing to computers. In the case of the Four-Color Theorem, the proof is simple to explain. For the most part, it is a proof by mathematical induction. There are three cases, one of which contains thousands of subcases. To handle the sheer number of subcases would be outside of the ability of any human. But, a computer could check each of these subcases. Moreover, the structure of the argument would resemble the DNA implementation algorithm in its gross structure. That is, the mathematician would prepare the computer in some way; then, by running the program, the mathematician would get an answer. If the answer were 'yes,' that would be a sign that subcase is four-colorable. If the answer were 'no,' that would be a sign that the subcase is not four-colorable. The program was executed and the answer was 'yes.' Hence, four colors suffice.

At least in these two cases,[17] the methodology of appraisal or assessment cannot be formal in the sense of formal logic—mathematicians didn't try to recast these arguments as formulas in a first-order language to determine whether the consequences follow as a matter of derivation or formal semantics. As I have shown, one can assess the reasoning using principles and techniques from informal logic.[18] Moreover, the techniques actually used by mathematicians to assess these arguments are precisely these informal logical techniques. One caveat is that no

---

[16]For an illuminating philosophical discussion of the use of computers in the proof of the Four-Color Theorem see (Detlefsen and Luker, 1980).

[17]Whereas DNA proofs will remain defeasible, computer proofs can have formal verifications.

[18]For a different assessment using the tools of Toulmin diagrams, see (Aberdein, 2007). There, Aberdein reconstructs (a part of) the proof of the four-color theorem as an explicit Toulmin diagram. The use of Argumentation Schemes vs. Toulmin Diagrams shouldn't be thought of as necessarily opposed. One may be able to capture all of the elements of a scheme in a diagram and vice versa.

mathematicians seems to have used an argumentation scheme like Argument from Sign explicitly.[19] One should focus on whether the reconstruction of mathematical reasoning using explicit argumentation schemes would shed any light on the practice. I think it will and does. That is, by reconstructing the DNA proofs as explicit Arguments from Sign, one simplifies the gross structure of the argument—the analysis—and one has recourse to general critical questions to guide the evaluation of the arguments.

15.3.2   Abduction in Mathematics

The previous subsection takes as its point of departure those mathematical practices under the purview of no axioms. Hence, those practices may be judged atypical. Yet, turning to axiomatic systems—the very heart of what some may call the *formal approach*—there is, even there, room for informal logic. Penelope Maddy argues that the acceptability of axioms is itself a matter of reason though not of direct or deductive proof (Maddy, 1988). She takes the development of set theory as a case study in this practice. For simplicity's sake, consider some arbitrary mathematical claim, C. Maddy distinguishes two ways of coming to believe C. On the one hand, C may be the consequence of some well established first principles, A, such that $A \vdash C$. This is usually the way mathematical claims are thought to garner acceptance. On the other hand, C may garner acceptability in terms of its 'fruitfulness.' Suppose $C \vdash P$ where P is some widely accepted mathematical claim. Does this implication make C any more acceptable? Maddy, quoting Gödel, thinks so.

> ...besides mathematical intuition, there exists another (though only probable) criterion of the truth of mathematical axioms, namely their fruitfulness in mathematics and, one may add, possibly also in physics (Gödel, 1947, 485, quoted in Maddy, 1992, 77).

Maddy runs through a considerable number of examples of what she terms *extrinsic* support. The important features of this practice, for my purposes, are that they are widespread and not themselves reducible to deductive inferences. The most widely discussed examples involve the acceptance of the axiom of choice.

> But the question that can be objectively decided, whether the principle is necessary for science, I should now like to submit to judgment by presenting a number of elementary and fundamental theorems and problems that, in my opinion, could not be dealt with at all without the principle of choice (Zermelo, 1908, 198–90, quoted in Maddy, 1988, 488).

---

[19]Below I suggest that proofs by Mathematical Induction have much in common with Argumentation Schemes. Perhaps, Mathematical Induction is, simply put, a mathematical argumentation scheme. Again, Poincaré's views on mathematical induction could importantly prefigure this idea, see (Detlefsen, 1992, 1993).

The point, obviously, is that the utility of the axiom of choice in dealing with open problems in mathematics provides reason to accept it, even if only hypothetically.

Let us now distinguish two questions about this practice. First, does the existence of fruitful consequences provide *any* support for a mathematical claim? Second, if so, how much? I take it that the widespread use of this practice suggests that practitioners would answer the first question affirmatively. Regarding the second question, I am much less sure what the correct answer is. But, I'm sure what it is not. Abduction provides some measure of support, I just don't know how much. This means that, at best, the conclusion of an abductive inference ought to be accepted *with reservation*. As I am not here arguing that abductive inferences are proofs, the (perhaps vague) reservations attached to conclusions supported abductively do not thereby undermine the use of informal logic.

Finally, to see that this practice both is even more widespread and has a long history, consider the oft discussed application of summation techniques by Leonhard Euler to divergent series, see (Pólya, 1968; Putnam, 1975; Sandifer, 2007). Put roughly, Euler applies techniques to divergent series that are only categorically valid for convergent series. In this way, Euler finds a value for the series $A = 1 - 1 + 2 - 6 + 24 - 120 + 720 - 5040 + \cdots$

> This is the key result of this paper, but Euler understands that some readers might not be convinced that he hasn't made any mistakes. So, he solves the same problem several other ways. For example, he finds diverging series for $1/A$ and $\log A$, and finds that similar methods also lead to a value of A near 0.59. He finds a way to write A and $1/A$ as continued fractions and evaluates those continued fractions to get still more estimates consistent with the ones before (Sandifer, 2007, 182).

Sandifer finds neither Euler's conclusion nor his method irrational. Still, Euler was obviously aware of the shaky footing of both result and method. So, to buoy the method, he shows that it is consistent with other results and accepted techniques.

> By the end of the article, Euler has estimated A at least six different ways, and every time he gets the same estimate. *When such different analyses all lead to the same conclusion, it is easy to understand why mathematicians of Euler's time believed in the utility of [divergent series]* (Sandifer, 2007, 183, emphasis added).

Sandifer's assessment of Euler's reasoning is simply an application of the techniques and principles of informal logic (though perhaps tacitly). Indeed, Sandifer's claims in the above quote suggest that one understand Euler's justification of the method abductively; and at the same time to understand the justification of the result as an accumulation of evidence. In both cases, there are argument schemes that accord with (or predict) Sandifer's assessment. This suggests a rather strong conclusion. Informal logic *is* the logic of mathematical reasoning. In a later section I consider this claim in the light of a debate between Jody Azzouni and Yehuda Rav concerning derivations and their relation to mathematical proofs.

15.3.3 A Preliminary Conclusion

Mathematicians already, though perhaps tacitly, use the techniques of informal logic. They use them when they appraise proofs, and they use them when they assess mathematical reasoning that isn't proof. This is not to say that mathematicians ought to pay more attention to informal logic or argumentation theory. Rather, this suggests that an accurate philosophy of mathematics ought to recognize this use. Hence, inasmuch as informal logic is already a part of mathematical practice, it makes sense to make the use explicit as part of a larger project to construct a philosophy of mathematics that takes practice seriously.

15.4 Proofs, Derivations and Algorithmic Systems

In this section, I turn to an account of mathematical proofs that has some affinities with the informal logical account I suggested above. However, whereas I've argued that mathematicians tacitly use informal logic as a theory of appraisal for mathematical reasoning, this account would seem to make much more use of formal technique. I argue that the appeal to formal techniques is unnecessary or illusory.

Jody Azzouni defends an account of proofs he calls the 'derivation indicator' view (Azzouni, 2004, 2006). Put roughly, this means that a mathematical 'proof' is a promissory note or an advertisement: either it promises that one can find a strict derivation of the result of the proof from first principles, or it advertises the existence of such a derivation. The proof is, in Azzouni's words, 'in the vernacular,' whereas the associated or promised derivation needn't be.

This account could make room for informal logic within mathematical practice. For, if proofs take place 'in the vernacular,' then, perhaps they ought to be assessed in the vernacular as well. But this doesn't seem to be what Azzouni intends.

> [A]lthough it's true (on [Azzouni's] view) that *proofs* in Rav's sense are indispensable to mathematical practice, this is compatible with the claim (which is also true on [Azzouni's] view) that it's *derivations*, derivations in one or another *algorithmic system*, which *underlie* what's characteristic of mathematical practice: in particular, the *social conformity* of mathematicians with respect to whether one or another proof is or isn't (should be, or shouldn't be) convincing (Azzouni, 2004, 83).

Azzouni defines *algorithmic system* as: 'one where the recognition procedure for proofs is mechanically implementable' (op. cit.). He allows for quite a broad understanding of what would count as such a system. For example, he thinks that Euclidean geometry would count on this view. But this complicates rather than clarifies the connection between proofs and derivations. Consider the proof of Euclidean Proposition I.19.

Proposition 19: In any triangle the greater angle is subtended by the greater side.

Proof. Let *ABC* be a triangle having angle *ABC* greater than angle *BCA*.; I say that the side *AC* is also greater than the side *AB*.

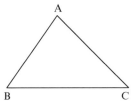

For, if not, *AC* is either equal to *AB* or less.
Now *AC* is not equal to *AB*; for then the angle *ABC* would also have been equal to angle *BCA* [Prop. I.5]; but it is not [by stipulation]; therefore *AC* is not equal to *AB*.
Neither is *AC* less than *AB*, for then the angle *ABC* would also have been less than angle *BCA* [by Prop. I.18], but it is not [by stipulation]; therefore *AC* is not less than *AB*.
And it was proved that it was not equal either.
Therefore *AC* is greater than *AB*.
Therefore [in any triangle the greater angle subtends the greater side.] (Euclid, 1956, 284).

The reason to choose this proposition and its *proof* is that it could be derived without recourse to any geometrical reasoning at all. Instead, it is a *logical* consequence of prior propositions by application of Aristotelian immediate inferences.[20] What is the proof and what is the derivation? Perhaps Azzouni would have the two identified in this case. He doesn't rule out this possibility, so it is open to him to take this option. Still, this would mean that all of Euclid's proofs are also derivations.[21] To identify these as such would, I think, take away some of the rhetorical force of Azzouni's otherwise tempting view.

What about proofs from other mathematical fields? Consider another example from graph theory (Aigner and Ziegler, 2002, 47)—this one from *The Book*!

Theorem 1. *In any configuration of n points in the plane, not all on a line, there is a line which contains exactly two of the points.*

Proof. Let $\mathscr{P}$ be the given set of points and consider the set $\mathscr{L}$ of all lines which pass through at least two points of $\mathscr{P}$. Among all pairs (P, l) with P not on l, choose a pair (P_0, l_0) such that P_0 has the smallest distance to l_0, with Q being the point on l_0 closest to P_0 (that is, on the line through P_0 vertical to l_0).

Claim. The line l_0 does it!
If not, then l_0 contains at least three points of $\mathscr{P}$, and thus two of them, say P_1 and P_2, lie on the same side of Q. Let us assume that P_1 lies between Q and P_2, where P_1 possibly

---

[20]Heath's footnote at (Euclid, 1956, 284–5).

[21]A distressing caveat: Euclid abbreviates his proofs, especially in the last step where the reasoning from a particular figure is generalized. This means that, strictly speaking, each of Euclid's proofs is a proof sketch. This is different from the usual complaint about the incompleteness of Euclidean proofs. The usual complaint is that the diagram fills in tacit assumptions, like continuity, that should be made explicit, as in Pasch's Axiom. I discuss proof sketches in more detail below.

coincides with Q. It follows that the distance of P_1 to the line l_1 determined by P_0 and P_2 is smaller than the distance of P_0 to l_0, and this contradicts our choice for l_0 and P_0.   □

Is this a proof or a derivation? It seems a prototypical proof. Yet, if Azzouni is correct, then one way to critique a proof is to undermine the underlying derivation. So, it matters whether this example is a proof, a derivation, or both. He asserts that proofs take place in natural language, but in what language do derivations occur? Here's a conjecture: (At least some) derivations occur in natural language too. There isn't a formal language in which proofs are recast as derivations. Instead, if the conjecture is correct, then the only proofs that aren't also derivations are proofs with gaps—intentional or not. Intentional gaps occur in 'proof sketches.' This raises two related questions, why isn't Azzouni's distinction simply between proofs and incomplete proofs? What gaps are sufficient to keep a proof sketch or an incomplete proof from being a proof-proper or a derivation?

To answer these questions, one must get a better sense for the prerequisites of Azzouni's algorithmic systems. Notice that they aren't called *logical systems* or *deductive systems*. Instead of being deductively or logically perfect, an algorithmic system's implementation must be 'mechanically recognizable.' This is one instance where Azzouni's account directly competes with Yehuda Rav's account (Rav, 1999). Rav asserts that proofs contain irreducibly semantic or intensional content that isn't captured by derivations. To see a non-mathematical example of what I think is at issue, consider the reasoning: Andrew is a bachelor; therefore he's unmarried. One might think that the 'therefore' indicates that the arguer has a (logical) derivation although the reasoning is, as stated, incomplete. On the other hand, one might think that the reasoning is complete—that it is complete because the meanings of 'bachelor' and 'unmarried' overlap in a way that allows one to infer unmarried-ness immediately from bachelorhood. On this view there is no intermediate step that connects bachelorhood and unmarried-ness. Compare with what Rav says about moves in proofs.

> [typical moves in a proof] bring to light the intensional components in a proof: they have no independent logical justification other than serving the purpose of constructing bridges between the initially given data, or between some intermediate steps, and subsequent parts of the argument. But the bridges are conceptual, not deductive in the sense of logic (Rav, 1999, 13), quoted in Azzouni, 2004, 101.

My guess is that Azzouni's profligate view on algorithmic systems would simply expand to countenance any 'proof' that Rav thinks is irreducibly intensional. That is, any proof that Rav would classify as containing intensional components would most likely be classified as algorithmically acceptable on Azzouni's account. Moreover, the assessments of either *proofs* for Rav or *derivations* for Azzouni will take place within informal logic. Again, this suggests a rather strong conclusion about which logic is the logic of appraisal in mathematics: it's informal logic.[22]

---

[22]There is nothing in the conception of informal logic I accept that would preclude the informal logician from partaking in the fruits of the formal logician's labor. The informal logician simply

15.5 Mathematical Discourse: Problems, Questions and Conclusions

When one applies informal logic to non-mathematical arguments in real contexts, there are many practical problems that don't occur in the mathematical cases. For example, it is quite difficult in practice to determine whether some (non-mathematical) discourse contains an argument. Indeed, in teaching the techniques of argument assessment, argument identification, although it is glossed in most textbook treatments, is where students have great difficulties.[23] Contrast this difficulty with one context in which proofs occur: mathematics texts including journals. In these contexts, mathematicians offer their proofs in a highly stylized manner. This includes clearly indicating the conclusion of the reasoning. Also, the starting point of the reasoning is, more often than not, indicated by the word, 'Proof.' Teaching argument assessment would be greatly simplified if authors generally indicated their conclusions and the start of their reasoning explicitly.

There are, of course, contexts in which even mathematicians are less explicit about indicating conclusions and reasons. But this serves to connect mathematical discourse more closely with other contexts rather than separating them. Even in the contexts in which both the conclusion and the reasoning are clearly indicated, this doesn't preclude the use of informal logical techniques and principles. Rather, it looks as if mathematicians use precisely the tools of informal logic, whether they realize and recognize this or not.

Insofar as the forgoing claim is correct and mathematicians are already informal logicians, is there anything to be gained by making this explicit—that is, are there mathematically important results that will accrue from making the practice explicit? I think not. Naming the practice of mathematicians 'informal logic' won't change the practice. But this isn't as negative a result as one might first imagine. Instead, if the focus is on philosophy of mathematics rather than mathematics proper, one can see that giving an accurate account of mathematics as it is practiced, as opposed to its mythic formal practice, is a gain over previous philosophies of mathematics.

This also means that there are a number of open research questions. For example, are there any paradigmatically mathematical argumentation schemes, fallacies, etc.? Above, I used the scheme *Argument from Sign* to assess a mathematical argument. But this scheme isn't mathematical, it is general. Perhaps mathematical induction is a quintessentially mathematical scheme.[24] As a scheme it is no more formal than *Argument from Sign*. Moreover, it is taught in much the same way that argument schemes are taught: the neophyte is presented with simple examples for illustration

---

denies that the method by which mathematical (or really any) reasoning gets appraised is translation into formal language and comparison with accepted formal results.

[23] See Michael Malone's argument that this is a theoretical as well as practical problem (Malone, 2003).

[24] Other possible mathematical argumentation schemes include: mathematical symmetry arguments and mathematical analogies, e.g., (Steiner, 1998, 48ff.).

purposes. The format of the argument is presented, though the actual format for any particular proof by mathematical induction may only approximate the canonical form—for example, how many cases, if any, to consider. Next, one tries out the technique on more complicated cases. Finally, after long practice, one learns when the base case is trivial, when there are multiple cases, when cases are treated *symmetrically*, etc. Mathematicians may make the analytic element of argument assessment much easier by labeling their reasoning explicitly. Indeed, one could take the practice of using mathematical induction to be one in which the arguer does all of the steps of argument analysis for the audience. An argument critic's only job is to assess the quality of the argument as presented.

Another open question regards whether, or to what extent, mathematics is dialectical. Imre Lakatos explicitly raised this issue (Lakatos, 1976). But there are hints at dialectical considerations much earlier in the history of mathematics.

> Let us first inquire why [Euclid] even includes in the theorem the equality of the angles under the base. He is never going to use this result for the construction or demonstration of any other problem or theorem. Since it will not be used later, why was it necessary to bring it into this theorem. To this question we must reply that, even if he was never intending to use 'and the angles under the base of an isosceles triangle are equal,' nevertheless it will be useful in *meeting objections* to them and *refuting* their adversaries. It is a mark of scientific and technical skill to arrange in advance for the undoing of those who attack what is going to be said and to *prepare the positions from which one can reply*, so that these previously demonstrated matters may later serve not only for establishing the truth, but also for *refuting error* (Proclus, 1992, 102–3).

The idea that Euclid presents the propositions in an order that allows for better strategic, rhetorical or dialectical positioning suggests that dialectics have long been essential in mathematics—perhaps going all the way to the very beginnings of mathematics. The question for informal logicians is how best to account for the dialectic elements of mathematics. For example, Erik Krabbe argues that proofs can be or ought to be recast as dialogues (Krabbe, 1991, 1997). The strategy of recasting these arguments as dialogue, i.e. of considering proofs to be implicit dialogues, falls out of a general theory of reasoning championed by Krabbe, see (Walton and Krabbe, 1995). The implicit dialogue strategy would account for the dialectical elements of proofs.

Whatever the answers to these questions are, it should be clear that mathematical reasoning is already in accord with principles and techniques from informal logic—even if this is unnoticed by the practitioners. Thus, it will be important, if one wants one's philosophy of mathematics to have an accurate account of mathematical practice, to include informal logic in the methodology of mathematics.

Acknowledgements A prior version of this paper was published in *Foundations of Science* (2009), 14(1–2):137–152. It received careful and helpful criticism from Andrew Aberdein and David Sherry. I thank them both. Previous versions of this paper were presented in Las Vegas and Amsterdam. I thank the audiences for their helpful questions and comments.

References

Aberdein, A. (2005). The uses of argument in mathematics. *Argumentation, 19*, 287–301.
Aberdein, A. (2007). The informal logic of mathematical proof. In J. P. Van Bendegam & B. Van Kerkove (Eds.), *Perspectives on mathematical practices* (pp. 135–151). Dordrecht: Springer.
Adleman, L. (1994). Molecular computation of solutions to combinatorial problems. *Science*, New Series, *266*(5187), 1021–1024.
Aigner, M., & Ziegler, G. M. (2002). *Proofs from THE BOOK* (2nd ed.). Berlin: Springer.
Allen, C., & Hand, M. (2001). *Primer in logic*. Cambridge, MA: MIT Press.
Azzouni, J. (2004). The derivation indicator view of mathematical practice. *Philosophia Mathematica (3)*, *12*, 81–105.
Azzouni, J. (2006). *Tracking reason*. Oxford: Oxford University Press.
Cáceres, I., Lozana M., & Saladié, P. (2007). Evidence for Bronze Age cannibalism in El Mirador Cave (Sierra de Atapuerca, Burgos, Spain). *American Journal of Physical Anthropology, 133*, 899–917.
Copi, I., & Cohen, C. (1994). *Introduction to logic* (9th ed.). New York: Macmillan.
Detlefsen, M., & Luker, M. (1980). The four-color theorem and mathematical proof. *Journal of Philosophy, 77*, 803–820.
Detlefsen, M. (1992). Poincaré against the logicians. *Synthese, 90*, 349–378.
Detlefsen, M. (1993). Poincaré vs. Russell on the role of logic in mathematics. *Philosophia Mathematica (3)*, *1*, 24–49.
van Eemeren, F. H., & Grootendorst, R. (2003). *A systematic theory of argumentation: The pragma-dialectical approach*. Cambridge: Cambridge University Press.
Euclid. (1956). *The thirteen books of Euclid's Elements* (Vol. I). New York: Dover.
Fallis, D. (1996). Mathematical proof and the reliability of DNA evidence. *The American Mathematical Monthly, 103*(6), 191–197.
Fallis, D. (1997). The epistemic status of probabilistic proofs. *Journal of Philosophy, 94*(4), 165–186.
Fallis, D. (2003). Intentional gaps in mathematical proofs. *Synthese, 134*, 45–69.
Finocchiaro, M. (1996). Critical thinking, critical reasoning and methodological reflection. *Inquiry: Critical Thinking Across the Disciplines, 15*, 66–79. (Reprinted in Finocchiaro, 2005, pp. 92–105).
Finocchiaro, M. (2003a). Dialectic, evaluation and argument: Goldman and Johnson on the concept of argument. *Informal Logic, 23*, 19–49. (Reprinted in Finocchiaro, 2005, pp. 292–326).
Finocchiaro, M. (2003b). Physical-mathematical reasoning: Galileo on the extruding power of terrestrial rotation. *Synthese, 134*, 217–244.
Finocchiaro, M. (2005). *Arguments about arguments: Systematic, critical and historical essays in logical theory*. Cambridge: Cambridge University Press.
Franklin, J. (1987). Non-deductive logic in mathematics. *British Journal for Philosophy of Science, 38*(1), 1–18.
Gödel, K. (1983). What is Cantor's continuum problem? In P. Benacerraf & H. Putnam (Eds.), *Selected readings in philosophy of mathematics* (pp. 470–485). Cambridge: Cambridge University Press. (Originally published 1947).
Johnson, R. H. (2000). *Manifest rationality: A pragmatic theory of argument*. Mahwah, NJ: Lawrence Erlbaum Associates.
Karp, R. M. (1972). Reducibility among combinatorial problems. In R. E. Miller & J. W. Thatcher (Eds.), *Complexity of computer computations* (pp. 85–103). New York: Plenum.

Krabbe, E. (1991). Quod erat demonstrandum: Wat kan en mag een argumentatietheorie zeggen over bewijzen? In M. M. H. Bax & W. Vuijk (Eds.), *Thema's in de Taalbeheersing: Lezingen van het VIOT-taalbeheersingscongres gehouden op 19, 20 en 21 december 1990 aan de Rijksuniversiteit Groningen* (pp. 8–16). Dordrecht: ICG.

Krabbe, E. (1997). Arguments, proofs and dialogues. In M. Astroh, D. Gerhardus & G. Heinzmann (Eds.), *Dialogisches Handeln: Eine Festschrift für Kuno Lorenz* (pp. 63–75). Heidelberg: Spektrum Akademischer Verlag. (This is an updated translation of (Krabbe 1991) and is reprinted in this volume)

Lakatos, I. (1976). *Proofs and refutations: The logic of mathematical discovery* (edited by J. Worrall & E. Zahar). Cambridge: Cambridge University Press.

Maddy, P. (1988). Believing the axioms, I and II. *Journal of Symbolic Logic, 53*(2), 482–511 and *53*(3), 736–764.

Maddy, P. (1992). *Realism in mathematics.* Oxford: Clarendon Press.

Malone, M. (2003). Three recalcitrant problems of argument identification. *Informal Logic, 23*(3), 237–261.

Mates, B. (1972). *Elementary logic* (2nd ed.). Oxford: Oxford University Press.

Maxwell, E. A. (1959). *Fallacies in mathematics.* Cambridge: Cambridge University Press.

Pólya, G. (1968). *Mathematics and plausible reasoning* (Vols. I and II). Princeton, NJ: Princeton University Press.

Proclus. (1992). *Commentary on the first book of Euclid's Elements* (Trans. by Glenn Morrow). Princeton, NJ: Princeton University Press.

Putnam, H. (1975). What is mathematical truth? In H. Putnam (Ed.), *Mathematics, matter and method: Philosophical papers* (Vol. 1, pp. 60–78). Cambridge: Cambridge University Press.

Rav, Y. (1999). Why do we prove theorems? *Philosophia Mathematica (3), 7,* 5–41.

Sandifer, C. E. (2007). *Divergent series. How Euler did it* (pp. 177–184). Washington, DC: Mathematical Association of America.

Scriven, M. (1976). *Reasoning.* New York: McGraw-Hill.

Steiner, M. (1998). *The applicability of mathematics as a philosophical problem.* Cambridge, MA: Harvard University Press.

Toulmin, S. E. (2003). *The uses of argument.* Updated edition (of 1958). Cambridge: Cambridge University Press.

Walton, D. (2006). *Fundamentals of critical argumentation.* Cambridge: Cambridge University Press.

Walton, D., & Krabbe, E. (1995). *Commitment in dialogue: Basic concepts of interpersonal reasoning.* Albany, NY: State University of New York Press.

Zermelo, E. (1908). A new proof of the possibility of well-ordering. In J. van Heijenoort (Ed.). (1967). *From Frege to Godel: A source book in mathematical logic, 1879–1931* (pp. 183–198). Cambridge, MA: Harvard University Press. (Originally published in *Mathematische Annalen, 65,* 107–128)

Chapter 16
Bridging the Gap Between Argumentation Theory and the Philosophy of Mathematics

Alison Pease, Alan Smaill, Simon Colton, and John Lee

16.1 Introduction

Mathematicians have traditionally attributed great importance to proofs, since Euclid's attempts to deduce principles in geometry from a small set of axioms (despite insufficiencies in certain proofs in Euclid's *Elements*, in terms of reliance on diagrams and physical constructions which were not formally defined). Experience of the fallibility of proofs led many mathematicians to change their view of the principal role that proof plays from a guarantee of truth; to new ideas such as an aid to understanding a theorem (Hardy, 1928), a way of evaluating a theorem by appealing to intuition (Wilder, 1944), or a memory aid (Pólya, 1945). Lakatos, via the voice of the teacher in (Lakatos, 1976), suggested that we see a proof as a thought-experiment which "suggests the decomposition of a conjecture into subconjectures or lemmas, thus *embedding it* in a quite possibly distant body of knowledge" (Lakatos, 1976, 9). This change in the perception of the role of proof in mathematics, from its lofty pedestal of infallible knowledge to the more familiar level of flawed and informal thought, suggests that work in argumentation theory, commonly inspired by practical argument such as legal reasoning, may be relevant

A. Pease (✉)
Department of Computing, Imperial College London, 180 Queens Gate, London, SW7 2RH, UK

School of Electronic Engineering and Computer Science, Queen Mary,
University of London, London, E1 4NS, UK
e-mail: apease@doc.ic.ac.uk

A. Smaill • J. Lee
School of Informatics, University of Edinburgh, Informatics Forum, Crichton Street,
Edinburgh, EH8 9LE, UK
e-mail: A.Smaill@ed.ac.uk; j.lee@ed.ac.uk

S. Colton
Department of Computing, Imperial College London, 180 Queens Gate, London, SW7 2RH, UK
e-mail: sgc@doc.ic.ac.uk

A. Aberdein and I.J. Dove (eds.), *The Argument of Mathematics*, Logic, Epistemology, and the Unity of Science 30, DOI 10.1007/978-94-007-6534-4__16,
© Springer Science+Business Media Dordrecht 2013

to the philosophy of mathematics. Conversely, the fertile domain of mathematical reasoning can be used to evaluate and extend general argumentation structures.

The relationship between the philosophy of mathematics and argumentation theory has already borne fruit. In Toulmin's well-known model of argumentation (Toulmin, 1958), written as a critique of formal logic, he argued that practical arguments focus on justification rather than inference. His layout comprises six interrelated components: a claim (the conclusion of the argument), data (facts we appeal to as the foundation of the claim), warrant (the statement authorising the move from the data to the claim), backing (further reason to believe the warrant), rebuttal (any restrictions placed on the claim), and a qualifier (such as "probably", "certainly" or "necessarily", which expresses the force of the claim). While Toulmin did consider mathematical arguments, for example, (Toulmin, 1958, 135–136), he initially developed his layout to describe non-mathematical argument.[1] However, he later applied the layout to Theaetetus's proof that there are exactly five platonic solids (Toulmin et al., 1979). Aberdein has shown that Toulmin's argumentation structure can represent more complex mathematical proofs; such as the proof that there are irrational numbers α and β such that α^β is rational (Aberdein, 2005), and the classical proof of the Intermediate Value Theorem (Aberdein, 2006). Alcolea Banegas (1998) has shown that Toulmin's argumentation structure can also be used to represent meta-level mathematical argument, modelling Zermelo's argument for adopting the axiom of choice in set theory (described in Aberdein, 2005). Alcolea Banegas also presents a case study of Appel and Haken's computer assisted (object level) proof of the four colour theorem (Aberdein suggests, in 2005, an alternative representation of this theorem, which also uses Toulmin's layout). In his (2006), Aberdein describes different ways of combining Toulmin's layout, and uses his embedded layout to represent the proof that every natural number greater than one has a prime factorisation.

Lakatos (1976) championed the informal nature of mathematics, presenting a fallibilist approach to mathematics, in which proofs, conjectures and concepts are fluid and open to negotiation. He saw mathematics as an adventure in which, via patterns of analysis, conjectures and proofs can be gradually refined. Lakatos demonstrated his argument by presenting rational reconstructions of the development of Euler's conjecture that for any polyhedron, the number of vertices (V) minus the number of edges (E) plus the number of faces (F) is equal to two, and Cauchy's proof of the conjecture that the limit of any convergent series of continuous functions is itself continuous. He also presented a rational reconstruction of the history of ideas in the philosophy of mathematics. Lakatos's work in the philosophy of mathematics had three major sources of influence: firstly, Hegel's dialectic, in which

---

[1] For instance, Toulmin argues that "mathematical arguments alone seem entirely safe" from time and the flux of change, adding that "this unique character of mathematics is significant. Pure mathematics is possibly the only intellectual activity whose problems and solutions are 'above time'. A mathematical problem is not a quandary; its solution has no time-limit; it involves no steps of substance. As a model argument for formal logicians to analyse, it may be seducingly elegant, but it could hardly be less representative" (Toulmin, 1958, 127).

the *thesis* corresponds to a naïve mathematical conjecture and proof; the *antithesis* to a mathematical counterexample; and the *synthesis* to a refined theorem and proof (described in these terms in Lakatos, 1976, 144–145); secondly, Popper's ideas on the impossibility of certainty in science and the importance of finding anomalies (Lakatos argued that Hegel and Popper "represent the only fallibilist traditions in modern philosophy, but even they both made the mistake of preserving a privileged infallible status for mathematics", Lakatos, 1976, 139); and thirdly, Pólya's (1954) work on mathematical heuristic and study of rules of discovery and invention, in particular defining an initial problem and finding a conjecture to develop (Lakatos claimed that the discussion in his (1976) starts where Pólya's stops (Lakatos, 1976, 7, footnote 1)). Lakatos held an essentially optimistic view of mathematics, seeing the process of mathematical discovery in a rationalist light. He challenged Popper's view that philosophers can form theories about how to evaluate conjectures, but not how to generate them[2] in two ways: *(i)* he argued that there *is* a logic of discovery, the process of generating conjectures and proof ideas or sketches *is* subject to rational laws; and *(ii)* he argued that the distinction between discovery and justification is misleading as each affects the other; *i.e.*, the way in which we discover a conjecture affects our proof (justification) of it, and proof ideas affect what it is that we are trying to prove (see Larvor, 1998). This happens to such an extent that the boundaries of each are blurred.[3]

There are a few explicit and implicit overlaps between Lakatos's work and argumentation theories. Aberdein (2006) uses Toulmin's layout to describe and extend Lakatos's method of lemma-incorporation, where a rebuttal (*R*) to a claim (*C*) is a global counterexample, and the argument is repaired by adding a lemma (*L*),

---

[2]Popper (1959) argued that the question of generation should be left to psychologists and sociologists: "The question of how it happens that a new idea occurs to man... may be of great interest to empirical psychology; but it is irrelevant to the logical analysis of scientific knowledge" (*ibid.*, 31), and shortly after emphasised again, that: "there is no such thing as a logical method of having new ideas" (*ibid.*, 32).

[3]The question of whether mathematical claims and proofs are socially constructed is clearly pertinent to our own thesis that argumentation theory is relevant in a mathematical context. Whether Lakatos held a social-constructivist philosophy of mathematics is controversial (but perhaps irrelevant for our current thesis). However, there are certainly social aspects in Lakatos's theory: in particular his emphasis on the influence that Hegel's dialectic had on his thinking, and his presentation of mathematical development as a social process of concept, conjecture and proof refinement, presented in dialogue form. Ernest (1997) develops Lakatos's fallibilism and ideas on negotiation and acceptance of mathematical concepts, conjectures and proofs (together with Wittgenstein's socially situated linguistic practices, rules and conventions) into a social-constructivist philosophy of mathematics. Goguen also presents a defence of the social-constructivist position, arguing that although mathematicians talk of proofs as real things, "all we can ever actually find in the real world of actual experience are proof events, or "provings", each of which is a social interaction occurring at a particular time and place, involving particular people, who have particular skills as members of an appropriate mathematical social community." (Goguen, 1999, 288). He continues in a very Lakatosian vein to criticise the way mathematicians often hide obstacles and difficulties when presenting their proofs, advocating that the drama be reintroduced.

which the counterexample refutes, as a new item of data ($R \rightarrow \neg L$) and incorporating the lemma into the claim as a precondition ($L \rightarrow C$). Pedemonte (2001) discusses the relationship between the production of a mathematical conjecture and the construction of its proof-object, referred to as *cognitive unity*. This can be compared to Lakatos's *logic of discovery and justification*. While Lakatos viewed the two processes as circular, with changes in the proof suggesting changes to the conjecture and vice versa, as opposed to Pedemonte's assumption of a chronological order, the two theories can be seen as cognitive and philosophical counterparts. Naess's (1953) work on argumentation can also be compared to Lakatos's theory: he argues that discussion can be about interpretation of terms, during which a process of precization takes place. If this fails to lead to agreement then evidence is weighed up to see which of two interpretations is more acceptable. There are very strong parallels between this and Lakatos's method of monster-barring in particular, but also his other methods of conjecture and proof refinement. Another similarity is the *degree* of precization required: both Lakatos and Naess argued that we should make our expressions sufficiently precise for the purposes at hand, rather than aim to resolve all ambiguity (Naess, 1947).[4]

The rest of this paper is structured as follows: in Sect. 16.2 we describe our computational model of Lakatos's theory, HRL, which has suggested new connections between argumentation theory and the philosophy of mathematics. In Sect. 16.3 we discuss how we have represented Cauchy's proof of Euler's conjecture by using work by Haggith on argumentation representation and structures. Section 16.4 contains a discussion of aspects of Lakatos's method of lemma-incorporation and how they have affected our algorithmic realisation: we also describe our algorithms for each type of lemma-incorporation and for determining which type to perform. In Sect. 16.5 we outline some connections between Haggith's argumentation structures and Lakatos's methods and show how other mathematical examples can be described in this way. We conclude in Sect. 16.6.

16.2   A Computational Model of Lakatos's Theory

Lakatos outlined various methods by which mathematical discovery and justification can occur. These methods suggest ways in which concepts, conjectures and proofs gradually evolve via interaction between mathematicians, and include surrender, monster-barring, exception-barring, monster-adjusting, lemma-incorporation, and proofs and refutations. Of these, the three main methods of

---

[4]We thank one of our reviewers for pointing out the relevance of precization. U is more precise than T if any interpretations of U are also interpretations of T, but there are interpretations of T which are not interpretations of U. Both Naess and Lakatos see agreeing on meaning of terms as a stage of a discussion (as opposed to a pre-requisite to it as, for example, Crawshay-Williams, 1957). Naess also shares with Lakatos an approach to philosophy which is mainly based on descriptive, rather than normative aspects.

theorem formation are monster-barring, exception-barring, and the method of proofs and refutations (Lakatos, 1976, 83). Crudely speaking, monster-barring is concerned with concept development, exception-barring with conjecture development, and the method of proofs and refutations with proof development. However, these are not independent processes; much of Lakatos's work stressed the interdependence of these three aspects of theory formation. We hypothesise that *(i)* it is possible to provide a computational reading of Lakatos's theory, and *(ii)* it is useful to do so. To test these two hypotheses we have developed a computational model of Lakatos's theory, HRL.[5] Running the model has provided a means of testing hypotheses about the methods; for instance that they generalise to scientific thinking, or that one method is more useful than another. Additionally, the process of having to write an algorithm for the methods has forced us to interpret, clarify and extend Lakatos's theory, for instance identifying areas in which he was vague, or omitted details.[6]

In keeping with the dialectical aspect of (Lakatos, 1976), our model is a multi-agent dialogue system, consisting of a number of student agents and a teacher agent. Each agent has a copy of the theory formation system, HR (Colton, 2002), which starts with objects of interest (*e.g.*, integers) and initial concepts (*e.g.*, division, multiplication and addition) and uses production rules to transform either one or two existing concepts into new ones. HR also makes conjectures which empirically hold for the objects of interest supplied. Distributing the objects of interest between agents means that they form different theories, which they communicate to each other. Agents then find counterexamples and use methods identified by Lakatos to suggest modifications to conjectures, concept definitions and proofs.

In this paper we are concerned with our algorithmic realisation of the method of proofs and refutations, and its mutually beneficial association with work from argumentation theory. Other aspects of the project are described in (Pease et al., 2004).

16.3 A Computational Representation of Cauchy's Proof

The method of *lemma-incorporation*, developed via the dialectic into the method of *proofs and refutations*, is considered by Lakatos to be the most sophisticated in his (Lakatos, 1976). Commentators and critics, for instance (Corfield, 1997) or (Feferman, 1978), usually share this view, often seeing the rest of the book as a

---

[5]The name incorporates HR (Colton, 2002), which is a system named after mathematicians Godfrey Harold Hardy and Srinivasa Aiyangar Ramanujan and forms a key part of our model, and the letter "L", which reflects the deep influence of Lakatos's work on our model.

[6]Since Lakatos's work (1976) was the first attempt to characterise informal mathematics (see Corfield, 1997; Feferman, 1978), it is likely to be incomplete, and hence be open to criticism and extension. Lakatos himself neither considered the methods complete nor definitive, arguing only that they provide a more realistic and helpful portrayal of mathematical discovery than Euclidean (deductive) methodology.

Fig. 16.1 Given the cube, after removing a face and stretching it flat, we are left with the network in **part 1**. After triangulating, we get **part 2**. When removing a triangle, we either remove one edge and one face, or two edges, one vertex and a face—shown in **parts 3** (*a*) and (*b*) respectively

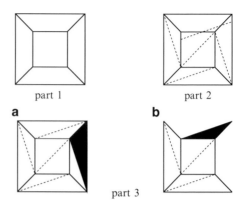

part 1   part 2

a   b

part 3

prelude to this method.[7] The method works on a putative proof of a conjecture: the main example in (Lakatos, 1976) is Cauchy's (1813) proof of Euler's conjecture that for all polyhedra, $V - E + F$ is 2. Therefore, in order to model *lemma-incorporation* in HRL, we need a way of representing an informal mathematical proof.

16.3.1   Cauchy's Proof

Lakatos argued that nineteenth century mathematicians viewed Cauchy's proof of Euler's conjecture in (Cauchy, 1813) as establishing the truth of the 'theorem' beyond doubt (Lakatos, 1976, 8, cites Crelle, 1827, Jonquières, 1890 and Matthiessen, 1863 as examples). For a diagrammatic representation of these steps, carried out on the cube, see Fig. 16.1, taken from (Lakatos, 1976, 8).

> *Step 1:* Let us imagine the polyhedron to be hollow, with a surface made of thin rubber. If we cut one of the faces, we can stretch the remaining surfaces flat on the blackboard, without tearing it. The faces and edges will be deformed, the edges may become curved, and V and E will not alter, so that if and only if $V - E + F = 2$ for the original polyhedron, $V - E + F = 1$ for this flat network - remember that we have removed one face. *Step 2:* Now we triangulate our map - it does indeed look like a geographical map. We draw (possibly curvilinear) diagonals in those (possibly curvilinear) polygons which are not already (possibly curvilinear) triangles. By drawing each diagonal we increase both E and F by one, so that the total $V - E + F$ will not be altered. *Step 3:* From the triangulated map we now remove the triangles one by one. To remove a triangle we either remove an edge - upon which one face and one edge disappear, or we remove two edges and a vertex - upon which one face, two edges and one vertex disappear. Thus, if we had $V - E + F = 1$ before a triangle is removed, it remains so after the triangle is removed. At the end of this procedure we get a single triangle. For this $V - E + F = 1$ holds true. (Lakatos, 1976, 7–8)

---

[7]This is possibly because, despite different perspectives on the role of proof in mathematics, the idea that it is an important one is generally accepted, and these are the only methods to consider the 'proof' of a conjecture.

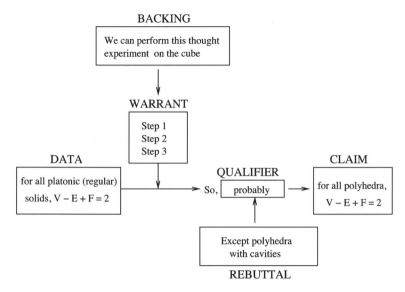

Fig. 16.2   A representation of Cauchy's proof of Euler's conjecture, using Toulmin's layout, where steps 1–3 are as described above, taking into consideration the first counterexample. The data are the facts which initially inspire the conjecture in Lakatos's (1976)

16.3.2   Representing Informal Mathematical Proofs

Cauchy's proof can be represented in a variety of ways using Toulmin's layout. We show one of the simplest ways in Fig. 16.2; more sophisticated versions might include multiple-linked, or nested layouts (as in Aberdein, 2005, 2006) where individual proof steps each form a *claim* in one argument, which then (possibly combined with other premises) forms the *data* in a subsequent argument.[8] However, while Toulmin picked apart argumentation structures and showed how the traditional "Minor Premise, Major Premise, so Conclusion" was too crude to represent the way in which people actually argue, he mainly identified different types of statement which in some way *support* a claim. There is only one type of statement in Toulmin's layout which *opposes* a claim: a rebuttal. This can be interpreted in different ways: as a rebuttal to the claim, a rebuttal to the warrant, a rebuttal to an implicit premise, or as a statement which supports a refutation of the claim, and its function is still under debate (see Reed and Rowe, 2005, 15–19). Pollock (1995) defines a *rebutting defeater R* as a reason for denying a claim *P* which is supported by prima facie reason *Q*. He claims (*ibid.*, 41) to have been the first to explicitly point out defeaters other than a rebuttal, in (Pollock, 1970), and identifies the *undercutting defeater*. This defeater attacks the connection between

---

[8]It would be interesting to investigate how many of the arguments described in (Lakatos, 1976) can be represented in this layout.

a prima facie reason and the conclusion, rather than attacking the conclusion directly. For the purposes of our computational model we have adopted a meta-level argumentation framework (Haggith, 1996), consisting of a catalogue of argument structures which give a very fine-grained representation of arguments, in which both arguments and counter-arguments can be represented. While this framework may lack Toulmin's analysis of statements which support a claim (and has no way of representing a qualifier), it is clear which part of an argument a rebutter rebuts. Given Lakatos's emphasis on both the importance and the different types of counterexample, we deemed this framework appropriate for our needs.

16.3.3 Haggith's Argumentation Structures

Haggith (1996) starts from the viewpoint that if a domain is controversial, then there may be more than one answer to a question and therefore disagreements may be useful, rather than an obstacle to be overcome. The primary goal of the system described in Haggith's (1996), therefore, is to *explore* rather than resolve conflicts. In order to incorporate a high degree of flexibility, Haggith represents arguments at the meta-level which is independent of logic or any specific representation language or domain.

Haggith's representation language describes three categories of meta-level object: proposition names, arguments and sets. The symbols used in the alphabet are: $A_1, A_2, \ldots$ to denote proposition names, "$\Leftarrow$" the argument constructor, the standard set and logical symbols, some relation names (such as disagree) and brackets and commas. An argument A in which C is the conclusion, derived from premises $P1$ and $P2$, where $P2$ is itself derived from premise $P3$, is represented as:

$$A = \{C \Leftarrow \{P1, P2 \Leftarrow \{P3\}\}\}$$

There are two destructor relations for looking inside argument and set terms: *set membership*, $\in$, which is a two-place, infix relation where $P \in S$ if, for some $S1$, $S = S1 \cup \{P\}$; and *argument* which is a three-place relation, where argument(A, P, S) if A is the argument $P \Leftarrow S$ and P is called the conclusion and S is called the premise set. A further destructor, the support relation can be defined from the other two. This holds between two propositions, the second of which occurs in an argument for the first: supports(P, Q) holds if there exists A, argument(A, P, S) and, either Q is a member of S, or there exists $A1$, a member of S, such that argument$(A1, P1, S1)$ and supports$(P1, Q)$. Haggith has defined four primitive relations at the meta-level which express links and contrasts between object level propositions: *equivalent*(P, Q), where P and Q are names of propositions which mean the same; *disagreement*(P, Q), where P and Q are names of propositions which disagree or express a conflict; *elaboration*(P, S), where P is a proposition name and S is a set of names of propositions which elaborate or embellish upon P; and *justification*(P, S), where P is a proposition name and S is a set of names of propositions which are a justification of P. Haggith provides some properties of these relations which restrain their possible applicability, for instance:

- $disagreement(P,Q) \rightarrow disagreement(Q,P)$;
- $disagreement(P,Q) \rightarrow not(equivalent(P,Q))$; and
- $(equivalent(P,Q) \,\&\, elaboration(P,S)) \rightarrow elaboration(Q,S)$.

Haggith then constructs higher order, meta-level relations defined in terms of the four primitive relations. It is Haggith's development of these argumentation structures that distinguishes her work from standard box-arrow systems. We give two of the structures in detail here and sketch the rest below. We use the letters "X" and "Y" to represent anonymous variables. *Rebuttal*, inspired by (Elvang-Goransson et al., 1993), is a relation between arguments whose conclusions disagree. The meta-level definition is:

- $rebuttal(P)$ is the set of arguments, A, such that $disagreement(P,Q) \,\&\, argument(A,Q,X)$.
- $rebuts(A,B)$ if $argument(A,P,X) \,\&\, member(B,rebuttal(P))$.

That is, the rebuttal of a proposition P is the set of arguments for any propositions which disagree with P. Two arguments rebut each other if one is a member of the rebuttal of the conclusion of the other. Haggith's notion of rebuttal fits with an interpretation of the Toulminian rebutter as one which rebuts the claim. It also coincides with Pollock's definition of rebuttal.

Undercutting is inspired by (Toulmin, 1958) and defined as follows: an argument A, with conclusion Q, undercuts an argument B if for some premise P of B, P disagrees with Q. That is, an argument undercuts another, if the first rebuts a premise of the second. The meta-level definition is:

- $undercutting(P)$ is the set of arguments, A, such that $argument(X,P,S)$ $\&\, member(P1,S) \,\&\, disagreement(P1,Q) \,\&\, argument(A,Q,Y)$.[9]
- $undercuts(A,B)$ if $argument(A,P,X) \,\&\, member(B,undercutting(P))$.

Haggith's notion of undercutting fits with an interpretation of the Toulminian rebutter as one which rebuts the warrant, and with Pollock's definition of undercutting. The only difference is that both the interpretation of Toulmin and Pollock specify the *type* of supporting premise which a statement must rebut (the warrant and the prima facie, respectively), while Haggith does not make that distinction.

Given a proposition P and an argument A for P, possible argument moves which provide support for P include:

- *corroboration*: an argument for a proposition which is equivalent to (or is) P;
- *enlargement*: an argument for an elaboration of P, and
- *consequence*: an argument in which P is a premise.

---

[9]Another way of saying this is $argument(X,P,S) \,\&\, member(P1,S) \,\&\, member(A,rebuttal(P1))$.

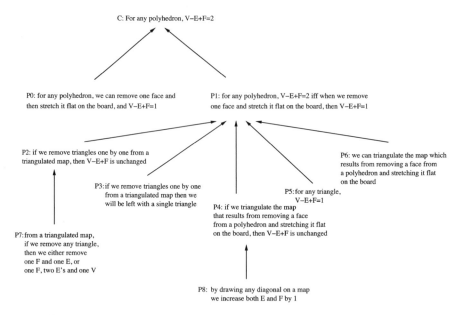

Fig. 16.3 The original proof of Euler's conjecture, represented in Haggith's terms. The *arrows* represent the justification relation, where the set of premises taken together on any line supports the proposition directly above it. Propositions without any *arrows* leading into them are unsupported assumptions (thus particularly open to counter-argument)

Argument moves which oppose *A* include:

- *rebuttal*: an argument for a proposition which disagrees with *P*;
- *undermining*: an argument for a proposition which disagrees with a proposition which is an elaboration of, or is equivalent to, *P*;
- *undercutting*: an argument for a proposition which disagrees with a premise of *P*;
- *target*: an argument which contains a premise which disagrees with *P*, and
- *counter-consequence*: an argument which contains a premise which disagrees with the conclusion of another argument in which *P* is a premise (inspired by Sartor, 1993).

16.3.4 Using Haggith's Argumentation Structures to Represent Mathematical Proofs

We have expressed Cauchy's proof in Haggith's terms by writing it as a series of propositions and showing the relationships between them. This is shown in Fig. 16.3, where the proof looks as follows:

$$A = \{C \Leftarrow \{P0, P1 \Leftarrow \{P2 \Leftarrow \{P7\}, P3, P4 \Leftarrow \{P8\}, P5, P6\}\}\}.$$

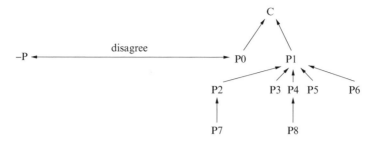

Fig. 16.4 The first counter-argument, represented in Haggith's terms. *Unmarked arrows* represent
the justification relation

The four initial counter-arguments (all questioning unsupported assumptions),
suggested by the ingenious students, to the three steps of Cauchy's proof on
(Lakatos, 1976, 7) are all examples of Haggith's *target* arguments, disagreeing
with a premise of C. We represent them below,[10] and show the diagrammatic
representation of the first in Fig. 16.4:

1. $-P0$: Some polyhedra, after having a face removed, cannot be stretched flat on a
   board (questioning the first step). This is shown diagrammatically below.
2. $-P8$: In triangulating the map, we will not always get a face for every new edge
   (questioning the second step).
3. $-P7$: There are more than two alternatives, when we remove the triangles one by
   one, that either one edge and a face; or two edges, a face and a vertex disappear
   (questioning the third step).
4. $-P3$: If we remove triangles one by one from a triangulated map, then we may
   not be left with a single triangle (also questioning the third step).

16.4 Lakatos's Method of Lemma-Incorporation

Lakatos's method of lemma-incorporation distinguishes *global* and *local* counterexamples, which refute the main conjecture or one of the proof steps (or lemmas),
respectively. When a counterexample is found, lemma-incorporation is performed
by determining which type of counterexample it is: if it is local but not global
(the conclusion may still be correct but the reasons for believing it are flawed)
then he proposes modifying the problematic proof step but leaving the conjecture unchanged; if it is both global and local (there is a problem both with
the argument and the conclusion) then Lakatos proposes modifying the conjecture by incorporating the problematic proof step as a condition; and if it is

---

[10]We state the counter-arguments as propositions, whereas in (Lakatos, 1976) they are questions,
i.e., "are you sure that..." rather than "it is not possible...".

Fig. 16.5 Given the network
which results from taking the
cube, removing a face and
stretching it flat, and
triangulating, we can remove
a triangle (shown in *black*)
which results in removing one
face, no edges and no vertices

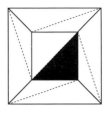

part 3 (c)

global but not local (there is a problem with the conclusion but no obvious
flaw in the reasoning which led to the conclusion) then he proposes looking
for a hidden assumption in the proof step, then modifying the proof and the
conjecture by making the assumption an explicit condition. The method of proofs
and refutations consists of setting out to prove and refute a conjecture, looking
for counterexamples both to the conjecture and the lemmas, determining which
type of counterexample it is, and then performing lemma-incorporation. In the
discussion below, we follow Lakatos's convention of using students with names
from the Greek alphabet to present different mathematical and philosophical
viewpoints.

16.4.1 Three Types of Counterexample

The cube is a local but not global counterexample since it violates the third lemma
in Cauchy's proof, but not the conjecture. That is, it is possible to remove a triangle
without causing the disappearance of one edge or else of two edges and a vertex,
by removing one of the inner triangles (see Fig. 16.5); in this case, we remove a
face but no edges or vertices. In this case we want to modify the proof but leave the
conjecture unchanged. We do this by generalising from a single counterexample to
a class of counterexamples, and modifying the problem lemma to exclude that class.
In this example, lemma three becomes "when one drops the *boundary* triangles one
by one, there are only two alternatives – the disappearance of one edge or else of
two edges and a vertex". The corresponding method is local, but not global lemma-
incorporation.

The hollow cube (a cube with a cube shaped hole in it), is a both a global
counterexample, since $V - E + F = 16 - 24 + 12 = 4$, and local, since it cannot
be stretched flat on the blackboard having had a face removed. In this case
we need to identify the faulty lemma, lemma one, and then make that step a
condition of the conjecture. The proof is left unchanged. Given the hollow cube,
we should incorporate the first lemma into the conjecture; this then becomes
"for any polyhedron which, by removing one face can be stretched flat onto a
blackboard, $V - E + F = 2$". The corresponding method is global and local lemma-
incorporation.

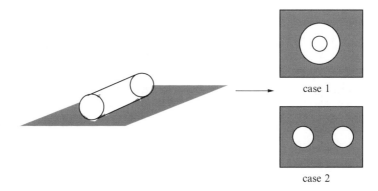

Fig. 16.6 If we remove a face from the cylinder and stretch it flat, then we either get *case 1* if we remove an end face, or *case 2* if we remove the jacket. Either way, we have satisfied the first lemma

The cylinder is a global counterexample, as $V - E + F = 0 - 2 + 3 = 1$ but not local, since it does not violate any of the proof steps. We can remove a face and stretch it flat, resulting in two circles which are either disjoint or concentric (see Fig. 16.6). In order to falsify the second lemma, we would have to draw an edge which joins two non-adjacent vertices, but does not create a new face. Clearly we cannot do this as there are no vertices on the map. Similarly, in order to falsify the third lemma, we would have to be able to remove a triangle and not remove either one edge and one face, or two edges, a vertex and a face, and since there are no triangles on the map, we cannot fail at this stage either. Thus it suggests that the conjecture is flawed and yet the proof of it is upheld. Lakatos argues that this strange situation occurs because of 'hidden assumptions' in the proof, which the counterexample *does* violate. Faced with such a counterexample, he suggests that we retrace our progress through the proof, until we come to a step which is in some way surprising, *i.e.*, one which violates some hidden assumption in the mathematician's mind. Once this has been identified, we should make it explicit in the proof. The counterexample then becomes one of the second type. For instance, one hidden assumption is that having performed lemma one, we are left with a connected network. Therefore we should add this into the proof explicitly, and modify lemma one to 'any polyhedron, after having a face removed, can be stretched flat on the blackboard, and the result is a connected network'. The cylinder clearly violates this, so we incorporate the new, explicit lemma, into the conjecture statement, which then becomes "for any polyhedron which, after having a face removed, can be stretched flat on the blackboard leaving a connected network, $V - E + F = 2$". The principle of turning a counterexample which is global but not local into one which is both is called the *principle of retransmission of falsity* in (Lakatos, 1976). This requires that falsehood should be retransmitted from a global conjecture to the local lemmas. Thus, any entity which is a counterexample to the conjecture must also be a counterexample to one of the lemmas. Lakatos called this *hidden lemma-incorporation.*[11]

16.4.2 Discussion of Lemma-Incorporation

In this section, we discuss various aspects of lemma-incorporation and how they have affected our algorithmic realisation.

16.4.2.1 Combining the Methods

Exception-barring can be used in lemma-incorporation to get a 'very fine delineation of the prohibited area' (Lakatos, 1976, 37). This means that we were able to reuse our exception-barring algorithms within our computational representation of the method of lemma-incorporation.

16.4.2.2 The Type of Entity in Hidden Lemma-Incorporation

The *type* of entity that we are discussing is important for computational purposes. This may change as we step through a proof. For instance, Cauchy's proof begins by referring to *polyhedra* and, once a face has been removed and a polyhedron stretched onto a board, then discusses *graphs*. Since a polyhedron can only be a counterexample to conjectures about polyhedra, not to conjectures about graphs, it may appear that we have a counterexample which is global and not local. It is necessary to look for the corresponding graph and determine whether this entity is a counterexample to those conjectures about graphs. In this case, the disconnected circles in lemma two correspond to the cylinder. This is glossed over in (Lakatos, 1976), as the following quote shows:

---

[11]Note that even if we disregard the different interpretations of the second lemma, and hence disagreement about whether the cylinder is a local as well as global counterexample, *Gamma's* argument that it is only global is not convincing. In the initial proof given (see Sect. 16.3.1) it *does* say explicitly that at the end of the process there is a single triangle: *if we drop the triangles one by one from a triangulated map, we will end up with a single triangle.* This lemma is violated by the cylinder, making it both a global and local counterexample. This would allow for the usual modification of making the lemma a precondition, *i.e.*, the conjecture would become:

'for any polyhedra which, after having a face removed, and then stretched flat, triangulated and the triangles removed one by one, *leaves a single remaining triangle, $V - E + F = 2$'.

Gamma is able to make his argument because the students get distracted by his claim that the cylinder can be triangulated. The discussion then turns to the meaning of statements which are vacuously true. If they had not disputed this point, *Gamma* would not have been able to uphold his argument. However, the cylinder is still an important example, in that it highlights hidden assumptions in the proof, which should be explicit.

> *Gamma:* The cylinder *can* be pumped into a ball—so according to *your* interpretation it does comply with the first lemma.

> *Alpha:* Well… But you have to agree that it does *not* satisfy the *second* lemma, namely that '*any face dissected by a diagonal fall into two pieces*'. How will you triangulate the circle or the jacket? Are these faces simply connected? (Lakatos, 1976, 44).

When *Alpha* uses the word 'it', he refers to the cylinder. However, he then moves on to talking about the associated graph. While for humans this leap may be acceptable, when implementing this in a program we need to be explicit about the types of entity to which we are referring.

16.4.2.3   Identifying a Problem Lemma in Hidden Lemma-Incorporation

In a proof where the lemmas chain together as a sequence of implications, the problem of identifying lemmas involving hidden assumptions, as presented by Lakatos, is not difficult for humans. This is because of the element of surprise which people feel when an entity does not "behave" in the expected way, where the 'expected way' has been learned from previous examples. Modelling this feeling of surprise, however, is a difficult task. To help us, we considered what caused the surprise and produced a simple model of that.[12] In Lakatos's example, hidden assumptions are found in two lemmas and cause surprise in different ways. In what follows, proof-schemes are a series of subconjectures constructed by the user, which fit together in a way determined by the user. These are input to an agent, who can identify counterexamples and use them to revise the proof. These revised proofs are then output and the proof is not currently incorporated into an agent's theory.

1. Surprise caused by unexpected behaviour

Lemma one states that any polyhedron, after having a face removed, can be stretched flat onto a blackboard. Although the cylinder is a supporting example of this conjecture, it is surprising: when we remove the jacket from the cylinder, it falls into two parts, leaving two disconnected circles. This is surprising since all previous examples resulted in connected networks. Therefore, we needed to capture the idea of an entity being surprising *with respect to a given conjecture*. We have defined this as follows:

- surprisingness (type 1): an entity m is surprising with respect to a conjecture C, $\forall x P(x) \rightarrow Q(x)$, if there exists another conjecture C' of the form $\forall x P(x) \rightarrow (Q(x) \wedge R(x))$, for some concept R, where m is the only known counterexample to C'.

Given a proof-scheme and an entity which is a global but not local counterexample, our algorithm for surprise caused by unexpected behaviour is to go through each lemma in the proof-scheme and, if possible, generate a further

---

[12]Note that we would not claim that our model itself is surprised, simply that the model can identify those lemmas which cause surprise to humans, and the hidden assumptions within the lemmas.

conjecture C' of the form above. In order to identify the 'hidden assumption' in a conjecture, we have to break down the concepts in it, in particular the concept Q in the conjecture $P \rightarrow Q$. This is made easy for our purposes since for each of its concepts, HR records the construction path, and in particular the concepts to which production rules were applied to get a current concept. This ancestor list allows HRL to gradually dissect a concept until a suitable further concept, R, is found. This R is the hidden condition.

2. *Surprise caused by non-meaningful terms*

The discussion of the second lemma, that *any face dissected by a diagonal falls into two pieces*, with respect to the two disconnected circles, is related to work on meaning and denotation, for instance (Russell, 1971, 496–504). The problem is that although there are no diagonals on a circle, we are making a claim about the properties that they have. Lakatos's character *Gamma* argues that it is correct to say that 'every new diagonal we draw on two disconnected circles results in a new face' (P), since the negation, that 'there is a diagonal of the two circles which does *not* create a new face' ($\neg P$), is false. This argument uses the law of excluded middle in classical logic, $P \vee \neg P$, *i.e.*, $\neg(\neg P) \rightarrow P$. According to this argument, the cylinder is a global but not local counterexample. *Alpha* disagrees, arguing that if we say that P is true then we must be able to construct at least one instance of it, *i.e.*, there must be an existential clause in the lemma. The statement 'a face is simply connected' means 'for all x, if x is diagonal then x cuts the face into two; *and there is at least one x that is a diagonal*' (Lakatos, 1976, 45). Under *Alpha*'s interpretation, the cylinder *is* a counterexample to this lemma, as there are no diagonals on the circle. Therefore the cylinder is a local as well as global counterexample, and the problem is no longer a case of hidden lemma-incorporation.

Although it would be difficult to model the surprise that a human feels when they attempt to triangulate a circle, the emphasis on vacuously true statements gave us an insight into how to automate this method. We defined the second type of surprise as follows:

- surprisingness (type 2): an entity m is surprising with respect to a conjecture C, $\forall x P(x) \rightarrow Q(x)$, if $\neg P(m)$.

Given a proof-scheme and an entity m which is a global but not local counterexample, our algorithm goes through each lemma C_i in the proof-scheme. If C_i is of the form $\forall x P(x) \rightarrow Q(x)$, and $\neg P(m)$, then HRL performs two steps: *(i)* it generates the conjecture $C'_i = \forall x (P(x) \rightarrow Q(x)) \wedge P(m)$ (the entity m is now a counterexample to C'_i), and *(ii)* it returns C_i as the hidden faulty lemma, and C'_i as the explicit lemma.

16.4.2.4 Multiple Applications of Lemma-Incorporation

In the discussion of lemma-incorporation in (Lakatos, 1976), the method is applied to the same conjecture (and proof) at least three times (thus enabling the description of different types of lemma-incorporation). This is like previous methods; one counterexample is found and dealt with and then more counterexamples to the modified conjecture and proof are sought. In HRL, proof-schemes and conjectures are passed around and modified and different students may consider them, or the same student may consider different versions of a proof and conjecture at different times.

16.4.3 Algorithms for Lemma-Incorporation

We have interpreted Lakatos's method of lemma-incorporation as the series of algorithms shown below. We define $P \leadsto Q$ to mean that it is *nearly* true that $P \to Q$, *i.e.*, there are lots of supporting examples and few counterexamples. We implemented these algorithms as a computer program: the teacher in HRL is given a proof-scheme and conjecture $P \to Q$ by the user, and asks the students to use Lakatos's methods to analyse both proof-scheme and conjecture. Further technical details are given in (Pease, 2007).

1. **Determine which type of lemma-incorporation to perform.** If there are any counterexamples to the global conjecture, then if either: *(i)* these are also counterexamples to any of the lemmas in the proof, or *(ii)* there is a counterexample to a local lemma, and there is a concept which links the local counterexample to the global counterexample and this concept appears in one of the local lemmas, then perform *global and local lemma-incorporation*. If there is a global counterexample but neither *(i)* nor *(ii)* hold then perform *hidden lemma-incorporation*. Otherwise, if there are counterexamples to any of the lemmas in the proof, then perform *local-only lemma-incorporation*.

2. **Perform local-only lemma-incorporation.** Given a conjecture to which there are no known counterexamples, and a proof tree which contains a faulty lemma, $P \leadsto Q$, to which there is at least one counterexample, then if there is a concept C in the theory which exactly covers the counterexamples (or such a concept can be formed), then make the concept $P \wedge \neg C$, replace the faulty lemma with the conjecture $P \wedge \neg C \to Q$, and return the improved proof-scheme.

3. **Perform global and local lemma-incorporation.** Given a proof-scheme where there are counterexamples to the global conjecture, $P \leadsto Q$, and these counterexamples are also counterexamples to a lemma in the proof, $R \leadsto S$; form the concept C 'objects which satisfy the faulty lemma', by merging the two concepts R and S (this is done by using a production rule to compose R and S), modify the global conjecture by making the new conjecture $C \to Q$, and replace the old global conjecture in the proof-scheme with the modified version.

4. **Perform global-only lemma-incorporation.** Given a proof-scheme where there are counterexamples to the global conjecture, but these counterexamples are not counterexamples to any of the lemmas in the proof, and none of the lemmas have counterexamples which are related to the global counterexamples; let the global conjecture be a near-implication $P \leadsto Q$. Then go through the proof-scheme and

take each lemma in turn, testing each to see whether the global counterexample is surprising in the first sense (type 1) with respect to the lemma. If it is, then return this lemma as the hidden faulty lemma and generate another conjecture, to which the global counterexample *is* a counterexample, as the explicit lemma. If not, then go through the proof-scheme and take each lemma in turn, testing each to see whether the global counterexample is surprising in the second sense (type 2) with respect to the lemma. If so, then return this lemma as the hidden faulty lemma, and generate another conjecture, to which the global counterexample *is* a counterexample, as the explicit lemma. If an explicit lemma has been found, then generate an intermediate proof-scheme in which the hidden faulty lemma is replaced by the explicit lemma, perform global and local lemma incorporation on the intermediate proof-scheme, and return the result. If no lemma is surprising in either sense, then return the proof-scheme unchanged.

Working in the polyhedra domain, and given the proof tree, conjecture and counterexample as input, HRL has replicated all three of Lakatos's types of lemma-incorporation (Pease, 2007).

16.5   Connections Between Haggith and Lakatos

We outline some connections between Haggith's argumentation structures and Lakatos's methods below, which have been suggested by our work in producing a computational reading of Lakatos's theory. In each example, P stands for the proposition "$\forall x, poly(x) \rightarrow euler(x, 2)$", *i.e.*, for all polyhedra, the number of vertices (V) minus the number of edges (E) plus the number of faces (F) is equal to two. Polyhedra for which $V - E + F$ is 2 are called Eulerian. Unless otherwise specified, page references refer to (Lakatos, 1976). In some examples we give alternative propositions for Q, which we express as Q'. We provide a diagram for each pattern where, again, unmarked arrows represent the justification relation between a proposition and a set of propositions. Given that Lakatos is commonly criticised (for example, Feferman, 1978) for claiming that his methods have general application despite only considering two case studies, we suggest how each argument pattern might describe other mathematical examples.

Corroboration: (Fig. 16.7)

Q: *All polyhedra in which circuits and bounding circuits coincide, are Eulerian* (Lakatos, 1976, 114). This is reformulated again to:

Q':   *If the circuit spaces and bounding circuit spaces coincide, the number of dimensions of the 0-chain space* minus *the number of dimensions of the 1-chain space* plus *the number of dimensions of the 2-chain space equals 2* (Lakatos, 1976, 116).

This is a reformulation of a problem, where a proposition P is reformulated as proposition Q, which is easier to prove. If a convincing proof of Q can be found

Fig. 16.7 Corroboration

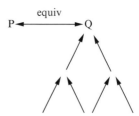

and it can be shown that Q is equivalent to P, then P has been proved. This is a common mathematical technique. Lakatos called this "the problem of translation" and devotes the second chapter of (Lakatos, 1976) to its description, discussion and accompanying problems. The roots of this method lie in Pólya's advice: "Could you restate the problem? Could you restate it again?" (Pólya, 1945). Analysis of the argument for Q and the premises it contains suggests insights into the original proposition P, *i.e.*, ways in which P, the concepts in P or the argument for P should be modified.

The trivial case, in which P is equivalent to itself, and multiple proofs are given for the same theorem, is common in mathematics. For instance, Pythagoras' theorem is proved using similar triangles, parallelograms (by Euclid), a trapezoid, similarity, by rearrangement, algebraically, with differential equations and using rational trigonometry. Examples of mathematical statements which are equivalent include: *(i)* Pythagoras' theorem and the parallel postulate; and *(ii)* Zorn's lemma, the well-ordering theorem and the axiom of choice. The second example is interesting since the equivalent statement to P, *i.e.*, Q, might also be a premise in the argument for P. For instance, the proof of the well-ordering theorem uses the axiom of choice. Although this satisfies Haggith's corroboration pattern it would clearly be circular to argue that the argument for the well-ordering theorem corroborates the axiom of choice.

There are many examples in mathematics where the relationship between P and Q is not equivalence, but the argument pattern is still relevant: for instance, where P is analogous to, similar to, weaker or more general than, or implies Q. Proving special cases of a theorem can be seen as corroborating the theorem: historically we see many examples, such as proving Fermat's Last Theorem for specific values of n (3,5,7), or for classes of number (all regular primes). This can also work the other way around, where a proof of a more general result is easier to find than the proof of the more specific version. For instance, Lagrange's Theorem, that for any finite group G the order (number of elements) of every subgroup H of G divides the order of G, is true under a suitable reformulation even for infinite groups G and H. In this case the proof of the more general statement is simpler than a proof that uses induction over finite structures (some historical details are in Roth, 2001).

Enlargement: (Fig. 16.8)

Q: A polyhedron is a solid whose surface consists of polygonal faces.
Q':   A polyhedron is a surface consisting of a system of polygons.

Fig. 16.8 Enlargement

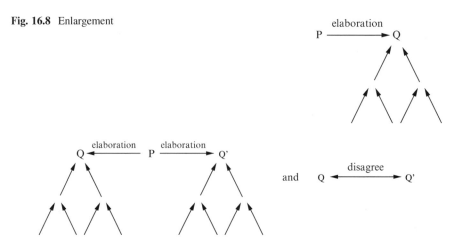

Fig. 16.9 Monster-barring and enlargement

Lakatos described both monster-barring, in which a narrow definition of a concept in a conjecture or proposition is proposed in order to exclude a counterexample and thus defend the conjecture, and concept-stretching, in which a wider definition of a concept which covers a counterexample and thus poses problems for the conjecture is proposed. Both methods, in which a concept definition that was previously vague or controversial is made more explicit, can be seen as examples of Haggith's *enlargement* structure. Arguments then given for *why* a particular concept definition should be accepted constitute the argument for *Q*. Lakatos demonstrated the intriguing situation in which we have two enlargement arguments for *P*, with conclusions Q_1 and Q_2, where Q_1 and Q_2 disagree. Thus, we might question whether the relationship between *P* and *Q* is really an elaboration or not. We show this in Fig. 16.9.

In emphasising the role of the argument for *Q*, Haggith stresses the need for *showing*, rather than stating that *Q* elaborates *P*. This aspect is not always made clear by Lakatos, who did not always clarify how one should select between competing definitions (relying on the literary device of a teacher to state a given definition as fact). This argumentation structure is similar to Walton's argumentation scheme for *argument from verbal classification* (Walton, 2006, 128–132). This is a scheme which takes the form: "$F(a)$, $\forall x, (F(x) \rightarrow G(x))$, therefore $G(a)$", where the classifications *F* and *G* may be vague or highly subjective.

Other examples of monster-barring or concept-stretching in mathematics include expanding the concept 'number' from natural numbers to include zero, negatives, irrationals, imaginary numbers, transfinite numbers and quaternions; narrowing the concept of 'set' (by limiting the types of sets which can be constructed), Cantor's expansion of our notion of 'size' and changing the definition of 'prime number' from 'a natural number which is only divisible by itself and 1' to 'a natural number with exactly two divisors', thus enabling many theorems about primes, where 1 would have been a counterexample (such as the Fundamental Theorem of Arithmetic), to be neatly stated.

Fig. 16.10 Consequence

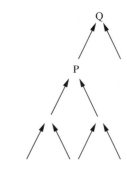

Fig. 16.11 Rebuttal

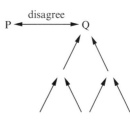

Consequence: (Fig. 16.10)

Q: A star polyhedron is Eulerian (Lakatos, 1976, 16).

   The proposition P, which states that for *all* polyhedra $V - E + F$ is 2, is a premise (the sole premise) in any argument for Q where Q is the proposition that $V - E + F$ is 2 for a particular type of polyhedron. Lakatos used this argumentation structure to great advantage, showing that analysing propositions which are implied by a conjecture is a fruitful way of analysing the conjecture itself. This pattern is similar to corroboration, where there is an implication relation between P and Q.

   Wiles's proof (with Taylor) for the statement that all rational semistable elliptic curves are modular (P) provides another example: this is famous because it implies Fermat's Last Theorem (Q). The implication relation was proved in the more general form (the Taniyama-Shimura-Weil conjecture) earlier on by Ribet. Clearly, proving a given relationship between two mathematical statements is just as important as the argument for one of them. Any examples of lemmas or corollaries in mathematics fit this pattern.

Rebuttal: (Fig. 16.11)

Q: $\exists x(poly(x) \wedge \neg euler(x, 2))$, or there is an x such that x is a polyhedron and it is *not* the case that the Euler characteristic of x is 2.

   An example is the cylinder (Lakatos, 1976, 22), for which $V - E + F$ is 1. This is a case of simple rebuttal, *i.e.* rebuttal without undercutting, which Lakatos addressed in his method of hidden lemma-incorporation. This method demonstrates that it is possible in mathematics to rebut without a known undercutter. This view is discussed (and supported) by Aberdein (2005, 298), who argues that the presence of a rebuttal is an *existence proof*, rather than a construction, for an undercutter.

Lakatos also describes how the method of hidden lemma-incorporation was used to fix Cauchy's faulty conjecture that 'the limit of any convergent series of continuous functions is itself continuous' (Lakatos, 1976, Appendix 1). The counterexample, found by Fourier, is:

$$cosx - \frac{1}{3}cos3x + \frac{1}{5}cos5x - \ldots$$

which converges to the step function. The most interesting aspect of this example is the timing of various discoveries. Fourier discovered the above series (see Fourier, 1808), and it was *after* this that Cauchy discovered the conjecture and proof (see Cauchy, 1821). One solution to this awkward situation was that the limit function was actually continuous, and therefore it was not a counterexample (Fourier held that it was continuous). However, Cauchy had provided a new interpretation of continuity, according to which the limit was *not* continuous (the existence of Fourier's example was considered by some to be evidence that the new interpretation should be rejected). Another possible solution was the argument that the series was not (pointwise) convergent, although this view was not accepted by most mathematicians, including Cauchy, who later proved that it did converge. There was then a long gap until 1847 when Seidel found the hidden assumption of uniform convergence in the proof. Indeed, it was Seidel who invented the method of proofs and refutations. Lakatos thought that the main reason for such a long gap, and the willingness of mathematicians to ignore the contradiction, was a commitment on the part of mathematicians to Euclidean methodology. Deductive argument was considered infallible and therefore there was no place for proof analysis.

A further example can be found in Hilbert's *Grundlagen der Geometrie*:[13]

Theorem 1. *For two points A and C there always exists at least one point D on the line AC that lies between A and C.*

Proof. (paraphrased from (Hilbert, 1902, 6) as a procedural proof)

lemma 1: draw a line *AC* between the two points
lemma 2: mark a point *E* outside the line *AC* (axiom (I,3))
lemma 3: mark another point *F* such that *F* lies on *AE* and *E* is a point of the segment *AF* (axiom (II,2))
lemma 4: mark on *FC* a point *G*, that does not lie on the segment *FC* (axiom (II,2) and axiom (II,3))
lemma 5: the line *EG* must then intersect the segment *AC* at a point *D* (axiom (II,4))

---

[13] Since Lakatos described his method in terms of a procedural proof, we paraphrase Hilbert's proof as a procedural proof, from the deductive proof which Hilbert gave. However, it is worth noting that the method of lemma-incorporation also applies to declarative proofs, which we demonstrate with respect to the first step of Hilbert's proof.

This proof is accompanied by a diagram containing the hidden assumption that the two points are different (as becomes obvious to humans when they try to draw the line which joins A and C). The counterexample comprises any two points which are identical, *i.e.*, (a,a). With the proof phrased as above, (a,a) is a global, but not local counterexample (note again that the *type* of counterexample changes as we step through a proof: in the global theorem the counterexample is a *point*, whereas in lemma one the counterexample is a *line*). This example is interesting, since, in Hilbert's original German edition of the Grundlagen in 1899, reprinted in (Hilbert, 2004), he does not exclude this example: there is nothing in the axioms to say that a line must join two *different* points (the relevant axiom is (I,1), which states that for every two points, A,B there exists a line that contains each of the points. The axiom does not specify that the points must be different), nor that a line which intersects a segment AC must be strictly between the points A and C, etc. However, in later editions, Hilbert has amended this: for instance, at the beginning of (Hilbert, 1902), which is a later, English translation, Hilbert states that in all of the theorems, he assumes that where he says two points, they will be considered to be two distinct points (the relevant omission in his first work is Hilbert, 2004, 437–438, Chap. 5). This example was highlighted by Meikle and Fleuriot's formalisation of (Hilbert, 1902), described in (Meikle and Fleuriot, 2003).

Just as mathematical concepts, conjectures and proofs do not emerge fully formed and perfect, neither does axiomatisation of mathematical domains. We believe that Lakatos's methods can be used to describe axiomatisation development in the same way as they describe other mathematical development. Hilbert's axiomatisation of geometry shown here (which itself built on previous work: see Pieri's 1895, 1897–1898) is one example. Another is the development of the axioms in set theory, where Frege's comprehension principle is modified to the axiom of subsets, in response to Burali-Forti, Cantor's and Russell's paradoxes.

Note that the teacher and students in (Lakatos, 1976) consider the possibility of an entity being a local only, global only, or a both local and global counterexample. They do not consider that a problem entity might be neither global nor local, assuming that such entities are positive examples of the theorem and proof, and therefore support rather than attack it. However, the admission that lemmas in the proof may contain hidden assumptions which mean that an entity satisfies the lemma, albeit it in a surprising way, raises the question of whether there could be an entity which satisfies the global conjecture in the same way, thus uncovering a hidden assumption in the global conjecture itself (one of us discusses this case in Pease and Aberdein, 2011, 38–39).

Undermining: (Fig. 16.12)

Q: A polyhedron is a solid whose surface consists of polygonal faces.
R: A polyhedron is a surface consisting of a system of polygons.

An example argument of this type is the dialectic over rival definitions in Lakatos's monster-barring method (for further mathematical examples, see the section on enlargement.)

Fig. 16.12 Undermining

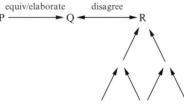

Fig. 16.13 Undercutting

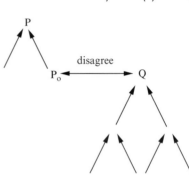

Undercutting: (Fig. 16.13)

P_0:   any polyhedron, after having a face removed, can be stretched flat on the blackboard.

Q: after removing a face from the hollow cube we cannot stretch it flat on the blackboard.

   Alternatively, let

P_0:   dropping a triangle from a triangulated map always results in either the disappearance of one edge or else of two edges and a vertex.

Q: we can drop a triangle from the triangulated map which results from removing a face from the cube and stretching it flat on the blackboard, without it resulting in either the disappearance of one edge or else of two edges and a vertex.

   This is the earliest argumentation structure to be found in (Lakatos, 1976), where the three steps (or premises) of Cauchy's proof are all questioned on page 8. The questions are later supported by counterexamples. Note that Lakatos's method of global and local lemma-incorporation is a combination of rebuttal and undercutting, and his local-only lemma-incorporation is just undercutting. Global only, or hidden lemma-incorporation is just rebuttal (as discussed above).

   Set theory provides further examples of local-only lemma-incorporation. For instance, Cantor's initial proof that the segment and the square are equivalent sets of points contained the premise that there is a one-to-one onto mapping: $f : \{(x, y) : x \in (0, 1], y \in (0, 1]\} \rightarrow (0, 1]$. Cantor identifies such a mapping: let (x, y) be co-ordinates of an arbitrary point in the unit square, where $x = 0.x_1x_2x_3\ldots$, and $y = 0.y_1y_2y_3\ldots$. Then they uniquely determine the point or the unit segment $z = 0x_1y_1x_2y_2x_3y_3\ldots$.

Fig. 16.14 Target

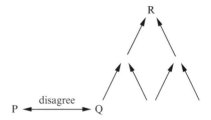

Conversely, every point in the unit segment can be expressed as an infinite decimal. Let z be an arbitrary point in the unit segment, where $z = z_1 z_2 z_3 z_4 z_5 z_6 \cdots$ Then this point uniquely determines two co-ordinates $x = z_1 z_3 z_5 \ldots$ and $y = z_2 z_4 z_6 \ldots$ which determines the points (x, y) of the unit square. Note that Cantor specified that where a number has two decimal expansions, for example $0.2299999999\ldots$ or 0.230000000, the expansions ending in an infinite set of zeros should be ruled out (so the real number 23/100 is identified as $0.2299999999\ldots$). Dedekind (at Cantor's request) checked this proof and found local (but not global) counterexamples. While to each (x, y) there corresponds a single z, there exist values of z that arise from no (x, y) in the above procedure. One such counterexample is $z = 0.13050706080\ldots$ which yields $x = 0.1$; another counterexample is $z = 0.513020109090\ldots$ which yields $y = 0.1$. In these cases neither x nor y is written in admissible decimal form, and if the trailing zeros are replaced by nines then the infinite decimal form does not correspond to the specified number z. The proof was then patched so that, instead of considering single digits, groups of digits are considered such that only the last digit of a group differs from 0; so for example $z = 0.1\ 2\ 05\ 0004\ 2\ 01\ 8\ldots$ would yield $x = 0.10528\ldots$ and $y = 0.2000401\ldots$ (Burton, 1985, 607-608).

Target: (Fig. 16.14)

Q: $\exists x(poly(x) \wedge \neg euler(x, 2))$; in particular, the hollow cube has this property, since $V - E + F$ is 4.
R For all polyhedra that have no cavities, $V - E + F$ is 2.

or,

R': $V - E + F = 2 - 2(n - 1) + \Sigma_{k=1}^{F} e_k$, for n-spheroid—or n-tuply connected—polyhedra with e_k edges deleted without reduction in the number of faces (Lakatos, 1976, 79).

Lakatos's exception-barring methods fit this pattern, where a counterexample is found and used, along with other premises, to modify P, just producing R_1. Another, later example is R_2 (Lakatos, 1976, 79), where many counterexamples have been found and a general repair, R_2 has been found which explains all of the positive and negative examples. Lakatos called this the problem of content, which deals with the problem that with every repair, the domain of application of the conjecture (originally all polyhedra, and then all polyhedra of an increasingly narrow type) has narrowed to the stage where the conjecture is no longer very interesting. Thus, math-

Fig. 16.15
Counter-consequence

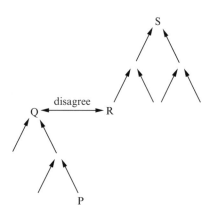

ematicians were "striving for truth at the expense of content" (Lakatos, 1976, 66). In this case, Lakatos suggested (again building on Pólya, 1945) distinguishing an initial problem in which a question is raised, from an initial conjecture which poses a first answer to the initial problem. If this answer is refined almost to the point of a tautology (approaching the 'conjecture' that any Eulerian polyhedron is Eulerian) then it may be preferable to return to the initial problem and pose another answer.

Historical conjectures about perfect numbers provide an example in number theory. For instance, using the (scientific) inductive argument that because the first four perfect numbers, 6, 28, 496 and 8128 each contain n digits, it was conjectured that the nth perfect number P_n contains exactly n digits (P). The fifth perfect number, 33,550,336 provides a counterexample (Q), the discovery of which led to the new conjecture that perfect numbers end alternately in 6 and 8 (R). Again, the argument pattern repeated itself with the discovery of the sixth perfect number, 8,589,869,056, which is a counterexample to R. This led to further statements and arguments, including the theorem that the last digit of any even perfect number must be 6 or 8 and the (open) conjecture that all perfect numbers are even.

Counter-consequence: (Fig. 16.15)

Q: $\neg\exists x(poly(x) \wedge \neg euler(x,2))$.
R: $\exists x(poly(x) \wedge \neg euler(x,2))$; in particular, the twin tetrahedra polyhedron has this
     property, since it has an Euler characteristic of 3 (Lakatos, 1976, 15).
S: for any polyhedron that has no 'multiple structure', $V - E + F = 2$.

This is very similar to the *target* structure described above. Again, Lakatos's exception-barring methods present an example of this type of structure. The perfect number examples described above also fit this pattern. Other examples of monster-adjusting seem rare in mathematics. However, we can see monster-adjusting as a type of monster-barring, where the concept in question may be the right hand concept in an implication or equivalence conjecture, rather than the domain. Let us formalise monster-barring as follows: from conjecture $\forall x, P(x) \rightarrow Q(x)$, and (known) counterexample m such that $P(m)$ and $\neg Q(m)$, (re)define either P or Q

Fig. 16.16 A new
argumentation structure for
monster-adjusting

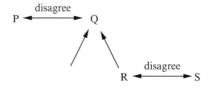

so that for the *m* in question, either ¬*P*(*m*) or *Q*(*m*) is true. Monster-adjusting can
now be seen as a case of this formalisation, where the concept under debate is *Q*
rather than *P*.

It is interesting that the structures which Haggith has identified can be used to
express most of Lakatos's methods. A method which has so far been neglected is
monster-adjusting. As with monster-barring, this method also exploits ambiguity
in concepts, but reinterprets an object in such a way that it is no longer a
counterexample. The example in (Lakatos, 1976) concerns the star polyhedron. This
entity is raised as a counterexample since, it is claimed, it has 12 faces, 12 vertices
and 30 edges (where a single face is seen as a star polygon), and thus $V - E + F$ is
−6. This is contested, and it is argued that it has 60 faces, 32 vertices and 90 edges
(where a single face is seen as a triangle), and thus $V - E + F$ is 2. The argument
then turns to the definition of 'face'. A new structure which corresponds to this
method might look like the one shown in Fig. 16.16, where:

Q: $\exists x(poly(x) \wedge \neg euler(x, 2))$; in particular, the star polyhedron has this property,
since $V - E + F$ is −6 (Lakatos, 1976, 16).

R: The star polyhedron has 12 (pentagonal) faces. (Other premises in the argument
for *Q* are that the star polyhedron has 12 vertices and 30 edges.)

S: The star polyhedron has 60 (triangular) faces (Lakatos, 1976, 31).

16.5.1 Mutually Beneficial Associations Between Lakatos and Haggith

16.5.1.1 Using Haggith's Work to Extend Lakatos's Theory

One shortcoming of Lakatos's representation of an initial proof, as a list of lemmas,
or conjectures, is that it fails to show how the conjectures in the proof fit together.
In both of his case studies, the proof is represented by Lakatos as a series of
local conjectures which, taken together, imply the global conjecture. In Haggith's
notation, this would be simply written as $A = \{C \Leftarrow \{P1, P2, P3\}\}$. This prevents the
expression of any dependencies between the conjectures and thus would be an extreme oversimplification of most mathematical proofs, even at an initial stage. Using
Haggith's notation makes explicit, rather than hides, the structure of a proof. Seeing
how a proof fits together may make it easier to identify flaws and their solution.

16.5.1.2   Using Lakatos's Work to Extend Haggith's Theory

Using Lakatos's case studies, we have shown that Haggith's argumentation structures, which were inspired by the need to represent different perspectives in natural resource management, can be usefully applied to mathematical examples. This is a new domain for Haggith, and thus supports her claim that the structures are of general use.

Combining Lakatos's conjecture-based and Haggith's proposition-based representations has the advantage of highlighting weak areas in a proof. This might be in the relationships between sets of conjectures, such as justification or elaboration, or in the claims asserted by the conjectures. Propositions are no longer black boxes, thus enabling new areas for flaws to be found and repaired. Lakatos's methods, for example his "problem of content", also consider the interestingness of a proposition.

Lakatos's methods suggest new structures for Haggith. Although she made no claim to have identified all structures, adding new examples to the catalogue is a valuable contribution to Haggith's work.

16.6   Conclusion

Many argumentation theorists have assumed, either implicitly or explicitly, that mathematics is about formal reasoning and is therefore not a suitable domain for argumentation. We hope that the ideas in this paper, in addition to other work such as (Aberdein, 2005), will show that this assumption is not justified and that mathematics is a rich and fertile domain for informal reasoning techniques. In accordance with the computational philosophy paradigm, we believe that implementing Lakatos's theory has provided a new perspective on it. One insight is discovering the utility of argumentation structures for proof-schemes, which has allowed us to represent them, as well as suggesting ways in which Lakatos's methods for repairing faulty conjectures and proofs can be implemented within our model. Thus, the process of producing computational models of theories can both suggest and exploit new connections.

Just as Lakatos suggested that embedding a conjecture in a different body of knowledge can lead to further insights into the conjecture, so we can see the value of embedding philosophy of mathematics in argumentation theory and vice versa. These two normally unconnected domains have much to offer each other. The computational modelling domain adds a further dimension of interest and will play an essential role in making these intriguing connections richly detailed and explicit.

Acknowledgements A prior version of this paper was published in *Foundations of Science* (2009), 14(1–2):111–135. We would like to thank our three reviewers who all gave comprehensive, constructive and thought-provoking reviews. One of the editors of this collection, Andrew Aberdein, made particularly helpful comments. This work was partly funded by EPSRC grant EP/F035594/1.

References

Aberdein, A. (2005). The uses of argument in mathematics. *Argumentation, 19*, 287–301.

Aberdein, A. (2006). Managing informal mathematical knowledge: Techniques from informal logic. In J. M. Borwein & W. M. Farmer (Eds.), *MKM 2006, LNAI 4108* (pp. 208–221). Berlin: Springer.

Alcolea Banegas, J. (1998). L'argumentació en matemàtiques. In E. Casaban i Moya (Ed.), *XIIè Congrés Valencià de Filosofia* (pp. 135–147). Valencià. [Trans.: A. Aberdein & I.J. Dove (Eds.), The argument of mathematics (pp. 47–60). Dordrecht: Springer].

Burton, D. (1985). *The history of mathematics*. Boston, MA: Allyn and Bacon.

Cauchy, A. L. (1813). Recherches sur les polyèdres. *Journal de l'École Polytechnique, 9*, 68–86.

Cauchy, A. L. (1821). *Cours d'Analyse de l'École Royale Polytechnique*. Paris: de Bure.

Colton, S. (2002). *Automated theory formation in pure mathematics*. Berlin: Springer.

Corfield, D. (1997). Assaying Lakatos's philosophy of mathematics. *Studies in History and Philosophy of Science, 28*(1), 99–121.

Crawshay-Williams, R. (1957). *Methods of criteria of reasoning: An inquiry into the structure of controversy*. London: Routledge and Kegan Paul.

Crelle, A. L. (1826–1827). *Lehrbuch der Elemente der Geometrie* (Vols. 1, 2). Berlin: Reimer.

Elvang-Goransson, M., Krause, P., & Fox, J. (1993). Dialectical reasoning with inconsistent information. In *Proceedings of the ninth conference annual conference on uncertainty in artificial intelligence (UAI-93)* (pp. 114–121). San Francisco, CA: Morgan Kaufmann.

Ernest, P. (1997). The legacy of Lakatos: Reconceptualising the philosophy of mathematics. *Philosophia Mathematica, 5*(3), 116–134.

Feferman, S. (1978). The logic of mathematical discovery vs. the logical structure of mathematics. In P. D. Asquith & I. Hacking (Eds.), *Proceedings of the 1978 biennial meeting of the Philosophy of Science Association* (Vol. 2, pp. 309–327). East Lansing, MI: Philosophy of Science Association.

Fourier, J. (1808). Mémoire sur la propagation de la chaleur dans les corps solides (extrait). *Nouveau Bulletin des Sciences, par la Société Philomathique de Paris, 1*, 112–16.

Goguen, J. (1999). An introduction to algebraic semiotics, with application to user interface design. In C. L. Nehaniv (Ed.), *Computation for metaphors, analogy, and agents: Vol. 1562, Lecture notes in artificial intelligence* (pp. 242–291). Berlin: Springer.

Haggith, M. (1996). *A meta-level argumentation framework for representing and reasoning about disagreement*. PhD thesis, Department of Artificial Intelligence, University of Edinburgh, Edinburgh, UK.

Hardy, G. H. (1928). Mathematical proof. *Mind, 38*, 11–25.

Hilbert, D. (1902). *The foundations of geometry* (1st ed.) (E. J. Townsend, Trans.). Chicago, IL: Open Court.

Hilbert, D. (2004). *David Hilbert's lectures on the foundations of geometry, 1891–1902: Vol. 1 of David Hilbert's lectures on the foundations of mathematics and physics, 1891–1933*. M. Hallett & U. Majer (Eds.). Berlin: Springer.

de Jonquières, E. (1890). Note sur un Point Fondamental de la Théorie des Polyèdres. *Comptes Rendus des Séances de l'Académie des Sciences, 110*, 110–115.

Lakatos, I. (1976). *Proofs and refutations: The logic of mathematical discovery* (edited by J. Worrall & E. Zahar). Cambridge: Cambridge University Press.

Larvor, B. (1998). *Lakatos: An introduction*. London: Routledge.

Matthiessen, L. (1863). Über die Scheinbaren Einschränkungen des Euler'schen Satzes von den Polyedern. *Zeitschrift für Mathematik und Physik, 8*, 1449–1450.

Meikle, L., & Fleuriot, J. (2003). Formalizing Hilbert's Grundlagen in Isabelle/Isar. In *Theorem proving in higher order logics* (Vol. 2758, pp. 319–334). Berlin: Springer.

Naess, A. (1953). *Interpretation and preciseness: A contribution to the theory of communication*. Oslo, Norway: Skrifter utgitt ar der norske videnskaps academie.

Naess, A. (1966 [1947]). *Communication and argument: Elements of applied semantics*. London: Allen and Unwin. (Translation of *En del Elementaere Logiske Emner*. Oslo: Universitetsforlaget, 1947).

Pease, A. (2007). *A computational model of Lakatos-style reasoning*. PhD thesis, School of Informatics, University of Edinburgh, Edinburgh, UK. Online at http://hdl.handle.net/1842/2113

Pease, A., & Aberdein, A. (2011). Five theories of reasoning: Interconnections and applications to mathematics. *Logic and Logical Philosophy, 20*(1–2), 7–57.

Pease, A., Colton, S., Smaill, A., & Lee, J. (2004). A model of Lakatos's philosophy of mathematics. *Computing, philosophy and cognition: Proceedings of the European Computing and Philosophy Conference (ECAP 2004)*. London: College Publications.

Pedemonte, B. (2001). Some cognitive aspects of the relationship between argumentation and proof in mathematics. In M. van den Heuvel-Panhuizen (Ed.), *Proceedings of the 25th conference of the international group for the Psychology of Mathematics Education PME-25, Utrecht, The Netherlands* (Vol. 4, pp. 3–40). Nottingham: PME Proceedings.

Pieri, M. (1895). Sui principi che reggiono la geometria di posizione. *Atti della Reale Accademia delle scienze di Torino*, 30:54–108.

Pieri, M. (1897–1898). I principii della geometria di posizione composti in sistema logico deduttivo. *Memorie della Reale Accademia delle Scienze di Torino 2, 48*, 1–62.

Pollock, J. (1970). The structure of epistemic justification. *American Philosophical Quarterly, Monograph Series, 4*, 62–78.

Pollock, J. (1995). *Cognitive carpentry*. Cambridge, MA: MIT Press.

Pólya, G. (1945). *How to solve it*. Princeton, NJ: Princeton University Press.

Pólya, G. (1954). *Mathematics and plausible reasoning: Induction and analogy in mathematics* (Vol. I). Princeton, NJ: Princeton University Press.

Popper, K. R. (1959). *The logic of scientific discovery*. New York: Basic Books.

Reed, C., & Rowe, G. (2005). Translating Toulmin diagrams: Theory neutrality in argument representation. *Argumentation, 19*(3), 267–286.

Roth, R. L. (2001). A history of Lagrange's theorem on groups. *Mathematics Magazine, 74*(1), 99–108.

Russell, B. (1971). *Logic and knowledge: Essays 1901–1950*. London: George Allen and Unwin.

Sartor, G. (1993). A simple computational model for nonmonotonic and adversarial legal reasoning. In *Proceedings of the fourth international conference on artificial intelligence and law, Amsterdam, The Netherlands* (pp. 19–201). New York: ACM.

Toulmin, S. (1958). *The uses of argument*. Cambridge: Cambridge University Press.

Toulmin, S., Rieke, R., & Janik, A. (1979). *An introduction to reasoning*. London: Macmillan.

Walton, D. (2006). *Fundamentals of critical argumentation*. Cambridge: Cambridge University Press.

Wilder, R. L. (1944). The nature of mathematical proof. *The American Mathematical Monthly, 51*(6), 309–323.

Chapter 17
Mathematical Arguments and Distributed Knowledge

Patrick Allo, Jean Paul Van Bendegem, and Bart Van Kerkhove

Because the conclusion of a correct proof follows by necessity from its premises, and is thus independent of the mathematician's beliefs about that conclusion, understanding how different pieces of mathematical knowledge can be distributed within a larger community is rarely considered an issue in the epistemology of mathematical proofs. In the present chapter, we set out to question the received view expressed by the previous sentence. To that end, we study a prime example of collaborative mathematics, namely the *Polymath Project*, and propose a simple formal model based on epistemic logics to bring out some of the core features of this case-study.

17.1 Introduction

In his *Objective Knowledge*, Karl Popper famously claimed that the beliefs of the individual are irrelevant for epistemological purposes; especially when that individual is understood as the individual scientist, and the field of epistemology is restricted to scientific knowledge (Popper, 1968, 1972). Even if, in view of the present state of the philosophy of science, this account has become untenable in general,[1] one might still be tempted to believe that, at least when it comes to mathematical knowledge, Popper's conception of objective knowledge remains a sensible

---

[1]Especially in the light of work carried out at the intersection of social epistemology and philosophy of science. See, e.g., (Kitcher, 1990) for a general defence, or (Zollman, 2007) for a specific study.

P. Allo • J.P. Van Bendegem • B. Van Kerkhove (✉)
Centre for Logic and Philosophy of Science, Vrije Universiteit Brussel, Pleinlaan 2, Brussels, B-1050, Belgium
e-mail: pallo@vub.ac.be; jpvbende@vub.ac.be; bvkerkho@vub.ac.be

A. Aberdein and I.J. Dove (eds.), *The Argument of Mathematics*, Logic, Epistemology, and the Unity of Science 30, DOI 10.1007/978-94-007-6534-4_17, © Springer Science+Business Media Dordrecht 2013

position. This, one could argue, follows from the fact that mathematical knowledge is obtained through proof. For when truth is exclusively arrived at via proof, we only need to be concerned with the correctness of that proof, and this does not depend on what any individual mathematician believes. Certainly there are many deep issues in the epistemology of mathematical proofs.[2] Yet, because the conclusion of a correct proof follows by necessity from its premises and is thus independent of the mathematician's beliefs about that conclusion, understanding how different pieces of mathematical knowledge can be distributed within a larger community is rarely considered an issue in the epistemology of mathematical proofs.

The above description is particularly compelling if one thinks that mathematics is just about correct proofs. By contrast, if we accept that mathematics is also about proofs that are *recognised* to be correct, or that are deemed to be *acceptable*, or widely thought to be *convincing*, knowledge and acceptance in different mathematical communities becomes relevant. This is particularly so when proofs are seen as arguments. On that conception, correctness is plainly insufficient. What arguments aim at is to be convincing, and this is something that depends on the audience that needs to be convinced of the truth (or provability) of a given theorem. This, for sure, depends on what the members of that audience know, but even that isn't enough. In addition, one would also like to know (or at least believe) that a certain argument will be convincing, which requires one to know what the members of one's audience know. These considerations not only show that what the individual mathematician knows matters for the epistemology of proofs-as-arguments, but also that an individual mathematician's beliefs about the knowledge of his audience (sometimes his peers, but equally often members of other communities or lay-people) is equally relevant. In sum: Not only do beliefs of the individual matter, but also beliefs about beliefs of others matter.

When we look at the received view about how proofs are conceived (discovered, found, designed, …), we find the same individualistic bias that is also present in mainstream epistemology. Proofs are the work of the individual mathematician, or at most the work of a small number of individuals. Again, one might be tempted to believe that traditional arguments for a social or interactive conception of knowledge need not apply to mathematics. Because individuals can come up with new results, and as the whole enterprise of mathematics is traditionally seen as cumulative, no real interaction is needed beyond the obvious reliance on prior results. Yet, this description no longer covers the totality of mathematical inquiry. Some proofs don't just happen to be the result of massive collaboration, but are actually results that are well beyond the reach of individuals and even smaller communities. This means that proofs cannot only be seen as arguments once they are completed (and indeed need to be used to convince the broader community), but that the process of looking for a proof can equally well be studied from an argumentation-theoretic perspective, and more generally from the perspective of social and interactive epistemology.

---

[2]What is the formal nature of proofs? Are proofs out there (in "The Book") or constructed by us? What is a surveyable proof and what not? How does one grasp (the content of) a proof? Are there more and less beautiful proofs? (etc.)

Formal models of scientific knowledge are often based on *Bayesian Models* (Bovens and Hartmann, 2003), or on *Signalling Games* (Zollman, 2007). Here, we pursue a different path and propose to use the epistemic logics that have been developed by computer scientists to reason about knowledge in *multi-agent systems* (Fagin et al., 1995). The not so obvious choice of a logical model is deliberate: We believe logic is equally relevant to the study how information flows in epistemic communities—our present aim—as it is to the study of validity—the traditional focus of mathematical proofs. This motivates the choice for a logical, as opposed to another formal framework.

Within the broader landscape of logical approaches, the choice for epistemic logics still needs further motivation. One element has already been advanced: We want to be able to reason about what different agents know and believe. But, as mentioned above, we also want to reason about knowledge in groups, and crucially also about higher order epistemic states (knowledge and belief about other knowledge or belief). This is a task that we think is best carried out within the framework of epistemic logics. Given the focus on a conception of proofs as arguments, formal models that directly deal with arguments like the logics for defeasible argumentation described in (Prakken and Vreeswijk, 2002) might be thought to be more obvious candidates. One issue with the latter type of formalism is that it primarily provides a formal model of *defeasible inference*, and this is a type of reasoning we do not readily associate with proofs *in* mathematics (Aberdein, 2005, 289–90). Still, as shown by Aberdein (2005) via the contrast between classical and constructive validity, Toulmin's model of argumentation (itself a model for defeasible argumentation or inference) can be used to model mathematical proofs, and the result of doing so can be informative (see also Aberdein, 2007). The main argument in favour of the doxastic/epistemic approach then, is that it allows one to reason explicitly about how agents and communities of agents interact; a perspective that is not available to more abstract approaches to argumentation.

No one will doubt that argumentation theory has a long and respectable history within philosophy, starting with Aristotle, who, at the same time, initiated the distinction, if not opposition, between the predominantly informal realm of argumentation (to convince an audience) and the more formally oriented field of logic (to establish necessary truths). In the twentieth century this situation of course drastically changed, as new developments were initiated that aimed at a *formal* argumentation theory. Among the major contributions, we find dialogue logic, as designed by Lorenzen and Lorenz (1978), the core idea of which is that a logical system, axiomatically and/or deductively presented, can be reformulated in terms of rules for a discussion between two parties, thus showing that the distinction between logic and argumentation is partially fictitious. A similar attempt, in terms of game theory, was made by Hintikka (1985), and related models were developed by Barth and Krabbe (1982).

Of major importance in the field of *informal* argumentation theory is the work of Grootendorst and Van Eemeren (see e.g. Van Eemeren et al., 2004). The Amsterdam school, which has been the driving force behind the International Society for the Study of Argumentation (ISSA), is mainly interested in specific case studies, with an eye on determining how participants in actual dialogues or discussions deal with arguments and their evaluation. There is moreover a direct connection with the Canadian school, with well known contributions like (Walton, 1998) and (Woods, 2003). Here too there is a strong focus on concrete situations and, above all, a careful investigation of whether fallacies are always to be rejected in a discussion or can sometimes be acceptable forms of reasoning. For example, classically, the argument *ad auctoritatem* is considered to be a fallacy, while expertise in a court of law is clearly acceptable.

The third development to be noted is that within the field of Artificial Intelligence (AI). Indeed there has been a strong interest for some time now in the modelling of argumentations (see e.g. Toulmin, 1958; Pollock, 1994; Vreeswijk, 1997). The main purpose of this research is to develop applications in decision-guided systems and intelligent agents (Reed, 1998). Recently, argumentation networks have been proposed as simple but very powerful models to represent static structures of arguments in competition with one another (Dung, 1995). However, the models remain rather abstract: arguments are viewed solely in relation to others, the basic scheme being one argument "attacking" another. Several semantics have been proposed for such networks, all taking a declarative and monological approach as a starting point (Bondarenko et al., 1997; Dung, 1995; Jakobovits and Vermeir, 1996, 1999b). In a further stage so-called dialectical semantics have been investigated (Jakobovits and Vermeir, 1999a; Prakken and Sartor, 1996), e.g. searching for applications in judicial reasoning (Verheij, 1995). Connections have already been shown between some of these semantics and the "well-founded", "stable" ones of logic programs (see e.g. Kakas et al., 1994). More generally speaking, such argumentation networks can be applied in areas where it is important to have a *motivation* for a result or a decision, e.g., in medicine (Atkinson et al., 2005) or in bioinformatics (Jeffreys et al., 2006). As far as we are aware, applications to mathematical contexts are absent and, as stated, it is our main purpose to start filling in this gap.

As mentioned, the abstract approach to argumentation lacks a notion of players or agents; it is solely concerned with the arguments themselves. While the main logical approach to agency is to be found in the field of modal epistemic logics, further connections with argumentation remain under-explored.[3] Because agency and interaction are central to our present enterprise, we privilege the framework of epistemic logics, and try to emphasise throughout this chapter to what extent it is relevant to the study of arguments in and about proofs. The upshot of the present article is to illustrate how epistemic logics that allow us to reason about the knowledge of individuals and the knowledge available within communities can be used to model proofs-as-arguments from a perspective that emphasises interaction and collaboration. This approach expands on a proposal described in Van Bendegem (1985b), but does so in a more flexible manner by modelling the structure of communities in terms of group and sub-group membership rather than with the

---

[3]For a first connection between abstract argumentation and modal logic, see (Grossi, 2010).

accessibility relation of a Kripke-model. In the following sections, we first present and analyse the Polymath Project as a potentially interesting case-study (Sect. 17.2), then on the basis of previous work by the second author (Van Bendegem, 1982, 1985a) and a number of other, more recent results develop a formalism (Sect. 17.3), and then also make it fit to capture cases plucked from contemporary mathematical practice, like the previously mentioned Polymath Project (Sect. 17.4).

17.2 Mathematics in the Cloud: The Polymath Project

In Gowers and Nielsen (2009), a mathematician and a science writer/former physicist report their endeavours in online collective problem solving. The Polymath Project, as it has been called, has drawn quite some attention lately. Some claim that it is an example of a new method or procedure of proof search in mathematics, whereas others merely see it as old methods implemented through new media, such as the internet. It is our hypothesis that the formal considerations presented here, in this chapter, can help us to clarify the matter. As it happens, we do believe that there are a sufficient number of distinctive characteristics that allow one to consider this development as indeed new and philosophically relevant.

The clue to the story is provided by the mathematician who initiated the whole process, Timothy Gowers, who wrote on his blog: "I'm interested in the question of whether it is possible for lots of people to solve one single problem rather than lots of people to solve one problem each" (Gowers, 2009a). Justin Cranshaw and Aniket Kittur made a similar comment: "In the case of finite simple group classification,[4] the larger task being solved had hundreds if not thousands of natural and predefined subtasks whose solutions were largely independent of the solutions of the other subtasks. Gowers was instead proposing to collaboratively solve a single problem that does not naturally split up into a vast number of subtasks" (Cranshaw and Kittur, 2011). In terms of a problem-solving community, this means that a quite different structure is required and that is one of the focal points of this chapter.

Some historical background.[5] In January 2009, Timothy Gowers opened a website, accessible to everyone, mathematicians and non-mathematicians alike, announcing that he was searching for a proof of a particular mathematical statement, and inviting them to join him in that search. Anyone could post a message about almost everything, on the obvious condition that it was somehow related to the proof search. In fact, a set of ground rules was announced to avoid the whole enterprise becoming all too chaotic. The hope was to find a proof and, if that were to occur, to publish the proof through the usual, existing channels, namely mathematical journals. Presented thus, this in fact does not really sound novel, except perhaps for the fact that a large number of people could (and did) participate. But that

---

[4]See, e.g., (Van Bendegem and Van Kerkhove, 2009, Sect. 4).

[5]A more complete and highly accessible version of events can be found in (Nielsen, 2011).

only seems to be a matter of scale and, although changing the scale of a process can induce important changes, it does not therefore affect the underlying standard picture of proof search. However, closer inspection does reveal some interesting elements. But let us first have a look at the problem itself.

The problem Gowers launched on the website is known as the Density Hales-Jewett Theorem (*DHJ* henceforth), for $k = 3$ at first, but later generalised to arbitrary k. This problem comes under the heading of Ramsey Theory, that is to say it involves combinatorics and so-called colouring problems, where "unavoidable" properties appear. The typical format of such problems is that, given a structure of sufficiently large size, there will always be substructures that have a particular property, the unavoidable property. More specifically, *DHJ* for $k = 3$ states the following. Let there be:

– a set $K = \{1,2,3\}$, with $\#K = k$
– a set $N = K^n$, i.e., the set of all words of length n, on the basis of K.

Next we need two definitions:

– A *combinatorial line* is a subset of N such that, for at least one x, the elements are of the form: $k_1 k_2 \ldots k_i x k_{i+2} \ldots k_n$ for $x = 1, 2, \ldots, k$.
  Example Take $n = 6$, then: the subset $\{122132, 122232, 122332\}$ is a combinatorial line, as is the subset $\{112132, 212232, 312332\}$.
– Define the density d of a subset M of N by $d = \#M/\#N$.

DHJ for $k = 3$ says that, for every d, there exists an n such that every subset M of N with density at least d contains a combinatorial line. The "unavoidable" property here is the presence of a combinatorial line. So the theorem says that no matter how low the density of a particular subset, if the words made on the basis of the alphabet can be sufficiently long, there will always appear a combinatorial line.[6,7]

What happened after the opening of the website? First, after Gowers, Terence Tao joined the enterprise (see below). After 6 weeks, during which 39 contributors had posted 1,228 comments (after every 100 of these, Gowers made summaries to keep an overview), not only was a proof found, but it became immediately clear how it could be generalized for arbitrary k. The proof has meanwhile been published under the pseudonym *D. H. J. Polymath* (2010), with a second paper (Polymath, 2012) being under review, reminiscent of other fictitious names covering a collective in the history of mathematics, the most famous one no doubt that

---

[6]This highly compact formulation of the *DHJ* can be complemented by, especially for the non-mathematician, "The gentle introduction to the Polymath project" to be found at: http://numberwarrior.wordpress.com/2009/03/25/a-gentle-introduction-to-the-polymath-project/.

[7]Actually, a proof already existed. It was formulated in (Furstenberg and Katznelson, 1991). However, this original proof relied on methods and techniques from domains far away from combinatorics, such as ergodic theory. So, as often happens in mathematical research, although one has a proof of the theorem, nevertheless this does not prevent mathematicians from searching for an alternative and, more importantly, an elementary proof, i.e., a proof using the concepts, proof methods and techniques of the domain itself. For more on this topic, see, e.g., (Rav, 1999).

of Nicolas Bourbaki. What can be concluded from this episode? Surely the most striking feature of the whole process is that "amateurs" could and did participate, although whether we should be as enthusiastic about this as Jacob Aron, claiming that this will "democratize the process of mathematical discovery" (see Aron, 2011), is another matter. A good reason for a restricted enthusiasm has to do with the history of amateur mathematics, probably known to all of us. Refutations of accepted results (see Hodges, 1998, for a rather depressing discussion of refutations of Cantor's diagonal argument or Dudley, 1992, for alleged proofs of almost everything, including the parallel postulate) and easy to understand, short proofs of famous conjectures such as Fermat's Last Theorem (even after Andrew Wiles settled the matter) or Goldbach's conjecture are by far the most common. However, in this case the situation is different. The amateur contributions are most of the time not in the form of proofs, but in the form of suggestions, ideas, outlines of possible proofs, and so forth (see Sect. 17.2.3 below). Perhaps more importantly, the "classic" amateur is an isolated individual who sends material to mathematicians or editors of journals,[8] preferably on a 1-1 basis, whereas here they contribute bits and pieces in an open social environment, where everything is noted and listed, for all to see. In short, your Polymath amateur is not your average "crank" amateur. Apart from that, are there any other, special features? We think there are at least three.

17.2.1 Feature 1

Combinatorial problems are rather specific mathematical problems that have two important properties: (i) the problem is usually quantified over the natural numbers, so, instead of having a direct go at the full proof, case-by-case proof search can be very revealing as to the general proof (as in the case of, e.g., Fermat's Last Theorem), and (ii) the search for counterexamples lends itself, at least for small numbers, to an exhaustive search[9] and, for larger numbers, to clever methods for reducing the cases to be examined. As it happened in the *DHJ* case, both research lines were pursued: next to the Gowers project, corresponding to (i), Terence Tao set up a project, complementary to the Gowers project and corresponding to (ii). More specifically, Tao asked the negative question: What is the size of the largest subset C of N that does *not* contain a combinatorial line as a function of k and n?

The question now becomes: given k and n, what is the lower limit of d? The reader can check for him- or herself that for $k = 3$ and $n = 2$, the size is 6. As k and n are finite, the set N will also be finite, so, in principle, all elements can be listed,

---

[8]One of the authors of this chapter is himself the editor of a logic journal, *Logique et Analyse*, and has a nice collection of, e.g., disproofs of Gödel's theorems.

[9]This type of research, much loved by some of the aforementioned amateurs can indeed be seen as "outsourcing drudge work, comparable to Amazon's Mechanical Turk" (https://www.mturk.com/mturk/welcome), as remarked by one of the referees.

all subsets can be listed and a step-by-step control of each subset can be checked for combinatorial lines, such that the size of C can be calculated. However, as one might expect, the concept of combinatorial explosion is definitely applicable here, as even for small k and n, the number of subsets grows exponentially and so one has to search, as mentioned, for clever tricks, techniques and methods to get sufficient insight in order to calculate the size of C.

This leads to the following interesting open question: are Polymath problems restricted to combinatorial problems (or to problems satisfying conditions (i) and (ii))? One might be tempted to answer positively, especially given the further development of the project after the success of the *DHJ* theorem (see below at the end of this section), but matters are not that simple. Note first of all that not all combinatorial (or similar) problems are good candidates. Take Fermat's Last Theorem as an example. The problem satisfies both conditions and similar to the *DHJ* theorem one could suggest the search for an elementary proof. There is however the commonly shared belief that such an elementary proof is highly unlikely, hence the risk is too high to meet failure and hence no such project will ever be launched (and the search for such a proof is definitely left to the "crank" amateurs). And it is true of course that Riemann's Hypothesis is a very unlikely candidate to be taken up as a project—in Tao's words on his blog: "I would imagine that a polymath to solve the Riemann Hypothesis will be a spectacular and frustrating fiasco; we should focus on problems that look like some progress can be made"[10]—but on the very same blog an important suggestion is made by Gowers, that can be seen as a generalisation of conditions (i) and (ii): "A lesson from the *DHJ* experience was that there were very different ways that people could contribute." So that seems to be the key: in a Polymath environment a methodological plurality is essential.

17.2.2   Feature 2

This feature can be described in sloganesque terms as "Lakatos visualised", in honour of Imre Lakatos, who has introduced historical and social elements in mathematics itself in his seminal study on Euler's Conjecture, entitled *Proofs and Refutations* (see Lakatos, 1976). What we mean by the "visualisation" is that, with Polymath, a register is made of all such social exchanges. Of course, we still do not know what happens in the individual minds of the participating mathematicians but, as soon as an item of information is exchanged, it is recorded. We thus obtain an extremely detailed record of the proceedings. This is obviously not only of interest for historians of mathematics, but more importantly so for the participating mathematicians. Surely this must lead to a different problem-solving practice in terms of shared knowledge in the community.

---

[10]See http://polymathprojects.org/2009/07/27/selecting-the-next-polymath-project/.

A first dimension is the role of authority in a problem-solving community. Anyone who had a glance at the Polymath blog will realise very quickly that Gowers and Tao are the authorities guiding the process. Not merely because both are internationally reputed mathematicians, but also because they formulated the minimal rules that any participant should respect, implying that any transgression of these rules could, in principle, lead to exclusion (although, as far as we know, this has never happened). In short they set up the framework wherein the game will be played. Their being in charge is also reflected in the number of contributions by Gowers and Tao, in comparison with other participants. Additionally, they continually made comments about the progress of the project, what suggestions and ideas were interesting and/or important to pursue, in short they both acted as filters. Another way to describe this situation is that we have here a network with two strongly connected "attractors" and no high density subnetworks, all nodes being mainly directed towards the two central ones.

A second dimension concerns the meta-level. For a similarly detailed recording of events and exchanges during the search process itself creates a "conscious" and shared meta-level, by which we mean all arguments, thoughts, ideas, etc., *about* the process itself, not so much as about the actual problem they are trying to solve. These comments are grouped together on a special blog.[11]

17.2.3 Feature 3

What kind of contributions were made during the process? One might expect that, since the overall aim was to find a proof, everyone would have directly contributed to it. In terms of the ideal picture of a proof—a labelled list of statements, justified at each step, starting with the premises and ending with the statement to be proved as conclusion—one could imagine Gowers writing down the first few lines, someone else contributing the next line, until it is realised that the road chosen is a dead-end, hence some backtracking is done and someone else proposes a new line to explore a different road, eventually arriving at the last line. Needless to say, that is not what happened. Actually, a stronger statement can be made: many contributions were of a quite different nature. What follows is not to be seen as an overview but as a selection of some elements that deviate from the picture sketched above:

– Looking at other domains: Very often analogies with other problems were proposed, both on the level of statements ("This problem looks very much like …"), where the analogical problem often comes from a different domain than combinatorics, and on the level of proof procedures ("Could you make use of this or that technique that turned out to be useful for this or that problem?"), that were successful in other domains.

---

[11] See http://polymathprojects.org/category/discussion/, for more specific comments and on a related blog, http://polymathprojects.org/general-discussion/, for more general comments.

- The same problem in different words: Equally often, concepts were proposed to reformulate the problem so that other problems could be related but also to get a better understanding of the problem. This is a technique that is well-known in mathematical practice: find equivalent forms of the statement you are trying to prove, hoping that the reformulation will be easier to prove. Think, e.g., about the four-colour theorem. Replace a region by a dot and, if two regions are neighbours, connect them with an edge. Now you obtain a graph and the theorem translates equivalently into: using four colours, colour the dots so that two connected dots do not get the same colour.
- Keep it simple(r): Not so much parts of possible proofs were proposed, but rather suggestions and quite vague ideas about possible routes to find the proof itself and what the proof could look like. Sometimes proofs of simpler problems, directly related to the original problem, were presented as a source of inspiration, reminding us of one of the strategies discussed in Pólya (1945, 75–85), namely, if the solution of a problem must satisfy a number of conditions, drop one of the conditions and see whether a solution can then be found, a special case of the general heuristic strategy, labeled "Decomposing and recombining".

Consequences of this high diversity in the contributions are, at least the following:

- The community, both in terms of members and in terms of the exchanges they make, is definitely not homogeneous. Different members play quite different parts.
- Different members will come up with different ideas and concepts. What goes on in the mind of the professional high-ranked mathematician is definitely not what goes on in the mind of the high school teacher or dedicated amateur. The one is not a "light" version of the other, they do have their proper characteristics and have their own specific contributions to make.
- Room is created for chance elements to play a part. Crazy ideas get a place in such networks. The upshot seems to be that all these diverse elements create a rather special and specific problem-solving context. In this respect, Gowers has remarked: "Reading the discussion provides some kind of strange random stimulus that causes your brain to go in to fruitful places where it might not have gone otherwise" (as quoted in Aron, 2011).

What happened after *DHJ*? At the time of writing, five more problems have already been tackled. It is rather striking, although of course the specialties of Gowers and Tao have to be taken into account, that most of the problems, either directly or slightly less so, deal with combinatorics and the search for or the improvement of upper and lower bounds. Polymath 2 deals with Banach spaces and although that seems sufficiently far away from combinatorics, Gowers on the blog has remarked: "I have given definition 3 above because I think it has the potential to 'combinatorialize' the problem" (Gowers, 2009b). Polymath 3, the Polynomial Diameter Conjecture, states that, if G is the graph of a d-polytope with n facets, then the diameter of G is bounded above by a polynomial of d and n, and needs no further discussion. The same goes for Polymath 4, the statement that there exists a deterministic algorithm which, given an integer k, is guaranteed to find a prime of at least k digits in length of time polynomial in k. (A proof of the theorem is under review.) Polymath 5 deals with Erdős's discrepancy problem, and Polymath 6 with improving bounds in Roth's theorem, both fitting nicely in the combinatorial domain. The question remains, as already mentioned above, whether there are problems *not* suited for this type of approach.

The case thus being laid out, it is now time to pass to the constructive part of our chapter, and propose a formalism that can capture instances of essentially collaborative mathematical research like the one met in this very example.

17.3 Formalising Shared Knowledge

Reasoning about how knowledge and belief can be shared by individuals is what epistemic logics, extended with operators for group knowledge, do best. Reasoning about how the distribution of shared knowledge and belief can be altered through interaction belongs to the domain of dynamic epistemic logic. In the present section the basic notions and insights from these fields are briefly described. We refer the reader to the Appendix for further details.

The epistemic and doxastic logics are built up in two steps. We start with the introduction of operators for single-agent knowledge (or belief) by extending the propositional language with a modal operator $[c]$ for each individual c we'd like to reason about. As is standard in the literature, we assume that these individuals are extremely powerful reasoners by stipulating that (a) they are logically and deductively omniscient, (b) they are infallible when it comes to their knowledge (they satisfy positive introspection), and (c) they are infallible when it comes to their ignorance (they satisfy negative introspection). When complemented with the standard view that knowledge is factive, this leads to a formalisation of individual knowledge that is based on the modal logic **S5**. Analogously, if we assume that beliefs can be false, but should still be consistent, the same assumptions lead to a formalisation of individual belief that is based on the modal logic **KD45**. In the remainder, we only focus on knowledge.

The operators for individual knowledge are sufficient to define two of the four main notions of group knowledge, namely *particular* and *general knowledge*. By particular knowledge in a group of individuals G, we mean knowledge by at least one of the members of G. We write $S_G\varphi$ to express that φ is particular knowledge in G. By general knowledge in a group of individuals G, we mean knowledge by all of the members of G. We write $E_G\varphi$ to express that φ is general knowledge in G. As is clear from the above, $S_G\varphi$ and $E_G\varphi$ are equivalent to, respectively, the (generalised) disjunction and conjunction of the respective knowledge claims for all individuals in G. No such straightforward reduction to claims about what the respective individuals know is available for the two remaining notions of group-knowledge. Once more, we refer the reader to the Appendix for the relevant details, and stick to an informal description of these notions. By *distributed knowledge* in G, we mean the knowledge

Table 17.1 Logical properties of group-knowledge

	$[c]\varphi$ for some $c \in G$	$[c]\varphi$ for all $c \in G$	Deductive closure	Pos & neg introspection
$D_G\varphi$			✓	✓
$S_G\varphi$	✓			(✓)
$E_G\varphi$	✓	✓	✓	
$C_G\varphi$	✓	✓	✓	✓

that can be obtained by pooling together the knowledge of all the members of G. We write $D_G\varphi$ to express that φ is distributed knowledge in G. To explain what we mean by *common knowledge* in G, we need to revert to the notion of general knowledge. Indeed, when we say that φ is common knowledge in G, we mean that φ is general knowledge in G, that it is general knowledge that it is general knowledge that φ etc. Common knowledge in a group not only amounts to general knowledge that cannot be doubted by any member of that group, but also implies that no member can doubt the fact that any other member is in such a position.

Every type of group knowledge has different logical properties, and therefore provides a model for how knowledge can be present in or available to a community with different strengths. For instance, φ being distributed knowledge within G is compatible with every member being ignorant about φ. That is, φ can be distributed knowledge without being particular knowledge. In more common terms: φ can be implicitly known without being explicitly known. Elsewhere (Allo, 2013), the first author proposed the following interpretation of each type of group knowledge: (a) distributed knowledge is implicitly available knowledge; (b) particular knowledge is explicitly available knowledge; (c) general knowledge is readily available (explicit) knowledge; and (d) common knowledge is transparently available (explicit) knowledge. This interpretation agrees with the logical properties of each type of group-knowledge as summarised in Table 17.1.

The static properties of each type of group-knowledge already give us the ability to discriminate between different ways in which something can be known or accepted within a community, but do not yet exhaust the possibilities. Consider for that matter the following principles

$$(S_G\varphi \wedge S_G(\varphi \rightarrow \psi)) \rightarrow D_G\psi$$
$$(S_G\varphi \wedge E_G(\varphi \rightarrow \psi)) \rightarrow S_G\psi$$

which express weaker types of deductive closure for particular knowledge. From the first principle it follows that the consequences of all particular knowledge are distributed knowledge. In other words, any deductive consequences of what is explicitly known are implicitly known, and to make merely implicit knowledge explicit the different agents will have to communicate.

Extensions of the standard epistemic logic known as dynamic epistemic logic (Baltag and Moss, 2004; Baltag et al., 2008; van Ditmarsch et al., 2007) bring in the required resources to express facts about what is known after a certain announcement. Crucially, different types of announcements will alter the way in which knowledge is distributed in a different manner. The best known example is that of so-called public announcements, whose effect is modelled as the removal of all alternatives that disagree with the content of what is announced. Thus, after the public announcement of p to a group of agents G, the model will no longer contain any alternative where p is true, and hence no p-world will be a G^k-alternative (see Appendix). That is, p will be common knowledge in G.

A first alternative to public announcements that we need to consider are *fully private announcements*. These are announcements where one or more agents learn the content of what is announced, while the remaining agents are not even aware of the fact that something was announced. When, for instance, p is privately announced to an agent c while the remaining members of the group G do not notice this, the resulting model will be one where c knows that p while the remaining members will be under the false impression that the knowledge available in the group didn't change at all. Thus, for instance, if prior to the announcement it was commonly known that c didn't know whether p was the case, then after the announcement the remaining members will falsely believe that c is ignorant with respect to p.

A second alternative are the so-called *fair game announcements*, where some agents receive new information while the remaining members of the group do notice that some information is being shared with others, but remain ignorant with regard to the specific content that is being learned. That is, after such an announcement of p it could be the case that c knows that p while the remaining agents only know that c knows whether p (i.e. they do not know as much as c does, but they do know that the state of c's knowledge has changed because they know that either c knows that p or c knows that $\neg p$).

Using these ideas, we can start to make further combinations. For instance, we can take the union of different fully private announcements to model the action where each member of a group G comes to know that p, and yet falsely believes that no other member of G comes to know that p. What is distinctive of such announcements is that after the announcement it is the case that p is generally known in G, but not commonly known in G. This type of result can be related to a well-known limitation about our means to achieve common knowledge in a group, namely the fact that common knowledge can only be attained through public announcements. That is, for p to become common knowledge in a group G, p must be publicly announced within G; it must be transparent to all members of G that p is announced and that all members of G know about this announcement. If communication is not transparent, as is the case with the other types of announcements we've described (or as is the case with unreliable communication, as we find it in the coordinated attack problem; see, e.g., Fagin et al., 1995), then common knowledge is absolutely beyond reach.

17.4 A Model for Collaborative Mathematics?

The main reason why common knowledge is such a valuable good is that it is a prerequisite for coordinated action. This feature is traditionally associated with the coordinated attack problem: To win the battle, two generals need to ensure that they will attack the enemy at the same moment. The latter can only be achieved when it is common knowledge between them that, say, the plan is to attack at dawn. As the problem is traditionally set up, common knowledge and thus agreement can never be achieved by these generals. Since all their means of communication are unreliable (e.g. messengers that might be captured by the enemy), the generals can always doubt that the message was delivered and hence that an agreement was reached. This situation is analogous to a situation where information is exchanged with (semi-) private announcements.

> Let $agree(\psi)$ be a formula that is true at states in which the players have agreed on ψ. (...) [W]e expect that if Alice and Bob agree on ψ, then each of them knows that they have agreed on ψ. This is a key property of agreement: in order for there to be agreement, every participant in the agreement must know that there is agreement. Thus, we expect $agree(\psi) \implies E(agree(\psi))$ to be valid. The Induction Rule for Common knowledge tells us that if this is the case, then $agree(\psi) \implies C(agree(\psi))$ is also valid. Hence, agreement implies common knowledge (Fagin et al., 1995, 189–90).

Assuming that collaborative mathematics requires coordinated action, one would presume that common knowledge would be as important for the success of the Polymath Project as it is for the victory of the two Byzantine generals. And hence, given the previously established connection with public announcements, one would expect that all knowledge should be shared in an entirely transparent manner. At first blush it would seem that this is exactly what is the case in the example from Sect. 17.2: The totality of what is communicated is publicly available on a website. And yet, if that were the case the whole case-study would turn out to be a rather boring example of pooling together resources (both in terms of computational capacity and in terms of mathematical knowledge). To reveal what makes the Polymath case an interesting one, we precisely need to focus on how the absence of common knowledge guarantees that the community remains sufficiently diverse, while the process of limiting this diversity ensures that the whole enterprise remains sufficiently goal-oriented to ultimately find a solution and end up with a proof.

When put in terms of announcements, the above description suggests that we need to look at the whole communication-process on the website as a mixture of both public and not so public announcements; with the public announcements enforcing a certain amount of coordination, and with the private and semi-private announcements to maintain a certain degree of diversity. Our aim in this section is to show that, first, the structure of the polymath community—with a core of two agents, and a highly diverse periphery—is indeed well-suited to maintain both diversity and a goal-oriented enterprise, and, second, to show that the structure of this type of community can be modelled with the tools described in the previous section.

Let us begin with a few methodological remarks about formal modelling. First, to design a formal model of a given phenomenon we need to decide on which features we're interested in, and thus need to be retained by the model, and which features we would like to abstract away. Here, the upshot is to focus on how information is distributed, and how that distribution can be altered by means of communication. This will mean that our model will abstract from many features of the case-study from Sect. 17.2, and in particular that a fine-grained rendering of the content of the specific mathematical problem will be absent from our model. As a result, the formal model that is further developed in this section ignores **Feature 1**, the fact that the polymath project dealt with a combinatorial problem. Such abstraction is typical of formal models of science.

A second type of abstraction follows from the fact that our formal model assumes all agents to be logically omniscient. This is a common type of idealisation in logical models. It entails that the model cannot reveal specific features that are related to the resources that are required to complete certain computations. As a result, the formal model we use will also ignore aspects that are related to the fact that by delegating the task of checking a number of limited cases computational resources can be shared (i.e. if all agents have unlimited resources, there's no point in delegating purely computational tasks).

Given these abstractions, the model we're after is primarily a model of the structure of a community understood in terms of how information can flow within that community. How can, given the formalism we've settled on, the latter be modelled? Here, we will need to make a few assumptions that strictly speaking go beyond the formalism described in Sect. 17.3. This is because the framework of dynamic epistemic logic lacks the expressive resources to make all the relevant features of this structure explicit. A *first* requirement is that we need to tie announcements to specific agents who make that announcement. To that end, we follow the standard practice of modelling the announcement of φ by some agent c as the announcement that c knows (or believes) that φ. This introduces a first type of distinction between otherwise similar announcements.

The *second* and more important requirement is that we need to account for the structure of a community. This we achieve in two ways. By picking out a number of *distinguished sub-groups* of that community, and by listing (or otherwise defining) the types of announcements that are available. By restricting the number of sub-groups we can make claims about,[12] we settle on a certain level of abstraction.[13] In doing so, we put a limit on how finely we can discriminate between different

---

[12] Again, this is something that the standard formalism does not allow for since the notions of general and particular knowledge are simply definable in terms of, respectively, the conjunction and disjunction of individual knowledge claims for each member of the group. Nevertheless, such restrictions on the available groups can be introduced syntactically with, what amounts to, awareness filters. We do not pursue the details of this additional modification, but stick to a rather informal approach.

[13] An intuitively plausible way to think about such levels of abstraction is this: Sometimes we like to consider the (or a) mathematical community as a monolithic bloc (in terms of consensus in that

factions in, say, the periphery of the community in terms of what is known and/or believed in such factions. A second effect of this type of expressive limitation is that it also precludes us from indiscriminately lumping together the knowledge and belief of arbitrary agents. For instance, while it obviously makes sense to talk about the knowledge that is distributed in the community as a whole, it isn't so obvious that for any two agents c_1 and c_2 we should equally likely be able to talk about the distributed knowledge (or lack thereof) in the two-agent group $\{c_1, c_2\}$.

The latter type of limitation on the types of claims we should be able to make about different sub-groups finds a natural companion in the restrictions we impose on the available announcements. For instance, if we are not allowed to make claims about the distributed knowledge in the group $\{c_1, c_2\}$, we should also not allow these two agents to privately communicate (for this would mean that they could in principle privately aggregate their knowledge). This is just one type of limitation on the available announcements. Another very natural type of limitation is on who is allowed to make public announcements. Here too, it often makes sense to assume that not all agents can make such announcements.

Depending on the structure of a community, that is, depending on the types of announcements that are available and that can alter the distribution of information in that community, the process of communication will determine the balance between coordination and diversity. For instance, only those agents who can make public announcements will be able to enforce the agreement required for coordination (and then only relative to their own information), while the other agents will at best be able to bring in new information and thereby enhance the diversity within the community. In the last part of this section we shall illustrate these processes by looking at three broad types of interaction. This will not only illustrate the connection between available actions and the balance between agreements and diversity, but will also shed a light on some distinctive features of the Polymath community.

17.4.1 The Inference Network

The main purpose of an inference network, conceived as a model of distributed computing, is to make the information that is implicitly present in the network explicit. In our terms, its aim is to move from merely distributed knowledge to (at least) particular knowledge. This can be achieved in many ways. One way to do so, is to allow every agent to send his information to a unique designated agent; a so-called *wise man*.[14] In our terms, this would mean that the only available actions

---

community), but equally often we prefer to take a more refined look that allows us to focus on more local phenomena.

[14]This approach trivialises the idea of *distributed computing*, but the comments we make equally apply to more refined protocols for inference networks.

are announcements that have the wise man among their recipients. Crucially, this type of network is indifferent with respect to how such announcements are made; all that matters is that the wise man should come to know every piece of information in the network. This can equally well be achieved by private announcements as it can be by public announcements. Coordinated action is, at least in principle, not required for this network to succeed in its goal; it only has an impact on how efficient the network is. For instance, if the identity of the wise man is generally known, each agent will only need to send his information to that wise man. Likewise if the communication of φ between some agent c and the wise-man c_w is public within the sub-group $\{c, c_w\}$, then c will also know that the message that φ has arrived and thus prevent it from sending it again (though this won't prevent others from sending φ as well).

17.4.2 Contributing by Solving Predefined Subtasks

While the process of solving a big problem within a group by splitting it up into smaller tasks and distributing these tasks to different agents or sub-groups shares many features with an inference network, it also gives rise to further constraints on the structure of the community. To begin with, while the presence of a wise man isn't absolutely required for an inference network, the presence of some agent or sub-community who coordinates the process by distributing the different pieces is crucial here. In our terminology, this amounts to saying that we need bi-directional information-flow: to distribute the predefined tasks, and to send back the results for aggregation. As before, some types of coordination and common knowledge can be useful. We could, for instance, require that the overall aim be commonly known, that it be publicly announced when the final goal is reached, or that the process whereby tasks are distributed amounts to reaching an agreement that a certain task is accepted. Yet, other types of common knowledge like knowledge about the different subtasks, their distribution among the community, and the completion of certain subtasks need not be present. Given the wide range of situations where problems are solved in this manner, the question of what exactly needs to be commonly known cannot and should not be settled in general (e.g. sometimes we might prefer a situation where some sub-tasks are simultaneously tackled by competing agents, who may (or may not) suspect that others might have received the same task).

17.4.3 Balancing Between Coordination and Diversity

This is the type of network or community we associate with the Polymath Project. In Sect. 17.2, this network is described as "a network with two strongly connected *attractors*". Here, we reflect this structure by describing the types of announcements that can be made, and by listing the types of knowledge claims that can be made.

First, to capture the fact that we have a core consisting of two agents, we should be able to refer to each of these agents separately, and indeed we should require that everyone in the community be aware of their identity. Second, to reflect that these two agents are strongly connected we should be able to make claims about what they know as a group (and how they know it). Similarly, to extend these considerations to the level of available announcements, we should also stipulate that these two agents can have private interactions, and that they both (as individuals, but perhaps also as a group) should be able to make public announcements to the community as a whole. In virtue of the fact that they can interact privately, the distribution of their knowledge can be altered unbeknownst to the community as a whole; in virtue of their ability to make public announcements, they have the certainty that when they address the community as a whole, the others will listen.

Third, to capture the fact that the rest of the community is best seen as a wide periphery, we may restrict our ability to refer to each agent separately, and we almost certainly need to limit their awareness of each other. Fourth, to reflect that all members of the community can contribute, that all information is made publicly available, and yet that it is by no means certain that all members of the community actually access each and every piece of available information, we should preclude that these agents be able to make public announcements. In fact, we should model the announcements of these agents in such a way that they do not know whether they actually reach the whole community, or only some subset of that community. Finally, as already mentioned, we might also want to preclude the agents in the periphery from interacting privately.

How does such a setting contribute to the intended aim of keeping a balance between diversity and coordination? This is achieved by the following division of labour: Agents in the periphery, on the one hand, can launch new ideas, and introduce new information into the network, but do not have the means to impose these ideas. This ensures that new ideas can at all time be introduced, and that most agents cannot decisively reject these new ideas. Agents at the core, by contrast, can (as they do by introducing the problem, or by giving summaries) set and revise the agenda, and this agenda can be considered binding. This is done by picking up and promoting certain ideas that were previously introduced by agents in the community, while objecting to (or merely ignoring) others. As such, the ability of the core members to make public announcements lets them, given an already available rich supply of new information, either stimulate or constrain this diversity.

17.5 Conclusion

In the field of philosophy there is a well-entrenched research tradition on argumentation, both from an informal and a formal point of view. In the latter case, this tradition has been strongly intertwined with the development of non-classical logics. Recently, from within the field of computer science, a novel approach to argumentation has been launched, namely that of argumentation networks. However, to date, the models have remained rather abstract: arguments are viewed solely in relation to others, the basic scheme being that of one-on-one "attacks". Also, as far as we are aware, applications to mathematical contexts are as yet unavailable. Correspondingly, the aim of this chapter has been to demonstrate and develop the potential of this type of approach for the understanding of the construction of mathematical proofs. In particular, we have illustrated how epistemic logics that allow us to reason about the knowledge of individuals and the knowledge available within communities can be used to model proofs-as-arguments from a perspective that emphasises interaction and collaboration. In particular cases like the Polymath Project, the absence of full common knowledge seems to guarantee that the community remains sufficiently diverse, while limits to this diversity ensure that the enterprise remains sufficiently goal-oriented.

Acknowledgements The first author is a postdoctoral fellow of the Research Foundation–Flanders, which through project G.0431.09 also supported research for this chapter by the third author.

Appendix

Where $\mathscr{C}$ is the set of individual agents, we have a modal operator $[c]$ for every c in $\mathscr{C}$. Thus, we say that

$$[c]\varphi \text{ is true at a state } w \text{ iff } wR_cw' \text{ implies that } \varphi \text{ is true at } w'.$$

If wR_cw' is read as saying that w' is an epistemic alternative to w for c, and that its negation $\neg wR_cw'$ means that at w, c can exclude w', we can say that c knows that φ at w iff

$$c \text{ can exclude at } w \text{ all states where } \varphi \text{ is not true,}$$

or, equivalently, iff

$$\varphi \text{ is true at all epistemic alternatives to } w \text{ for } c.$$

Traditionally, epistemic operators are presumed to satisfy some further conditions. For present purposes we opt for the strongest possible set of conditions: We presuppose that knowledge is factive (only truths can be known), that it is positively introspective (knowing implies knowing that one knows), and that it is negatively introspective (one always knows about one's ignorance). It is a standard result in modal logic that $[c]$ has all these properties iff R_c is a reflexive, transitive and symmetric relation (see, for instance, van Ditmarsch et al., 2007, 2.2).

In addition to being able to talk about what individual agents know, we also want to say something about what is known in communities of agents. Crucially, knowledge can be available to communities in different guises. Where $G \subseteq C$ is a community of agents and φ is any formula, φ can be known at a state w by a community G because:

1. At w every state w' where φ is not true is excluded by at least one $c \in G$. In that case, we say that φ is *distributed knowledge* at w in G and write $w \models D_G\varphi$ to express this.
2. At w some $c \in G$ excludes every state w' where φ is not true. In that case, we say that φ is *particular knowledge* at w in G (informally, 'someone knows') and write $w \models S_G\varphi$ to express this.
3. At w every $c \in G$ excludes every state w' where φ is not true. In that case, we say that φ is *general knowledge* at w in G (informally, 'everybody knows') and write $w \models E_G\varphi$ to express this.

To explain a final way in which φ can be known in G, we first define a G^k alternative with the following inductive clauses:

- A world w is a G^1 alternative iff w is an epistemic alternative for some member of G.
- A world w is a G^{k+1} alternative iff at some G^k alternative the world w is an epistemic alternative for some member of G.

Using this notion, we can now stipulate that φ can also be known when

4. At w and for any finite k, no state w' where φ is not true is a G^k alternative. In that case, we say that φ is *common knowledge* in G and write $w \models C_G\varphi$ to express this.

References

Aberdein, A. (2005). The uses of argument in mathematics. *Argumentation*, *19*(3), 287–301.
Aberdein, A. (2007). The informal logic of mathematical proof. In J. P. Van Bendegem & B. Van Kerkhove (Eds.), *Perspectives on mathematical practices: Bringing together philosophy of mathematics, sociology of mathematics, and mathematics education* (pp. 135–151). Dordrecht: Springer.
Allo, P. (2013). The many faces of closure and introspection. *Journal of Philosophical Logic*, *42*(1), 91–124.
Atkinson, K., Bench-Capon T., & McBurney, P. (2005). A dialogue game protocol for multi-agent argument over proposals for action. *Journal of Autonomous Agents and Multi-Agent Systems*, *11*(2), 153–171.
Aron, J. (2011). Maths can be better together. *New Scientist*, *210*(2811), 10–11.
Baltag, A., & Moss, L. S. (2004). Logics for epistemic programs. *Synthese*, *139*(2), 165–224.
Baltag, A., van Ditmarsch, H. P., & Moss, L. S. (2008). Epistemic logic and information update. In P. Adriaans & J. van Benthem (Eds.), *Handbook on the philosophy of information* (pp. 361–456). Amsterdam: Elsevier Science Publishers.
Barth, E. M., & Krabbe, E. C. W. (1982). *From axiom to dialogue. A philosophical study of logics and argumentation.* Berlin: Walter de Gruyter.
Bondarenko, A., Dung, P. M., Kowalski, R. A., & Toni, F. (1997). An abstract, argumentation-theoretic approach to default reasoning. *Artificial Intelligence*, *93*(1–2), 63–101.

Bovens, L., & Hartmann, S. (2003). *Bayesian epistemology*. Oxford: Oxford University Press.

Cranshaw, J., & Kittur, A. (2011). The Polymath project: Lessons from a successful online collaboration in mathematics. In *Proceedings of the conference on human factors in computing systems (CHI-11)* (pp. 1–10). Vancouver, BC.

Dudley, U. (1992). *Mathematical cranks*. Washington, DC: MAA.

Dung, P. M. (1995). On the acceptability of arguments and its fundamental role in nonmonotonic reasoning, logic programming and n-person games. *Artificial Intelligence, 77*(2), 321–358.

Furstenberg, H., & Katznelson, Y. (1991). A density version of the Hales-Jewett Theorem. *Journal d'Analyse Mathématique, 57*, 64–119.

Gowers, T. (2009a). Why this particular problem? In Gowers's Weblog. Mathematics related discussions. http://gowers.wordpress.com/2009/02/01/why-this-particular-problem/. Cited 11 Oct 2011.

Gowers, T. (2009b). Must an "explicitly defined" Banach space contain c_0 or ell_p? In Gowers's Weblog. Mathematics related discussions. http://gowers.wordpress.com/2009/02/17/must-an-explicitly-defined-banach-space-contain-c_0-or-ell_p/. Cited 11 Oct 2011.

Gowers, T., & Nielsen, M. (2009). Massively collaborative mathematics. *Nature, 461*, 879–881.

Grossi, D. (2010). Doing argumentation theory in modal logic. *ILLC-Preprint*, 2009–24. Amsterdam: University of Amsterdam.

Fagin, R., Halpern, J. Y., Moses, Y., & Vardi, M. Y. (1995). *Reasoning about knowledge*. Cambridge, MA: MIT Press.

Hintikka, J. (1985). A spectrum of logics of questioning. *Philosophica, 35*, 135–150

Hodges, W. (1998). An editor recalls some hopeless papers. *The Bulletin of Symbolic Logic, 4*(1), 1–16.

Jakobovits, H., & Vermeir, D. (1996). Contradiction in argumentation frameworks. In *Proceedings of the IPMU conference* (pp. 821–826). Granada.

Jakobovits, H., & Vermeir, D. (1999a). Dialectic semantics for argumentation frameworks. In *Proceedings of the seventh international conference on artificial intelligence and law* (pp. 53–62). ACM: New York.

Jakobovits, H., & Vermeir, D. (1999b). Robust semantics for argumentation frameworks. *Journal of Logic and Computation, 6*(2), 215–261.

Jeffreys, B. R., Kelley, L. A., Sergot, M. J., Fox, J., & Sternberg, M. J. E. (2006). Capturing expert knowledge with argumentation: A case study in bioinformatics. *Bioinformatics, 22*, 924–933.

Kakas, A. C., Mancarella, P., & Dung, P. M. (1994). The acceptability semantics for logic programs. In P. Van Hentenrijck (Ed.), *Proceedings of the 11th international conference on logic programming* (pp. 504–519). Cambridge, MA: MIT Press.

Kitcher, P. (1990). The division of cognitive labor. *The Journal of Philosophy, 87*(1), 5–22.

Lakatos, I. (1976). *Proofs and refutations: The logic of mathematical discovery* (edited by J. Worrall & E. Zahar). Cambridge: Cambridge University Press.

Lorenzen, P., & Lorenz, K. (1978). *Dialogische logik*. Darmstadt: Wissenschaftliche Buchgesellschaft.

Nielsen, M. (2011). *Reinventing discovery: The new era of networked science*. Princeton, NJ: Princeton University Press.

Pollock, J. (1994). Justification and defeat. *Artificial Intelligence, 67*, 377–407.

Pólya, G. (1945). *How to solve it*. New York: Doubleday.

Polymath, D. H. J. (2010). Density Hales-Jewett and Moser numbers. In I. Bárány & J. Solymosi (Eds.), *An irregular mind (Szemerédi is 70): Vol. 21. Bolyai society mathematical studies* (pp. 689–753). Berlin: Springer.

Polymath, D. H. J. (2012). A new proof of the Density Hales-Jewett Theorem. *Annals of Mathematics, 175*(3), 1283–1327.

Popper, K. R. (1968). Epistemology without a knowing subject. In B. Van Rotselaar & J. F. Staal (Eds.), *Logic, methodology and philosophy of science III* (pp. 333–373). Amsterdam: North-Holland. (Reprinted as Chapter III of Popper 1972)

Popper, K. R. (1972). *Objective knowledge*. Oxford: Oxford University Press.

Prakken, H., & Sartor, G. (1996). A dialectical model of assessing conflicting arguments in legal reasoning. *Artificial Intelligence and Law*, *4*, 331–368.

Prakken, H., & Vreeswijk, G. (2002). Logics for defeasible argumentation. In D. M. Gabbay & F. Guenthner (Eds.), *Handbook of philosophical logic* (2nd ed., Vol. 4, pp. 219–318). Dordrecht: Kluwer.

Rav, Y. (1999). Why do we prove theorems? *Philosophia Mathematica (III)*, *7*(3), 5–41.

Reed, C. A. (1998). Dialogue frames in agent communication. In *Proceedings of the 3rd international conference on multi agent systems (ICMAS–98)* (pp. 246–253). Washington, DC: IEEE.

Toulmin, S. (1958). *The uses of argument*. Cambridge: Cambridge University Press.

Van Bendegem, J. P. (1982). Pragmatics and mathematics or how do mathematicians talk? *Philosophica*, *29*, 97–118.

Van Bendegem, J. P. (1985a). Dialogue logic and problem-solving. *Philosophica*, *35*, 113–134.

Van Bendegem, J. P. (1985b). A connection between modal logic and dynamic logic in a problem solving community. In F. Vandamme & J. Hintikka (Eds.), *Logic of discourse and logic of discovery* (pp. 249–262). New York: Plenum Press.

Van Bendegem, J. P., & Van Kerkhove, B. (2009). Mathematical arguments in context. *Foundations of Science*, *14*(1–2), 45–57.

Van Eemeren, F. H., Grootendorst, R., & Kruiger, T. (2004). *A systematic theory of argumentation: The pragma-dialectical approach*. Cambridge: Cambridge University Press.

Van Ditmarsch, H., van der Hoek, W., & Kooi, B. (2007). *Dynamic epistemic logic*. Dordrecht: Springer.

Verheij, B. (1995). Two approaches to dialectical argumentation: Admissible sets and argumentation stages. In J.-J. C. Meyer & L. C. van der Gaag (Eds.), *NAIC'96: Proceedings of the eighth Dutch conference on artificial intelligence* (pp. 357–368). Utrecht: Utrecht University.

Vreeswijk, G. A. W. (1997). Abstract argumentation systems. *Artificial Intelligence*, *90*(1–2), 225–279.

Walton, D. N. (1998). *The new dialectic: Conversational contexts of argument*. Toronto, ON: University of Toronto Press.

Woods, J. (2003). *Paradox and paraconsistency. Conflict resolution in the abstract sciences*. Cambridge: Cambridge University Press.

Zollman, K. J. S. (2007). The communication structure of epistemic communities. *Philosophy of Science*, *74*(5), 574–587.

Chapter 18
The Parallel Structure of Mathematical Reasoning

Andrew Aberdein

18.1 The Dance of Mathematical Practice

What is mathematics about? A standard answer for philosophers has long been that mathematics is concerned with the derivation of formal proofs. And yet, as many mathematicians point out, truly formal proof has little to do with actual mathematical practice. Here is one such mathematician, David Ruelle:

> Human mathematics consists in fact in talking about formal proofs, and not actually performing them. One argues quite convincingly that certain formal texts exist, and it would in fact not be impossible to write them down. But it is not done: it would be hard work, and useless because the human brain is not good at checking that a formal text is error-free. Human mathematics is a sort of dance around an unwritten formal text, which if written would be unreadable. This may not seem very promising, but human mathematics has in fact been prodigiously successful (Ruelle, 2000, 254).

Explaining that success poses a problem for philosophy of mathematics as traditionally conceived. If mathematical practice were ultimately reducible to formal proof, which has been analysed in great detail in mathematical logic, then actual practice would differ only in degree from the elementary and/or foundational work upon which most philosophers of mathematics concentrate. But if mathematical practice cannot be understood solely in such terms, then philosophy of mathematics needs to pay it much closer attention. My goal in this chapter is to address this shortcoming by taking Ruelle's metaphor seriously, and seeking to devise, as it were, a choreographic notation for the dance which he describes.

A. Aberdein (✉)
Department of Humanities and Communication, Florida Institute of Technology,
150 West University Boulevard, Melbourne, FL 32901-6975, USA
e-mail: aberdein@fit.edu

A. Aberdein and I.J. Dove (eds.), *The Argument of Mathematics*, Logic, Epistemology, and the Unity of Science 30, DOI 10.1007/978-94-007-6534-4_18,
© Springer Science+Business Media Dordrecht 2013

18.2 Structures of Mathematical Reasoning

Many mathematicians and philosophers of mathematics have observed the dual nature of mathematical proof: proofs must be both persuasive and rigorous. The passage from Ruelle quoted above is one example. Here is another from a more famous mathematician, G.H. Hardy:

> If we were to push it to its extreme we should be led to a rather paradoxical conclusion; that we can, in the last analysis, do nothing but *point*; that proofs are what Littlewood and I call *gas*, rhetorical flourishes designed to affect psychology, pictures on the board in the lecture, devices to stimulate the imagination of pupils. ... On the other hand it is not disputed that mathematics is full of proofs, of undeniable interest and importance, whose purpose is not in the least to secure conviction. Our interest in these proofs depends on their formal and aesthetic properties. Our object is *both* to exhibit the pattern and to obtain assent. (Hardy, 1928, 18, his emphasis)

It follows from this account that 'proof' is ambiguous between two different activities: 'exhibiting the pattern' and 'obtaining assent'. In most circumstances both activities must be satisfactorily performed for the proof to be a success. There are some special cases, such as proofs that have been fully formalized, or have been reified as mathematical objects, where only the first activity is attempted. But in the characteristic sense of 'proof' we need more than this; we need a dialectical interaction with the mathematical community.

For Richard Epstein, proofs intended to obtain assent are *arguments* by means of which mathematicians convince each other that the corresponding *inferences* are valid. He represents this situation schematically (Fig. 18.1). However, proofs are typically made up of many steps, not all of which are necessarily developed with the same rigour. So closer examination of proofs will represent them not as single arguments but as structures of arguments (technically trees, or directed acyclic graphs). Applying this detail to the broader picture suggested by Epstein requires the articulation of two parallel structures: an *inferential structure* of formal derivations linking formal statement to formal statement, and an *argumentational structure* of arguments by which mathematicians attempt to convince each other of the soundness of the inferential structure, and thereby of the acceptability of the informal counterparts of those statements. In Hardy's terms, it is the inferential

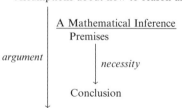

A Mathematical Proof
Assumptions about how to reason and communicate.

A Mathematical Inference
Premises

argument   *necessity*

Fig. 18.1 Epstein's picture   Conclusion
of mathematical proof
(Epstein, 2013, 267)   The mathematical inference is valid.

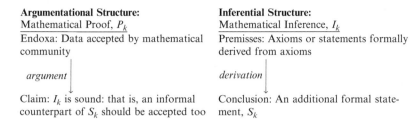

Argumentational Structure:	Inferential Structure:
Mathematical Proof, P_k	Mathematical Inference, I_k

Endoxa: Data accepted by mathematical community	Premises: Axioms or statements formally derived from axioms
argument	*derivation*
Claim: I_k is sound: that is, an informal counterpart of S_k should be accepted too	Conclusion: An additional formal statement, S_k

Fig. 18.2 The parallel structure of mathematical reasoning

structure which is responsible for 'exhibiting the pattern' while the argumentational structure is responsible for 'obtaining assent'. Figure 18.2 summarizes this picture.

This account both conserves and transcends the conventional view of mathematical proof. The inferential structure is held to strict standards of formal rigour, without which the proof would not qualify as mathematical. However, the step-by-step compliance of the proof with these standards is itself a matter of argument, and susceptible to challenge. Hence much actual mathematical practice takes place in the argumentational structure. Careful demarcation of these two levels is essential to the proper understanding of mathematics. If this account is correct, important concepts in the philosophy of mathematics, such as mathematical rigour and mathematical explanation, can only properly be addressed when *both* of the parallel structures are accounted for. In order to do so, we need to say more about the details of the individual components of the two structures. The formal nature of the inferential structure makes its characterization comparatively straightforward. It is a graph of vertices linked by edges, where the vertices are statements expressed in some formal system and the edges are derivations admissible in that system. The underlying formal system might, for example, be a natural deduction presentation of a particular system of logic, but more characteristically it will be a higher-level language, such as MIZAR (see Alama and Kahle, 2013, for further details of such systems). The argumentational structure poses more of a challenge, which I shall turn to in the next section.

18.3   The Argumentational Structure of Mathematics

Wilfrid Hodges offers a brief analysis of 'unformalised deductive argument[s]' that is of help here (although I shall argue that it is necessary to generalize his account beyond deductive argumentation). For Hodges, such arguments contain components of three sorts:

- There are the stated conclusion, the stated or implied starting assumptions, and the intermediate propositions used in getting from the assumptions to the conclusion. I shall call these the *object sentences*.

- There are stated or implied justifications for putting the object sentences in the places where they appear. For example if the argument says '*A*, therefore *B*', the arguer is claiming that *B* follows from *A*.
- There are instructions to do certain things which are needed for the proof. Thus 'Suppose *C*', 'Draw the following picture, and consider the circles *D* and *E*', 'Define *F* as follows' (Hodges, 1998, 6).

The argumentational structure must account for all three of these components.

The object sentences are the vertices of the argumentational structure. They are informal counterparts to formal statements of the inferential structure. Strictly speaking, the inferential structure must be grounded in axioms: propositions within the formal system whose truth must be granted if the system is to be employed. (As such the truth of the axioms cannot be established by formal proof; it is one area of mathematics where informal argument is the only option.) But practicing mathematicians seldom take things back that far. Instead, they begin as informal arguers do in any domain, with the informal counterpart of formal axioms, *endoxa*: 'propositions considered true by everybody, or by the majority, or by the wise' (Walton et al., 2008, 273). In mathematics, the endoxa comprise what Reuben Hersh has called 'established mathematics':

> Established mathematics is the body of mathematics that is accepted as the basis for mathematicians' proofs. It includes proved statements 'in the literature,' and also some simpler statements that are so well accepted that no literature reference is expected. The central core of established mathematics includes not only arithmetic, geometry, linear and polynomial algebra, and calculus, but also the elements of function theory, group theory, topology, measure theory, Banach and Hilbert spaces, and differential equations—the usual first two years of graduate study. And then to create new mathematics, one must also master major segments of the established knowledge in some special area (Hersh, 2013).

Thus, since informal mathematics need not always start with first principles, the argumentational structure will typically correspond only to a proper substructure of its inferential counterpart.

However, the correspondence will not be vertex to vertex: formal mathematics necessarily proceeds by very fine-grained increments, which the corresponding informal mathematical proofs typically elide. Thus many formal propositions of the inferential structure will have no counterpart object sentences in the argumentational structure. Conversely, some object sentences will have no counterpart formal propositions, since in some informal proofs there are object sentences representing intermediate statements which are not strictly needed for the validity of the proof, but only for its intelligibility (see Alama and Kahle, 2013, 168).

As Hodges observes, the argumentational structure is more complex than the inferential structure because it contains instructions as well as justifications. The edges of the argumentational structure must be defined loosely enough to articulate both components. Hersh has the following to say about the linkages between object sentences:

> Established mathematics is an intricately interconnected web of mutually supporting concepts, which are connected both by plausible and by deductive reasoning. ... Deductive

Fig. 18.3 (a) Basic Toulmin
layout. (b) Full Toulmin
layout

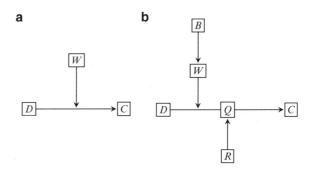

proof, mutually supporting interconnections, and close interaction with social life (commerce, technology, education) all serve to warrant the assertions of established mathematics. Deductive proof is the principal and most important warrant (Hersh, 2013).

Of course, this is still *unformalized* deductive proof, so it belongs in the argumentational structure not the inferential structure. Nonetheless, in many cases, the formalization is fully worked out, or more typically, it is clear how it could be. In this situation the corresponding steps of the argumentational structure can be very simple; they need do no more than point (as Hardy puts it) at the steps of the inferential structure.[1] But where the derivation is more complex or contested, much more of the burden of the proof rests on the argumentational structure. In those circumstances it becomes critical to track and provide responses to the objections that may be raised to the gaps in the inferential structure. As we shall see, exactly how this is to be achieved turns on which steps are admissible in the argumentational structure. Moreover, Hersh observes what I suggested above: even within established mathematics, not all the interconnections are deductive. Hence, some of the admissible steps will need to be framed in more permissive terms.

Further progress requires closer attention to the individual steps of the argumentational structure. One of the most influential attempts to analyze argumentational steps without appealing to logical form was developed in the 1950s by Stephen Toulmin. His 'layout' can represent deductive inference, but encompasses many other species of argument besides. In its simplest form, shown in Fig. 18.3a, the layout represents the derivation of a Claim (*C*), from Data (*D*), in accordance with a Warrant (*W*). This DWC pattern may appear to resemble a deductive inference rule, such as modus ponens, but it can be used to represent looser inferential steps.

The differences between the types of inference which the layout may represent are made explicit by the additional elements of the full layout shown in Fig. 18.3b. The warrant is justified by its dependence on Backing (*B*), possible exceptions or Rebuttals (*R*) are allowed for, and the resultant force of the argument is stated

---

[1] Such 'pointing' may bring to mind Jody Azzouni's 'derivation indicator view' of mathematical practice (Azzouni, 2004). However, Azzouni's 'indicating' describes a looser correspondence, closer to that holding in general between the two structures. Moreover, at least on some construals, such as that in (Dove, 2013, 304), Azzouni's conception of derivation is broader than mine.

in the Qualifier (Q). Hence the full layout may be understood as 'Given that D, we can Q claim that C, since W (on account of B), unless R'. In a frequently cited example (derived from Toulmin, 1958, 104), 'Given that HARRY WAS BORN IN BERMUDA, we can PRESUMABLY claim that HE IS BRITISH, since ANYONE BORN IN BERMUDA WILL GENERALLY BE BRITISH (on account of VARIOUS STATUTES . . .), unless HIS PARENTS WERE ALIENS, SAY.' In recent years a number of authors have demonstrated that informal deductive proofs may be represented using Toulmin layouts (for example Alcolea Banegas, 1998; Inglis et al., 2007). Elsewhere I have shown how the layout may be generalized to exhibit arguments of a greater degree of structural complexity (Aberdein, 2006, 213 ff.).

The Toulmin layout is well-adapted to display the object sentences and justifications of specific steps of the argumentational structure. However, instructions, the last of Hodges's three argument components, would in general be difficult to fit into a Toulmin layout (although it has been argued that the warrant should be understood as an instruction not a proposition, Hitchcock, 2003, 71). We also need a means to characterize generic steps, if structures are to be analalyzed in terms of the admissibility of steps. One way of tackling both issues is in terms of *argumentation schemes*: stereotypical reasoning patterns, often accompanied by *critical questions*, which itemize possible lines of response. Schemes are framed in generic terms and there is no obstacle to their inclusion of instructions as well as object sentences and justifications (see Pease and Aberdein, 2011, 49 f., for an exploration of how this might be done). Like the Toulmin layout, schemes can be deductive, although they were devised to describe a more diverse range of arguments, such as argument from consequences or argument from expert opinion. (Walton et al., 2008, Chap. 9, catalogues sixty such schemes, many with multiple subvariants.) However, despite their topic-neutral provenance, many of these schemes are applicable to mathematical argumentation (as I have argued elsewhere: Aberdein, 2010, 2013). Indeed, there is at least a family resemblance between argumentation schemes and Toulmin layouts. Most (instantiations of) schemes could be expressed as layouts: the data and warrant correspond to premises, the claim (suitably modified by the qualifier) to the conclusion, and the backing and rebuttal both comprise possible answers to critical questions (for further discussion of the relationship of schemes to layouts, see Pease and Aberdein, 2011, 28 ff.).

We shall divide the schemes which will be needed to describe mathematical reasoning into three groups, on the basis of how their instantiations are related to the corresponding steps (if any) of the inferential structure. *A-schemes* correspond directly to a derivation rule of the inferential structure. (Equivalently, we could think in terms of a single A-scheme, the 'pointing scheme' which picks out a derivation whose premises and conclusion are formal counterparts of its grounds and claim.) B-schemes are less directly tied to the inferential structure. Their instantiations correspond to substructures of derivations rather than individual derivations (and they may appeal to additional formally verified propositions). *B-schemes* may be thought of as exclusively mathematical arguments: high-level algorithms or macros that may in principle be formalized as multiple inferential steps. *C-schemes* are even looser in their relationship to the inferential structure, since the link between their grounds and claim need not be deductive. However, their instantiations may still correspond to substructures of the inferential structure, although there will be no guarantee that this is so and no procedure that will always yield the required structure even when it exists.

I shall make three initial observations about the relationships between the three types of scheme. Firstly, the classification is relative to the composition of the inferential structure. For example, if the inferential structure was a natural deduction presentation of classical logic, then only schemes corresponding to the rules of that system, such as modus ponens, would qualify as A-schemes; even fairly elementary mathematical deductive inferences would comprise B-schemes; and the C-schemes would cover inference patterns that were not in general deductive. Alternatively, if the inferential structure was a higher-level but constructive system, such as Nuprl (MacKenzie, 2001, 285), many of these B-schemes would be A-schemes since they corresponded to higher-level rules, but some of the A-schemes, the constructively inadmissible ones, would now be C-schemes. Secondly, in terms of the Toulmin layout, the difference between the three types of scheme is tracked by their qualifier. The A-schemes would all be qualified as 'by an immediate, formally valid derivation', whereas the B-schemes would all be qualified as 'by an in-principle formalizable valid deduction'. The C-schemes would have different qualifiers in different cases: typically something like 'by plausible mathematical reasoning', but sometimes weaker or stronger. This is a good reason to explicitly state the qualifier, at least of C-schemes, as Matthew Inglis has urged (Inglis et al., 2007). Thirdly, many schemes have subschemes: special cases, all of whose instantiations might also be seen as instantiations of the general scheme, but where additional constraints are imposed. But the classification of a scheme as belonging to a specific type need not transfer to its subschemes, which may well be more rigorous. For instance, the example mentioned above of a scheme containing instructions is a B-scheme, but a subscheme of a C-scheme for analogy (Pease and Aberdein, 2011, 49 f.).

18.4 Four Views of Mathematical Practice

Different views about which steps should be admitted to the inferential structure give rise to different accounts of the foundations of mathematics. Most conspicuously, the divergence between classical and constructive mathematics may be characterized in these terms. Correspondingly, different views about which steps should be admitted to the argumentational structure give rise to different accounts of mathematical practice. In this fashion, at least four such accounts may be distinguished:

0. Only A-schemes are admissible. There is no such thing as 'informal mathematical reasoning': only formalized reasoning can count as mathematical. All that the argumentational structure can do is 'point' at the inferential structure.

1. Only A- and B-schemes are admissible. Informal mathematical reasoning is possible, but the argumentational structure must employ exclusively mathematical steps, albeit ones characterized informally.
2. All three types of scheme are admissible. Informal mathematical reasoning is possible, and the argumentational structure employs both exclusively mathematical steps, and steps of more general application.
3. Only topic-neutral A- or C-schemes are admissible. Informal mathematical reasoning is possible, and must be understandable purely in terms of steps of general application. No argumentational structure need contain any exclusively mathematical steps; that is, all such steps must be reducible to instances of general steps.

Support for each account may be found in mathematical practice. This section explores their competing merits.

18.4.1 Option 0: No Such Thing as Informal Mathematics

The first question to ask about Option 0, the claim that all mathematics is formal, and hence that 'informal mathematics' is not mathematics, is whether it has ever been taken seriously. Both formalism and logicism require that mathematics be formaliz*able*, but neither thesis need insist that mathematics isn't mathematics until it has been formaliz*ed*. This broader claim is more familiar as a polemical exaggeration, as with Russell's insistence that the history of (pure) mathematics began in 1854 (Russell, 1901, 75), or as a straw man erected by critics of such programmes. Hence Lakatos informs us that 'according to logical positivism, informal mathematics, being neither analytical nor empirical, must be meaningless gibberish' (Lakatos, 1978b, 59). While this may be a reasonable inference about the party line of a movement now rather lacking in adherents, specific endorsements are harder to find. For example, Carnap might seem a plausible candidate, but closer inspection reveals this is not actually so (Lavers, 2008, 14). Looking beyond logical positivism, even the arch-formalists of Bourbaki do not consistently practice what they preach, stressing instead the importance of 'experience and mathematical flair' (Bourbaki, 1968, 8; see also Corry, 2009). As the example of Bourbaki demonstrates, whether or not this position is credible as a regulative ideal, even its most single-minded proponents have difficulty living up to it.

A more fundamental defence of Option 0 (or perhaps a narrow interpretation of Option 1) arises from the wholesale rejection of informal reasoning proposed by David Miller. He states that 'It cannot be denied that a complex sequence of interlocked blind guesses and cruel rejections may look much like directed thought, just as Darwinian evolution simulates orthogenesis or design. But we must not be hoodwinked into thinking that it is reasoning, or anything else that we know, that drives us forward to what is unknown. What reasoning does is pull us back' (Miller, 2005, 68). For Miller, the only legitimate task that arguments, deductive or otherwise, can perform is critical (and only deductive arguments are any use for this) (Miller, 2006, 65). He canvasses, but rejects, three other possible roles: persuasion, discovery, and justification. Justification, he states, must entail regress or circularity, because an argument can only justify if its premisses are themselves justified. Persuasion he dismisses as unrelated to the argument: if the argument is thought persuasive, it can only be because the premisses are themselves persuasive, but then the argument makes no contribution. His rejection of discovery as a property of argument is more tendentious: he regards any discoveries arising from valid inference as trivial, in so much as they would be already known to us, were we logically omniscient. But this leads to two important concessions: 'To be sure, [deductive arguments] can provide new subjective knowledge, in the way that mathematical proofs uncannily do. Arguments, it may be conceded, do have an exploratory function, even if what they explore is what is already known, or conjectured, about the world, and not the world itself' (Miller, 2005, 66). Already known, that is, to the logically omniscient. Moreover, 'inferences ... that resist deductive reconstruction,' while 'evidently indistinguishable from blind guesses, ... can indeed lead to an augmentation of knowledge (provided that knowledge is recognized, as it must be, to be conjectural through and through and through)' (Miller, 2005, 67). So, if our concern is with actual mathematical practice, in which logical omniscience is sadly unavailable, then informal deductive arguments, and even non-deductive arguments, can have a place in discovery.

In broader terms, we may observe that Miller's challenge is most effective against a position dual to his own: that argument only draws us forward and never pulls us back. This would indeed be a dangerous position, especially if the steps of such arguments need not be deductive. Even if individual steps carry high levels of confidence, providing that the doubts are independent and thereby cumulative, multistep arguments would swiftly grow less convincing as they lengthen. A system which encouraged us to pursue such arguments without correction could lead us far astray. But what of systems which permit both the forward propagation of confidence and the back propagation of doubt? Why must our guesses be blind? Certainly, prior experience can be a poor guide and even the most confident hunches can prove wide of the mark. But, if only for purposes of resource management, they are often the best place to start.

18.4.2 Option 1: All Steps Exclusively Mathematical

B-schemes, the sort of schemes which are admissible at Option 1 but not at Option 0, comprise mathematical argument patterns of more than purely local application. For example, diagonalization, or finding the determinant of a matrix or the adjoint of an operator. In more complicated cases, multiple B-schemes can act in concert as a transferable technique which may be applied to diverse mathematical problems. This is what Jody Azzouni calls an *inference package*:

a capacity to recognize the implications of several assumptions by means of the representations of objects wherein those several assumptions have been knit together (psychologically). I also claim that the employment of inference packages shows up everywhere in mathematical practice. What an inference package allows a mathematician to do is to reason about a subject matter compatibly with a formal mechanical proof. His reasoning, however, is topic-specific and the various assumptions operative in that reasoning function together in a way that makes them phenomenologically invisible to him (Azzouni, 2009, 20).

B-schemes are in principle formalizable, that is to say any instantiation of a given scheme in the argumentational structure should correspond to a substructure of the inferential structure in a predictable fashion. Nonetheless, actually working out what the latter structure should be in a given case may be forbiddingly laborious. Many B-schemes are intrinsic to established mathematics, and fluency in their use is a prerequisite for participation in mathematical practice. (For an attempt to construct an explicit B-scheme, see Pease and Aberdein, 2011, 49 ff.)

A lot of mathematics can be conducted using A- and B-schemes, so Option 1 is a much more reasonable characterization of mathematical practice than Option 0. However, as we shall see, it does have important limitations. To stick at Option 1, and thereby reject the admissibility of C-schemes, is to restrict mathematical practice to established mathematical techniques. Hence, Option 1 comprises a defence of the purity of proof method. This defence originates in Aristotle's rejection of 'metabasis', or 'kind crossing', the use of methods proper to one domain of reasoning within another. Indeed, some proponents of Option 1 appeal directly to Aristotle's arguments (for example, Krabbe, 2013, 183; see Cantù, 2010, for discussion of the significance of metabasis in the history of mathematics). Thus, although geometrical methods would be permitted in astronomy, which Aristotle regards as a science subordinate to geometry, they should not be used in arithmetic (Aristotle, 1947, 75a). Hence mathematics *are* genuinely plural: for Aristotle, they comprise several distinct domains whose methods overlap only by analogy (76a). In this respect he differs from Plato, for whom dialectic, that is informal reasoning, comprises a highest domain to which all others are subordinate (Detlefsen, 2008, 179).

As an example of Aristotle's approach, consider the following:

Bryson's method of squaring the circle, even if the circle is thereby squared, is still sophistical because it does not conform to the subject in hand. So, then, any merely apparent reasoning about these things is a contentious argument, and any reasoning that merely appears to conform to the subject in hand, even though it be genuine reasoning, is a contentious argument: for it is merely apparent in its conformity to the subject matter, so that it is deceptive and plays foul (Aristotle, 1995, 171b).

Some unpacking is required. Sources both ancient and modern disagree as to precisely what Bryson's method comprised (Mendell, n.d; Heath, 1949, 47 ff.; Hankinson, 2005, 35). But there is consensus that it began by both inscribing and circumscribing the circle with squares, perhaps as in Fig. 18.4. One school of thought, attributed to Proclus, represents Bryson's argument as merely an existence claim: some square with an area between the two must be equal in area to the circle. Another, questionably attributed to Alexander of Aphrodisias, has Bryson proceed

Fig. 18.4 Bryson's method
of squaring the circle

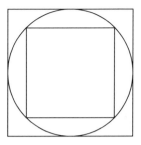

to inscribe and circumscribe pentagons, then hexagons, and so on. With each extra side the margin of error shrinks, so the method might charitably be interpreted as a partial anticipation of the method Archimedes used to estimate the area of a circle. However, in either case Aristotle would reject the result as fallacious. This has the uncomfortable corollary that, had he lived to see it, Aristotle would apparently have rejected Archimedes's method too.

Of course, in one sense, Aristotle would be right: Archimedes's method is not purely geometrical, and however far it is taken it does not constitute a solution to the ancient (and provably insoluble) problem of squaring the circle by a ruler and compass construction. The difficulty with Aristotle's approach is whether he can acknowledge an innovative method as mathematical if it does not fit into any of the kinds of mathematics which he recognizes. This is exacerbated by his rejection of a common kind to which all other mathematical kinds would be subordinate. Later purists have been less reticent: Descartes proposed a mathesis universalis based on algebra; in the twentieth century set theory and subsequently category theory have been likewise proposed as common mathematical kinds. However, such a relationship between a general system and specific mathematical inferences is easier to reconcile with formal than with informal mathematics.

We may ask whether purity is an appropriate, or even intelligible ideal for informal mathematics. Methodological innovation often originates in informal mathematics. When a new method is first proposed, it can be controversial whether it qualifies as mathematical. Typically, its early use is restricted to the heuristic pursuit of results subsequently confirmed by more conventional means. Euler's employment of alternating series to establish that $\sum_{n=1}^{\infty} 1/n^2 = \pi^2/6$ is one frequently discussed example (Pólya, 1954); the anticipation of the Laplace transform by Heaviside's operator methods another (Hunt, 1991). To insist on purity of method at this point may be appropriate if the informally derived results are (mis)represented as formally sound, but can otherwise only be an unnecessary brake on progress.

In conclusion, Aristotle may charitably be read as talking only of (mutatis mutandis) the inferential structure, and therefore not endorsing Option 1 after all. Latterday Aristoteleans, in so far as they insist on 'certain conventions to which an argument must conform to be an argument within the discipline' (Krabbe, 2013, 183), beg the question against methodological innovation. Requiring that the new method be shown to be mathematical before it can be admitted into informal mathematical reasoning is to require that it be shown to be either formally valid or heuristically useful. But the former is inappropriate for informal mathematics and the latter can only be demonstrated through extensive use.

18.4.3   Option 2: Not All Steps Exclusively Mathematical

The Victorian mathematician J.J. Sylvester, in defending his discipline from the ill-informed criticism of T.H. Huxley, stressed 'how much observation, divination, induction, experimental trial, and verification, causation, too (if that means, as I suppose it must, mounting from phenomena to their reasons or causes of being[2]) have to do with the work of the mathematician' (Sylvester, 1869, 1762). This perspective emphasizes the connexions between mathematical practice and ordinary reasoning, much of it comprising C-schemes, thereby leading to Option 2 or, when the ordinary reasoning is understood as underpinning the mathematical practice, Option 3.

One important challenge to Option 2 arises from an ambiguity between two senses of the phrase 'mathematical argument'. The ambiguity may be resolved by applying the terminology of Toulmin layouts. When 'mathematical argument' occurs in ordinary, that is non-mathematical, discourse it often serves to indicate that the argument has a mathematical warrant and/or backing, and thereby a mathematical qualifier. For example, in a discussion on the prospects of two competing political candidates it might be argued that one candidate is too far behind to win, even if all the uncounted ballots had been cast in his favour. The proponent of such an argument may stress that it was a 'purely mathematical argument', thereby emphasizing its difference from typical political arguments: a more robust warrant and qualifier. If that were the only context in which 'mathematical argument' were used it would appear that the mathematical nature of the warrant (or qualifier) was essential, thereby ruling out topic-neutral schemes, and thereby Option 2. But this would be to ignore the non-mathematical context—the warrant is singled out since it is what makes the argument different from other arguments to the same claim. In broader terms, we might also classify this as a political argument, since it derives a political claim from political data. In this second sense, a mathematical argument will be one which arises in mathematical discourse. Characteristically, this will entail the data and claim being mathematical, but not necessarily that the warrant should be so too.

If warrants in mathematical arguments are not mathematical what are they? The positive argument in defence of the mathematical use of C-schemes can best be made by answering this question. As discussed in Sect. 18.3, although argumentation schemes have been developed primarily for non-mathematical contexts, many

---

[2]This would seem to be an attempt to characterize what would later be described as abductive reasoning, or inference to the best explanation.

of them lend themselves readily to mathematical application. For example, consider the following scheme for argument from analogy:

ARGUMENTATION SCHEME FOR ARGUMENT FROM ANALOGY

Similarity Premise	Generally, case C_1 is similar to case C_2.
Base Premise	A is true (false) in case C_1.
Conclusion	A is true (false) in case C_2.

Critical Questions:

CQ1.   Are there differences between C_1 and C_2 that would tend to undermine the force of the similarity cited?

CQ2.   Is A true (false) in C_1?

CQ3.   Is there some other case C_3 that is also similar to C_1, but in which A is false (true)? (Walton et al., 2008, 315)

This is a C-scheme because it is topic-neutral and not necessarily deductive. Whether or not this scheme is deductively valid turns on the interpretation of the warrant, that is the similarity premise. If 'similar' is defined narrowly enough, providing an equivalence between ostensibly unrelated mathematical structures, the premise would be categorical, and the scheme valid. This can be a source of formidable insight. The graph theorist Bill Tutte gives an impressive example in describing how he and three colleagues came to falsify Lusin's Conjecture, that no square can be dissected into smaller squares, all of different sizes: 'Eventually we found that a squared rectangle was equivalent to an electrical network of unit resistances. … We now understood that the basic theory of squared rectangles was that of Kirchoff's Laws of electrical networks. That was something we could look up in textbooks' (Tutte, 1998, 3 ff.). The similarity premiss is the equivalence that Tutte and his colleagues found, which could be formally demonstrated. Hence the base premisses, which they looked up in textbooks, were known to have true analogues in the theory of squared rectangles. This suggests that this specific case of analogy may also instantiate some B-scheme which is a subscheme of the general C-scheme for analogy above. One such subscheme is discussed in (Pease and Aberdein, 2011, 49 f.), but a better fit may be found in the account of analogy in (Bartha, 2013), which may be construed as the articulation of several such subschemes.

But even if the similarity premise is defeasible, the scheme at least characterizes a versatile pattern-spotting procedure which can be an invaluable source of mathematical hypotheses. (This is not to prejudge the question whether defeasible analogies are *only* useful as heuristics.) A familiar example is the extension of the number concept, from natural numbers to rational numbers, irrational numbers, negative numbers, imaginary numbers, transfinite numbers, and so forth. Each of these steps proceeded by analogy, and was initially controversial. These moves also gave rise to further analogies, as properties known to apply to one sort of number were conjectured to apply to other sorts. For example, Wallis conjectured that $n^p \times n^q = n^{p+q}$, known to apply for integers p, q, also held when the indices were fractional or negative (Stedall, 2002, 158).

Informal analogies can be drawn from areas outside mathematics. For example, the optimization technique of 'simulated annealing' is motivated by an analogy with the metallurgical technique of annealing (Michalewicz and Fogel, 2004, 120). The mathematical problem is that many search procedures designed to find the global maximum of functions of multiple independent variables will halt at local optima, that is globally suboptimal solutions. The analogous problem in metallurgy is that metals (and other crystalline substances) if cooled too quickly from a molten state will 'freeze in' irregularities, producing a structure of higher energy level, and typically weaker structural integrity. The solution is gradual cooling, which finds an analogy in the incremental reduction of a control parameter in the simulated annealing algorithm. Here, the base premiss, that is the data, is drawn from metallurgy, not mathematics, while the warrant asserts the analogy between the two fields, and is thus not wholly mathematical either. Nonetheless, the claim is a substantial contribution to a mathematical problem. While more rigorous arguments have been made for the effectiveness of simulated annealing, as they must be if it is to have a permanent place in mathematics, this analogy is independently compelling, and was crucial to its discovery.

We have seen that the data, the warrant, and thereby the qualifier of a mathematical argument may be non-mathematical. However, arguments with non-mathematical claims would ipso facto not be contributions to mathematics, but rather mathematical contributions to whatever discourse the claim was drawn from. What of the other components of the Toulmin layout? The susceptibility of mathematical argument to what Lakatos describes as 'heuristic falsifiers' would suggest the admissibility of non-mathematical rebuttals (see Lakatos, 1978a, 36). (If a heuristic falsifier succeeds, the inferential structure is unaffected, since it is still formally valid, but the argumentational structure would be forced to follow a different path, bypassing the 'falsified' content.) And even ostensibly mathematical warrants may have non-mathematical backing, as the next section demonstrates.

18.4.4 Option 3: No Steps Exclusively Mathematical

The distinction between Option 2 and Option 3 turns on whether all the instantiations of mathematical schemes in the argumentational structure are reducible to instantiations of non-mathematical schemes, or more precisely, whether all instantiations of B-schemes are decomposable into sequences or combinations of instantiations of A- and C-schemes. Advocates of Option 2 hold that room must be left for the irreducibly mathematical, even in the argumentational structure; proponents of Option 3 disagree.

This dispute may be difficult to resolve. On the one hand, there are elementary topic-neutral A- and C-schemes in mathematics, and many instantiations of more complex B-schemes can be reduced to instantiations of such schemes in an obvious enough if tedious fashion. Such, after all, is one of the goals of proof. On the other hand, there are many B-schemes of formidable complexity, for example those invoking widely used but hard to prove results. Even if B-schemes of this character can be understood as ultimately depending on elementary, topic-neutral

A- or C-schemes, such schemes would seem so remote from the use to which the B-scheme was put, especially in informal contexts, as to be of no practical relevance. But notice that this ostensible counterexample to Option 3 makes a crucial concession: if the arguer is merely using the result, and not reestablishing it on the fly, then the backing for his argument is not wholly mathematical. Rather he is ultimately relying, quite properly, on an argument from expert opinion. If the argument is challenged, the arguer will typically make the backing more explicit to show how the contentious scheme is supported, thereby answering one of its critical questions. This may take the form of restating the argument in more elementary steps, thereby bringing it closer to topic-neutral schemes, or the arguer may be forced to acknowledge that he is using a result which he is unable to prove, at least on demand, but has on good authority, at which point he has clearly resorted to a non-mathematical scheme.

More direct motivation for Option 3 may be found in the widely held thesis that 'there are certain basic forms of thought and argument which are prior to the development of formal Mathematics' (Robinson, 1964, 237). Such intuitions have taken a variety of forms. For example, Jody Azzouni once developed (but subsequently repudiated) a view of informal mathematics as a subdoxastic process analogous to that governing grammar, from which he infers that 'all reasoning is topic-neutral in nature' (Azzouni, 2009, 18 f.). Keith Devlin reaches a similar conclusion from the grounds that mathematical 'thought processes are comprised of brain activation patterns that are associated with real world stimuli' (Devlin, 2008, 378). Penelope Maddy proposes basing mathematics on 'a characterization of classical logic as grounded in a rudimentary logic that's both true of the world and embedded in our most primitive modes of cognition' (Maddy, 2007, 288). As Ian Dove concludes, 'mathematical reasoning is already in accord with principles and techniques from informal logic—even if this is unnoticed by the practitioners' (Dove, 2013, 306).

We have seen that all four options have defenders amongst both mathematicians and philosophers. However, Option 0 relies on either an untenable idealization of mathematical practice, or an arbitrary restriction of 'mathematics' to exclude much of that practice. And Option 1 is hard to reconcile with methodological innovation. That leaves Option 2 and Option 3, both of which leave room for informal reasoning in mathematical practice. Resolving the debate between these positions turns on whether all inferences of informal mathematics may be reduced to combinations of topic-neutral inferences. A final answer to this question is beyond the scope of this chapter, but I would suggest that the onus is on proponents of Option 2 to provide examples of informal mathematical inferences for which no such reduction is possible.

18.5   Mathematical Reasoning in Context

We have so far observed that the admissibility of schemes in the argumentational structure gives rise to different views of mathematical practice. However, mathematical practice is itself highly diverse, a diversity which the relationship between

Table 18.1 Systematic survey of dialogue types

	Initial situation		
Main goal	Conflict	Open problem	Unsatisfactory spread of information
Stable agreement/resolution	*Persuasion*	*Inquiry*	*Information seeking*
Practical settlement/decision (not) to act	*Negotiation*	*Deliberation*	N/A
Reaching a (provisional) accommodation	*Eristic*	N/A	N/A

From Walton and Krabbe, 1995, 80, Table 3.2

schemes and structure should reflect. Elsewhere, I have argued that mathematical discourse is characteristically a dialogue, if only with oneself; that it exhibits a diversity of dialogue types; and that the analysis of mathematical reasoning should have regard to the type of dialogue in which the mathematical reasoning arises (Aberdein, 2007, 144 ff.). This account of mathematical dialogue types was developed from an account of dialogue types in non-mathematical discourse, including persuasion, negotiation, inquiry, deliberation, information-seeking, and quarrel (developed in Walton and Krabbe, 1995; see also Krabbe, 2013). Dialogue types may be distinguished in terms of their initial situation, and the shared and individual aims of their participants, as in Table 18.1 (Walton and Krabbe, 1995, 80, Table 3.2). Since different argumentational practices are legitimate in different types of dialogue, the evaluation of arguments and argumentation schemes must have regard to the type of dialogue in which they are advanced. Table 18.2 summarizes a variety of mathematical dialogue types, some of which are more appropriate for successful proof than others (cf. Walton and Krabbe, 1995, 66, Table 3.1).

The first three rows of Table 18.2 characterize contexts in which a stable resolution is sought—a traditional expectation of mathematical proof. But there is still significant diversity and competing expectations of what a proof should provide. To see this, observe that a proof of the same result may pass through all three types of dialogue at different stages in its career. It may originate in the context of open-minded inquiry, either as a purely internal dialogue conducted within its originator's head or in spoken or written communication between collaborators; but it must then convince a sceptical interlocutor, typically a journal referee; and, if it is sufficiently important, it may eventually figure in pedagogic dialogues with audiences of different levels of sophistication: peers in the subfield, the broader professional community, graduate or undergraduate students and so on. All of these contexts are consistent with the uncontroversial acceptance of the proof, but they impose different constraints on the admissibility of schemes in its argumentational structure. For example, highly specialized mathematical B-schemes might be appropriate in the inquiry dialogue and in a subsequent pedagogic dialogue with peers who are well-acquainted with such manoeuvres, but could reasonably be challenged by the referee, and would be inappropriate for an undergraduate audience.

The last three rows of Table 18.2 represent other forms of mathematical argumentation, allied to proof, but which fall short of the standards conventionally

Table 18.2 Some mathematical dialogue types

Dialogue type	Initial situation	Main goal	Goal of prover	Goal of interlocutor
Proof as Inquiry	Open-mindedness	Prove or disprove conjecture	Contribute to main goal	Obtain knowledge
Proof as Persuasion	Difference of opinion	Resolve difference of opinion with rigour	Persuade interlocutor	Persuade protagonist
Proof as Pedagogical Information-Seeking	Interlocutor lacks information	Transfer of knowledge	Disseminate knowledge of results and methods	Obtain knowledge
'Proof' as Oracular Information-Seeking (e.g. Tymoczko, 1979)	'Prover' lacks information	Transfer of knowledge	Obtain information	Inscrutable
'Proof' as Deliberation (e.g. Swart, 1980; Jaffe and Quinn, 1993)	Open-mindedness	Reach a provisional conclusion	Contribute to main goal	Obtain warranted belief
'Proof' as Negotiation (e.g. Zeilberger, 1993)	Difference of opinion	Exchange resources for a provisional conclusion	Contribute to main goal	Maximize value of exchange

Adapted from Aberdein, 2007, 148

expected of proof. Oracular information-seeking describes Thomas Tymoczko's hostile characterization of computer-assisted proof (Tymoczko, 1979, 71) as blatant appeal to authority. But it also describes other contexts in which the standards of admissibility for the scheme for argument from expert opinion are lower than might be expected; where, for example, there is no obvious means by which the prover or interlocutor could replicate the cited result. This context preserves the goal of a stable resolution, although it allows unorthodox strategies for its attainment. The next two contexts drop this goal in favour of a more provisional outcome. Deliberation encompasses the establishment of less rigorous results, such as Edward Swart's 'agnograms', theorem-like statements about which we are presently agnostic, and the conjectures or speculations of Arthur Jaffe and Frank Quinn's 'theoretical mathematics' (Swart, 1980; Jaffe and Quinn, 1993). Lastly, Doron Zeilberger's 'theorems for a price' combine a high but not absolute level of certainty with a cost estimate for the computational resources necessary to attain full certainty (Zeilberger, 1993, 980), a context requiring negotiation over the best allocation of those resources. It would be highly controversial to substitute such dialogues for rigorous proof, but they uncontroversially describe other forms of mathematical practice, including much applied mathematics.

18.6   Conclusion

This chapter has defended an account of mathematical reasoning as comprised of two parallel structures. The argumentational structure is composed of arguments by means of which mathematicians seek to persuade each other of their results or, more generally, to achieve goals appropriate for whatever dialogue they are having. The inferential structure is composed of derivations which offer a formal counterpart to these arguments. The precise relationship between the two structures may be understood in terms of the range of argumentation schemes which may be instantiated by steps of the argumentational structure. Just as different views about the foundations of mathematics may be characterized in terms of the admissibility of steps in the inferential structure, different views about mathematical practice may be characterized in terms of the admissibility of steps in the argumentational structure. I have made the case that a wide range of schemes should be admitted to the argumentational structure. Two distinct areas emerge for further exploration: firstly, the investigation of fine-tuned, specifically mathematical B-schemes; and secondly, the application to mathematics of topic-neutral C-schemes.

Acknowledgements Previous versions of parts of this chapter were delivered in Vienna, St Andrews, Gainesville, FL, Windsor, ON, and Birmingham. I am grateful to the audiences for helpful discussion. I am also grateful to Ian Dove for insightful comments and to Alison Pease for ideas developed during our collaboration on (Pease and Aberdein, 2011).

References

Aberdein, A. (2006). Managing informal mathematical knowledge: Techniques from informal logic. In J. M. Borwein & W. M. Farmer (Eds.), *MKM 2006*, Vol. 4108 in *LNAI* (pp. 208–221). Berlin: Springer.

Aberdein, A. (2007). The informal logic of mathematical proof. In B. Van Kerkhove & J. P. Van Bendegem (Eds.), *Perspectives on mathematical practices: Bringing together philosophy of mathematics, sociology of mathematics, and mathematics education* (pp. 135–151). Dordrecht: Springer.

Aberdein, A. (2010). Observations on sick mathematics. In B. Van Kerkhove, J. P. Van Bendegem & J. De Vuyst (Eds.), *Philosophical perspectives on mathematical practice* (pp. 269–300). London: College Publications.

Aberdein, A. (2013). Mathematical wit and mathematical cognition. *Topics in Cognitive Science, 5*(2), 231–250.

Alama, J., & Kahle, R. (2013). Checking proofs. In A. Aberdein & I.J. Dove (Eds.), *The argument of mathematics* (pp. 147–170). Dordrecht: Springer.

Alcolea Banegas, J. (1998). L'argumentació en matemàtiques. In E. Casaban i Moya (Ed.), *XIIè Congrés Valencià de Filosofia* (pp. 135–147). Valencià. [Trans.: A. Aberdein & I.J. Dove (Eds.), The argument of mathematics (pp. 47–60). Dordrecht: Springer].

Aristotle (1947). Posterior analytics (Trans. by G. R. G. Mure). In R. McKeon (Ed.), *Introduction to Aristotle* (pp. 9–109). New York: Random House.

Aristotle (1995). On sophistical refutations (Trans. by W. A. Pickard-Cambridge). In H. V. Hansen & R. C. Pinto (Eds.), *Fallacies: Classical and contemporary readings* (pp. 19–38). University Park, PA: Pennsylvania State University Press.

Azzouni, J. (2004). The derivation-indicator view of mathematical practice. *Philosophia Mathematica*, *12*(2), 81–105.

Azzouni, J. (2009). Why do informal proofs conform to formal norms? *Foundations of Science*, *14*(1–2), 9–26.

Bartha, P. (2013). Analogical arguments in mathematics. In A. Aberdein & I.J. Dove (Eds.), *The Argument of Mathematics* (pp. 197–236). Dordrecht: Springer.

Bourbaki, N. (1968). *Elements of mathematics: Theory of sets*. Berlin: Springer.

Cantù, P. (2010). Aristotle's prohibition rule on kind-crossing and the definition of mathematics as a science of quantities. *Synthese*, *174*(2), 225–235.

Corry, L. (2009). Writing the ultimate mathematical textbook: Nicolas Bourbaki's *Éléments de mathématique*. In E. Robson & J. Stedall (Eds.), *Oxford handbook of the history of mathematics* (pp. 565–587). Oxford: Oxford University Press.

Detlefsen, M. (2008). Purity as an ideal of proof. In P. Mancosu (Ed.), *The philosophy of mathematical practice* (pp. 179–197). Oxford: Oxford University Press.

Devlin, K. (2008). A mathematician reflects on the useful and reliable illusion of reality in mathematics. *Erkenntnis*, *68*(3), 359–379.

Dove, I. J. (2013). Towards a theory of mathematical argument. In A. Aberdein & I.J. Dove (Eds.), *The Argument of Mathematics* (pp. 291–308). Dordrecht: Springer. (Reprinted from *Foundations of Science*, 2009, *14*(1–2), 137–152).

Epstein, R. L. (2013). Mathematics as the art of abstraction. In A. Aberdein & I.J. Dove (Eds.), *The Argument of Mathematics* (pp. 257–289). Dordrecht: Springer.

Hankinson, R. J. (2005). Aristotle on kind-crossing. In R. Sharples (Ed.), *Philosophy and the sciences in antiquity* (pp. 23–54). Aldershot: Ashgate.

Hardy, G. H. (1928). Mathematical proof. *Mind*, *38*, 11–25.

Heath, T. L. (1949). *Mathematics in Aristotle*. Oxford: Clarendon.

Hersh, R. (2013). To establish new mathematics, we use our mental models and build on established mathematics. In C. Cozzo & E. Ippoliti (Eds.), *From an heuristic point of view: In honor of Carlo Cellucci*. Newcastle upon Tyne: Cambridge Scholars Publishing, forthcoming.

Hitchcock, D. (2003). Toulmin's warrants. In F. H. van Eemeren, J. Blair, C. Willard & A. F. Snoeck-Henkemans (Eds.), *Anyone who has a view: Theoretical contributions to the study of argumentation* (pp. 69–82). Dordrecht: Kluwer.

Hodges, W. (1998). An editor recalls some hopeless papers. *Bulletin of Symbolic Logic*, *4*(1), 1–16.

Hunt, B. J. (1991). Rigorous discipline: Oliver Heaviside versus the mathematicians. In P. Dear (Ed.), *The literary structure of scientific argument* (pp. 72–95). Philadelphia, PA: University of Pennsylvania Press.

Inglis, M., Mejía-Ramos, J. P., & Simpson, A. (2007). Modelling mathematical argumentation: The importance of qualification. *Educational Studies in Mathematics*, *66*(1), 3–21.

Jaffe, A., & Quinn, F. (1993). "Theoretical mathematics": Toward a cultural synthesis of mathematics and theoretical physics. *Bulletin of the American Mathematical Society*, *29*, 1–13.

Krabbe, E. C. W. (2013). Strategic maneuvering in mathematical proofs. In A. Aberdein & I.J. Dove (Eds.), *The Argument of Mathematics* (pp. 181–197). Dordrecht: Springer. (Reprinted from *Argumentation*, 2008, *22*(3), 453–468)

Lakatos, I. (1978a). A renaissance of empiricism in the recent philosophy of mathematics. In *Philosophical papers* (Vol. 2, pp. 24–42). Cambridge: Cambridge University Press.

Lakatos, I. (1978b). Cauchy and the continuum: The significance of non-standard analysis for the history and philosophy of mathematics. In *Philosophical papers* (Vol. 2, pp. 43–60). Cambridge: Cambridge University Press.

Lavers, G. (2008). Carnap, formalism, and informal rigour. *Philosophia Mathematica*, *16*(1), 4–24.

MacKenzie, D. (2001). *Mechanizing proof: Computing, risk, and trust*. Cambridge, MA: MIT Press.

Maddy, P. (2007). *Second philosophy: A naturalistic method*. Oxford: Oxford University Press.

Mendell, H. (n.d.). Bryson's squaring of the circle. http://www.calstatela.edu/faculty/hmendel/ Ancient%20Mathematics/Philosophical%20Texts/Bryson/Bryson.html

Michalewicz, Z., & Fogel, D. B. (2004). *How to solve it: Modern heuristics*. Berlin: Springer.

Miller, D. (2005). Do we reason when we think we reason, or do we think? *Learning for Democracy, 1*(3), 57–71.

Miller, D. (2006). *Out of error: Further essays on critical rationalism*. Aldershot: Ashgate.

Pease, A., & Aberdein, A. (2011). Five theories of reasoning: Interconnections and applications to mathematics. *Logic and Logical Philosophy, 20*(1–2), 7–57.

Pólya, G. (1954). *Mathematics and plausible reasoning* (Vol. 2). Princeton, NJ: Princeton University Press.

Robinson, A. (1964). Formalism 64. In Y. Bar-Hillel (Ed.), *Proceedings of the international congress for logic, methodology and philosophy of science, Jerusalem* (pp. 228–246). Amsterdam: North-Holland.

Ruelle, D. (2000). Conversations on mathematics with a visitor from outer space. In V. Arnold, M. Atiyah, P. Lax & B. Mazur (Eds.), *Mathematics: Frontiers and perspectives* (pp. 251–269). Providence, RI: American Mathematical Society.

Russell, B. (1986 [1901]). Mathematics and the metaphysicians. In *Mysticism and logic* (pp. 75–95). London: Unwin.

Stedall, J. A. (2002). *A discourse concerning algebra: English algebra to 1685*. Oxford: Oxford University Press.

Swart, E. (1980). The philosophical implications of the four-color problem. *The American Mathematical Monthly, 87*, 697–707.

Sylvester, J. J. (1956 [1869]). The study that knows nothing of observation. In J. R. Newman (Ed.), *The world of mathematics* (Vol. 3, pp. 1758–1766). New York: Simon & Schuster.

Toulmin, S. (1958). *The uses of argument*. Cambridge: Cambridge University Press.

Tutte, W. T. (1998). *Graph theory as I have known it*. Oxford: Oxford University Press.

Tymoczko, T. (1979). The four-color problem and its philosophical significance. *Journal of Philosophy, 76*, 57–83.

Walton, D. N., & Krabbe, E. C. W. (1995). *Commitment in dialogue: Basic concepts of interpersonal reasoning*. Albany, NY: State University of New York Press.

Walton, D. N., Reed, C., & Macagno, F. (2008). *Argumentation schemes*. Cambridge: Cambridge University Press.

Zeilberger, D. (1993). Theorems for a price: Tomorrow's semi-rigorous mathematical culture. *Notices of the American Mathematical Society, 46*, 978–981.

Index

A. Aberdein and I.J. Dove (eds.), *The Argument of Mathematics*, Logic, Epistemology,   381
and the Unity of Science 30, DOI 10.1007/978-94-007-6534-4,
© Springer Science+Business Media Dordrecht 2013

Printed by Printforce, the Netherlands